Conserving Soil Resources

European Perspectives

Conserving Soil Resources

European Perspectives

Selected papers from the First International Congress of the European Society for Soil Conservation

Edited by R.J. Rickson
Lecturer in Soil Erosion Control
Silsoe College, Cranfield University, UK

CAB INTERNATIONAL

CAB INTERNATIONAL
Wallingford
Oxon OX10 8DE
UK

Tel: +44 (0)1491 832111
Telex: 847964 (COMAGG G)
E-mail: cabi@cabi.org
Fax: +44 (0)1491 833508

A catalogue record for this book is available from the British Library.

ISBN   0 85198 948 9

Printed and bound in the UK at the University Press, Cambridge.

CONTENTS

Part II: Prediction and Modelling

Part III: Perspectives on Soil Conservation Measures

PREFACE

Conserving Soil Resources: European Perspectives is a collection of selected papers presented at the First International Congress of the European Society for Soil Conservation. The Society was founded in November 1988, by a number of European scientists, government officials and private individuals. They perceived an increasing awareness that soil degradation *is* a reality on the continent of Europe, and that there is an urgent need to develop and promote conservation policies not only over the next decade, but well into the next century. This cause for concern is now shared by over 600 members of the Society, representing 42 countries, primarily in Europe, from the Atlantic seaboard in the West to the Ural Mountains in the East.

One of the aims of the Society is to promote the exchange of scientific ideas and expertise at workshops, seminars and conferences. With this in mind, the Executive Council of the Society proposed that an inaugural International Congress should be organised in order to gauge the 'state of the art' as regards environmental degradation and protection in Europe. The Congress was held in April 1992 at Silsoe College, Bedfordshire in the U.K., with the specific aim of presenting research and development concerned with the current European status of soil degradation and soil conservation. Unlike other international meetings on these subjects, the Congress was unique in concentrating specifically on European issues, but with the intention that the outcome of the dialogue set up during the Congress would have wider reaching applications.

The ESSC also encourages the publication and dissemination of work concerned with soil erosion and conservation. As a result, a number of the best papers presented at the Congress were reviewed and selected for this volume. The papers fall into three distinct groups. The assessment of soil degradation (Part I) illustrates the continued emphasis on measuring and monitoring the problem of physical, biological and chemical degradation of the soil resource, and on the development of new methodologies for studying these problems. Research is reported at the national down to the field scale. A greater knowledge of the extent and degree of soil degradation in Europe enables the most susceptible areas to be identified and targeted for protection policies. This knowledge can also enhance understanding and explanation of the complex processes operating, so encouraging the development of protective measures to combat the loss of the soil resource. A preventative approach towards environmental protection is always more effective than attempts to cure the problem once the processes of degradation have been initiated.

Factors affecting erosion processes such as rainfall erosivity and soil erodibility are quantified for European conditions, and it is evident that the fundamental methodology used in soil degradation studies, primarily developed in the US (such as the Universal Soil Loss Equation) may have rather limited application and accuracy elsewhere. Land management can also seriously affect the degree and extent of soil degradation. This volume does not restrict coverage of such factors to agricultural land alone; industrial and engineered sites are also considered.

Emphasis is placed on both the on-site and off-site consequences of soil erosion. On-site, crop yields and soil productivity can be detrimentally affected by soil degradation. These consequences may have little impact given the overproduction of food in Europe as a whole.

Also, the loss of productivity can be masked by additions of chemical fertilizers, although this practice can aggravate some of the off-site consequences such as pollutants and nutrients being preferentially transported by eroded sediments, and concentrated where the sediments are deposited. Such off-site effects warrant considerable environmental concern, as shown in the papers concerned with nutrient losses, pollution and water quality.

The second area of interest was in prediction and modelling of soil degradation processes (Part II). There is still some scope to apply models such as the Universal Soil Loss Equation to European studies of soil erosion, providing accurate calibration is undertaken. This methodology is well established and suited to new technologies such as Geographical Information Systems and satellite imagery interpretation. However the papers at the Congress also reflected the current development of newer, alternative approaches to prediction and modelling of erosion processes, including movement away from empirical models towards more mathematical representations of the processes operating. Indeed the book contains the first collection of papers specifically addressing the development, validation and application of the European Soil Erosion Model (EUROSEM), which has been designed with the specific aim of creating an erosion model applicable to European conditions. Again different authors consider prediction and modelling of soil degradation processes at varying scales, from the catchment level down to individual experimental plots.

The final section of the book deals with the current perspectives on soil conservation measures in Europe (Part III). The chapter by G. Chisci illustrates the wealth and diversity of conservation measures already in use in Europe, including methods used over the centuries, which may have fallen out of favour. In some cases these measures are being resurrected to combat the accelerating trend of soil degradation. Newer technologies such as the use of geotextiles and soil conditioners for erosion control are also covered.

It is significant that very few papers presented at the Congress deal with the social and economic aspects of soil conservation. Whilst the papers reflect our understanding of soil degradation processes and the technical solutions available, the farmers viewpoint (such as the perception of the problem or the motivation to control the problem) attracts little research. Ironically, it is these non-technical aspects of soil conservation planning which are essential if we are to protect and maintain our soil resources. Consideration of these subjects must be given priority in the immediate future.

During the organization of the Congress and the call for papers, the range of the presentations to be made was unknown. From the papers presented at the Congress, and the chapters included in this volume, it can be concluded that the subject of soil degradation and soil conservation in Europe is extremely diverse and is of interest to a multidisciplinary audience. Membership of the European Society for Soil Conservation continues to grow, reflecting both a flourishing development of this area of study, and an increasing need to conserve soil resources before the trend of degradation becomes irreversible.

R.P.C. Morgan,
President, ESSC,
Chairman, Organizing Committee,
First ESSC Congress.

R.J. Rickson,
Proceedings Editor,
Secretary, Organizing Committee,
First ESSC Congress.

Silsoe, 1994

Part I:

ASSESSMENT OF
SOIL DEGRADATION

Chapter 1

SOIL EROSION IN BRITAIN: A REVIEW

J. Boardman[1] and R. Evans[2]

[1] School of Geography, University of Oxford,
Mansfield Road, Oxford OX1 3TB, UK
[2] Department of Geography, University of Cambridge,
Downing Place, Cambridge CB2 3EN, UK

Summary

Information on erosion in Britain is available from national and regional monitoring schemes, plot studies, observations of discrete incidents, river yield and reservoir surveys, and caesium-137 measurements. There is a need to integrate these results but this is difficult because of the different methods used.

Areas at risk of erosion are well known, and land use, farming practice, soil and climatic factors controlling erosion are reasonably well understood. Erosion by water is now far more important than erosion by wind. Both regional and national surveys of rates and frequencies of erosion cover short time periods. Measurements using ^{137}Cs attempt to fill this gap in knowledge by providing mean annual rates for 30-35 years. The highest rates of erosion are on intensively farmed arable land in the Midlands and in the south and east of England, and on peat covered uplands.

Over the long term, crop yields will decline due to erosion but in the short term most concern relates to off-farm impacts, particularly the flooding of residential areas by runoff from winter cereal fields.

Government response to the erosion problem has been minimal. A monitoring scheme has been discontinued and off-farm effects are dealt with on an *ad hoc* basis by local councils and affected householders. Changes in agriculture related to climate change may in the long term increase erosion rates. These challenges should be met with agricultural and environmental policies based on an awareness of the threat of erosion and the need for soil conservation.

1. INTRODUCTION

The potential for widespread soil damage in Britain was first noted in 1971, with the emphasis placed on water erosion of upland peat and mineral soils, and wind erosion on lowland arable fields (Evans, 1971). Eight years later Reed (1979) listed soils at risk of erosion, mainly from publications of the Soil Survey of England and Wales. He also demonstrated that in the West Midlands, erosion by water occurred frequently and caused spectacular gullying. A survey of erosion in Britain by Morgan (1980) pointed out that many cases had occurred in the past, and were recorded in local data sources such as engineer's reports and newspapers. These data sources have been largely ignored. Speirs and Frost (1985) review the reasons for a recent increase in the incidence of erosion in Scotland.

In 1982 an erosion monitoring scheme was set up by the Soil Survey of England and Wales and the Ministry of Agriculture, Fisheries and Food. Seventeen localities in England and Wales were monitored each year between 1982 and 1986, using aerial photography, ground checking and measurement (Evans, 1988). The localities were selected on the basis of known or suspected erosion and the possibility of land-use change from pasture to arable. Some of the study areas were assumed to be at low risk of erosion. In each locality aerial photographs were taken annually at a scale of 1:10,000 or 1:15,000, covering an area of 15 to 30km in length, and 2 to 3km wide. This survey forms the principal data source of any assessment of the extent, frequency and magnitude of erosion on agricultural land in England and Wales. Preliminary findings are discussed by Evans and Cook (1986) and results for the first three years of the five year monitoring period are presented in Evans (1988; 1990a). Since 1989, a scaled-down version of the scheme has operated but no results have been published.

Data from other sources can supplement those from the SSEW/MAFF monitoring scheme. The National Soil Map (1:250,000) shows soil associations that are at risk of erosion by water and wind (Mackney et al., 1983). The map was compiled from data available in 1981 but there are omissions. For example, Andover 1 Association soils on chalk have subsequently proved to be at risk of erosion. A more up-to-date listing of soils at risk is in Evans (1990b). Morgan's (1985a) assessment of erosion risk is based on the National Soil Map data, allied to a rainfall erosion index (KE>10) taken from Morgan (1980).

Suspended sediment yields from rivers and volumes of sediment trapped in reservoirs are frequently used to estimate rates of erosion in catchments. The main problem with this approach is that it takes no account of sediment storage between the site of erosion and the sampling point. A delivery ratio may be used but there are many theoretical and practical difficulties (Walling, 1983). The average suspended sediment yield from 56 catchments in Britain ranging in size from 0.1 to 6850km^2 is 0.63t ha^{-1}yr^{-1}. The longer term average from reservoir sedimentation studies is 0.3t ha^{-1}yr^{-1} (Walling and Webb, 1981). Erosion rates from catchments of mixed land use and varying size are therefore in the order of 0.5t ha^{-1} yr^{-1} (Boardman, 1986). This, however, is not a very useful figure in an agricultural context since fields within a catchment may experience high rates of erosion with little immediate impact on streams due to storage of eroded soil within the field or in flood plain sites (Trimble, 1983).

Other data sources include single storm measurements (e.g. Morgan 1985b); estimates based on caesium-137 (Quine and Walling, 1991); regional studies (e.g. Reed, 1979; Colborne and Staines, 1985; and Boardman, 1990a), and data from several studies using experimental plots (Morgan, 1985b; Fullen and Reed, 1986; Robinson and Boardman, 1988). However, there are many difficulties in integrating data from these different sources to the extent that apparently simple questions such as 'how serious is erosion on agricultural land in Britain?' are not easy to answer!

Table 1. Erosion rates per field in 17 monitored localities (data from Evans, 1988).

	1982			1983			1984		
	n	m	M	n	m	M	n	m	M
Cambridge/ Bedfordshire	21	0.32	2.33	14	0.25	2.36	2	0.20	0.25
Cumbria	3	0.31	3.62	-	-	-	1	0.01	0.01
Devonshire	1	1.29	-	6	1.67	5.94	-	-	-
Dorset	12	0.58	7.14	35	1.45	22.20	5	0.74	1.41
Gwent	19	0.95	4.90	26	0.66	15.62	4	0.70	1.15
Hampshire	3	0.81	4.90	7	0.66	21.22	6	0.61	1.82
Hereford	47	0.87	9.44	5	0.62	7.64	1	0.63	2.07
Isle of Wight	13	1.40	4.79	31	2.0	20.45	2	1.00	3.69
Kent	8	0.61	3.46	12	1.13	7.09	7	7.92	12.76
Norfolk East	3	0.90	1.46	48	0.80	6.75	9	1.47	4.77
Norfolk West	5	0.05	1.20	42	0.58	8.54	10	0.14	3.11
Nottingham	9	0.33	3.89	85	1.81	47.25	16	0.24	1.50
Shropshire	29	0.99	28.84	69	1.35	35.24	21	0.40	8.57
Somerset	19	4.30	39.74	37	2.75	29.52	8	3.42	7.85
Stafford	30	0.90	77.34	85	1.31	55.24	22	0.75	10.77
Sussex East	-	-	-	10	0.56	6.72	3	0.13	0.48
Sussex West	-	-	-	30	0.43	7.15	6	0.21	0.41

n = number of fields
m = median volume of eroded soil (m ^{3} ha^{-1})
M = maximum volume of eroded soil in each locality (m ^{3} ha^{-1})

2. AREAS AT RISK AND RATES OF EROSION

Areas of lowland arable farming most at risk of erosion are reasonably well known as a result of studies of individual erosion events, the SSEW/ MAFF monitoring scheme and information presented on the National Soil Map. They have been identified as:

1. Lower Greensand soils of southern England including the Isle of Wight;
2. Sandy and loamy soils in Nottinghamshire and the West Midlands;
3. Sandy and loamy soils in Somerset and Dorset;
4. Sandy and loamy soils in parts of East Anglia;
5. Chalk soils of the South Downs, Cambridgeshire, Yorkshire and Lincolnshire Wolds, Hampshire and Wiltshire;
6. Sandy soils in South Devon;
7. Loamy soils in eastern Scotland.

In most of these areas the main agent of erosion is runoff, but wind erosion occurs locally on sandy and peaty soils in the Vale of York, the Breckland of East Anglia and on peat mosses in Lancashire (Mackney et al., 1983; ADAS, 1985; Evans, 1990b). Sandy soils in the West Midlands and Nottinghamshire are at risk of both wind and water erosion. Evans (1990b) lists 144 soil associations as being at moderate to very high risk of water erosion and 20 at similar risk of wind erosion.

Fourteen soil associations are at moderate to very high risk of erosion in the uplands of England and Wales (Evans, 1990b). The limited amount of information on rates of upland erosion is reviewed by Evans (1989; 1990c). Soils at risk include deep peats, where gullying occurs, and mineral soils.

The SSEW/MAFF monitoring scheme quantified rates of erosion by measurement of volumes of soil lost from rills and gullies, or volumes of soil deposited in fans (Evans, 1988). The data for erosion rates is very left skewed, with many low and few high values. For this reason the median rather than the mean is a better reflection of central tendency. Median values for the years 1982-84 are generally below $1m^3ha^{-1}$ with higher values in a few localities (Table 1). Silty soils in Somerset generally have the highest median values, but the highest was in Kent in 1984 when all the eroded fields were sown to irrigated market garden crops (Boardman and Hazelden, 1986). Areas of sandy soils in the Isle of Wight, Nottinghamshire, Shropshire and Staffordshire also have high median rates of erosion. The maximum values of erosion show a wider range than the median values.

In an area of about $35km^2$ of agricultural land on the South Downs Boardman (1990a and unpublished) monitored erosion from 1982-92. Median annual erosion rates ranged from 0.5 to $5.0m^3ha^{-1}$ measured on the basis of contributing area rather than field area. For one year (1987-88) the median rates were $3.3m^3ha^{-1}$ on a field basis and $5.0m^3ha^{-1}$ on a contributing area basis (Boardman, 1988).

Rates of erosion estimated using caesium-137 for 13 sites in England and Wales are presented by Quine and Walling (1991) (Table 2). The locations were in areas where erosion is known to be a problem and covered a range of soil textures. Each sampling site is one field, which may not be representative of the wider area. Erosion rates are average annual rates for approximately the last 30 years. The ability to estimate erosion retrospectively makes the technique potentially very useful. It is however based on several assumptions and a calibration procedure. The net erosion rate represents the amount of soil which leaves the field. The gross rate is calculated by dividing the total mass of eroded sediment by the area of the field (Quine and Walling, 1991). The gross figure is therefore comparable to the figures quoted by Evans (1988 and Table 1), and Boardman (1990a). The gross erosion rates in Table 2 are higher than the median figures in Table 1, sometimes by a considerable amount. This may be because Quine and Walling sampled sites with unusually high rates of erosion, whereas Evans used the median value of all eroding fields in a locality. At the Lewes site (Table 2), a gross rate of erosion of $4.3t\ ha^{-1}yr^{-1}$ compares with a rate estimated by measurement of rills and gullies over eight years of approximately $1t\ ha^{-1}yr^{-1}$ (Boardman, unpublished). Walling and Quine (1991) recognise this problem but regard the values as showing reasonable agreement, considering the differences between the two approaches. It seems unlikely that the discrepancy is due to unmeasured sheet and wind erosion since these processes appear to be insignificant on the South Downs at present. It also seems unlikely that rates of erosion were higher before 1982 when Boardman began monitoring. All available evidence suggests the reverse. There is therefore an unexplained contrast in rates obtained by the two methods which would repay further investigation.

Table 2. Caesium-137 based soil erosion rates from Quine and Walling (1991) (t ha^{-1}yr^{-1}).

Site	Soil texture*	Net	Gross
Yendacott, Devon	Loamy	1.9	5.3
Mountfield, Somerset	Silty	2.2	4.6
Higher, Dorset	Clayey	3.1	5.2
Fishpool, Gwent	Silty	1.9	5.1
Wootton, Herefordshire	Silty	2.8	6.4
Dalicott, Shropshire	Sandy	6.5	10.2
Rufford, Nottingham	Sandy	10.5	12.2
Brook End, Bedford	Clayey	1.2	3.6
Keysoe Park, Bedford	Clayey	0.6	2.2
Manor House, Norfolk	Coarse loamy	2.4	6.3
Hole Farm, Norfolk	Sandy	3.0	6.3
West Street, Kent	Fine silty	4.3	7.7
Lewes, Sussex	Silty	1.4	4.3

* From description of soil association (Mackney et al., 1983)

In many localities winter cereals are most at risk of erosion because of the extensive area under this crop. However, rates of erosion are generally higher on sugar beet, potatoes, market garden crops and soft fruits (Evans, 1988).

3. CAUSES OF EROSION

There has been considerable increase in erosion on agricultural land in Britain in the last fifteen years (Evans and Cook, 1986; Boardman, 1990a). This is the result of the intensification of agriculture (Morgan, 1980; Speirs and Frost, 1985; Bullock, 1987; Boardman, 1990b), with heavier and more powerful vehicles leading to compaction problems (Fullen, 1985), and cultivation of steeper slopes. Contour working is not possible for reasons of safety. Fields have been enlarged by the removal of hedges, banks, walls and ditches, so that larger machinery is easier to use. There is some evidence that a decrease in the organic matter content of soils has made them more susceptible to degradation and erosion (Morgan, 1985c). Farmers have been encouraged to produce fine seedbeds by the use of the power harrow (Frost and Speirs, 1984). On some soils fields drilled to cereals, oil seed rape or grass are rolled after drilling to produce smooth surfaces with low micro-topography on which runoff readily occurs.

The association of erosion with winter cereal crops is very clear (Boardman, 1984; Colborne and Staines, 1985; Evans and Cook, 1986; Speirs and Frost, 1985). In England and Wales between 1969 and 1983 there was a threefold increase in the area of winter cereals (Evans and Cook, 1986) and in Scotland a fourfold increase between 1967 and 1986 (Evans, 1989). Winter cereals are susceptible to erosion following drilling until the fields have about 30% crop cover (Robinson and Boardman, 1988). This period is frequently the wettest of the year with rilling initiated on bare smooth surfaces (Boardman, 1992).

7

Furthermore, under winter cereals farmers are often unable to complete drilling operations on all their land before early November due to periods of wet weather or late harvesting of some previous crops (e.g. potatoes and sugar beet). This means that fields drilled late will not attain sufficient crop cover to inhibit erosion before the spring. Fields will also be drilled when soils are damp and compaction is more likely to occur. Large areas of a single crop increase the likelihood of adjacent fields being bare at the same time. In some parts of the South Downs winter cereals occupy 60% of the landscape and runoff is able to travel uninhibited from one bare field to the next. Under these conditions runoff and eroded soil have moved up to 5km along normally dry valleys and into a housing estate (Boardman and Evans, 1991).

There is no evidence that the recent increase in erosion can be explained by a change in the amount or character of rainfall. Wetter seasons undoubtedly occur, such as the autumns of 1982 and 1987 on the South Downs. However, in each autumn current farming practices and the area of land under arable crops means that the landscape is primed for erosion. It is a question of 'How much?' and 'Where?' rather than 'If?' and 'When?' In areas such as the West Midlands and Nottinghamshire, the intensification of agriculture has meant that much erosion occurs on sugar beet and potatoes as a result of spring and summer convective rain (Evans, 1988; Fullen, 1985; Fullen and Reed, 1986).

Data on the extent and rate of wind erosion on agricultural land in Britain is lacking. Soils at risk are known, but the magnitude and frequency of erosion events is unknown. Wilson and Cooke (1980) give a range of soil loss between 20 and 44t ha^{-1}yr^{-1}. Hepworth (1987) recorded widespread damage and significant costs in a study area south east of York in April 1986. Wind erosion is difficult to measure because much soil is redistributed within the field and fine components may be moved long distances. However, estimates based on material trapped in ditches and behind field boundaries could be made and are needed.

Erosion in the uplands is a result of water, wind, frost, fire, animal and human agencies (such as footpaths, ski impacts, and afforestation). In the peat areas of the southern Pennines a phase of gullying seems to have been initiated by pollution at the beginning of the Industrial Revolution which killed the protective cover of sphagnum moss. The gullying has continued to the present day (Burt and Labadz, 1990). In many upland areas a large increase in sheep numbers since the second World War has caused overgrazing problems.

4. IMPLICATIONS OF EROSION IN BRITAIN

Erosion affects crop yields due to the decrease in soil depth and loss of nutrients. The full economic impact may, however, be delayed for hundreds of years, depending on the rate of erosion, the depth of the soil and the economics of farming. Erosion damages current crops which may have to be redrilled or abandoned. Arden-Clarke and Evans (in press) review the impact of erosion on yields in Britain (Table 3). On thick soils the reduction in cereal yields with rates of erosion typical of Britain (e.g. 3t ha^{-1}yr^{-1}) is significant. On thin soils similar rates of erosion give more cause for concern (Table 4). Where topsoils are thin and subsoils are rapidly exposed by erosion, or are incorporated into the topsoil, a less workable seedbed results and yield losses from erosion quickly become evident. On eroding valley sides in Cambridgeshire, where chalky boulder clay had been brought to the surface, winter wheat yields were only 59% of those on the deeper soils of the valley floor (Evans and Catt, 1987).

Table 3. Estimated percentage reduction in yield of cereals over 100 years at various rates of erosion.

Rates of erosion (t ha^{-1}yr^{-1})	Reduction in yield (%)			
	Spring Barley		Winter Wheat	
	Hempstead	Maxey	Hempstead	Mean
1.0	0.3	0.5	1.4	0.7
3.0	1.0	1.8	3.6	2.1
12.0	5.3	7.9	13.1	8.8

Initial soil depth at Hempstead = 0.9m
Initial soil depth at Maxey = 1.2m
After Evans, 1981

Table 4. Calculated life-span of soils.

Soil Type	Sandy loam			Clay loam		
Bulk density (g.cm^{-3})	1.4			1.3		
Minimum soil depth (m)	0.2			0.15		
Available soil depths (m)	0.50	0.30	0.25	0.30	0.25	0.20
Life span (yr) at mean annual erosion rates of (t ha^{-1})						
2	7800	1400	700	1950	1300	650
5	700	350	175	488	325	163
10	311	156	78	217	144	72
20	147	74	37	103	68	34
50	57	29	14	40	27	13
100	28	14	7	20	13	7

After Morgan (1987).

Off-farm costs associated with erosion incidents are far higher than on-farm costs (Crosson, 1984). In Britain many costs are unquantified. Most attention has focused on the flooding of properties by soil-laden water from farmers' fields, mostly in autumn and winter from cereal fields. Recently, there appears to have been a substantial increase in flooding, although the phenomenon is not new (e.g. at Shepton Beauchamp in Somerset, June 1966 and July 1968, in Morgan, 1980). On the South Downs prior to the 1980s there are few records of erosion and associated flooding - even in the wet autumn of 1976 (Boardman, 1990a). In the 1980s there have been many incidents. In 1982, Lewes District Council spent about £12,000 in emergency and protective measures at Highdown (Stammers and Boardman, 1984). In 1987, total off-farm costs attributable to flooding at four major sites around Brighton were in excess of £660,000 (Robinson and Blackman, 1990).

The impact of erosion on rivers in Britain is reviewed by Evans (1989). This includes the impact on fish stocks and the reasons for increased turbidity in many chalk streams of southern England. Discolouration of water by inwashed peat or soil has caused severe difficulties with water filtration at the Cray Reservoir in South Wales (Stretton, 1984), near Huddersfield, South Yorkshire (Austin and Brown, 1982), and in Nidderdale, North Yorkshire (Edwards, 1986). The annual loss of water storage capacity due to sediment deposition is reviewed by Butcher (1992) for six reservoirs in the southern Pennines. The loss of capacity per century ranges from 3 to 75% for individual reservoirs, varying with erosion rate and reservoir size.

5. CONCLUSION

The data from the SSEW/MAFF monitoring scheme have been related to soil texture, topography, climate and farming practices. The causes of erosion under current conditions are now reasonably well understood (Evans, 1988, 1990a). Unfortunately the scheme was abandoned prior to 1987 when severe erosion occurred in parts of southern England (e.g. Boardman, 1988). Boardman et al. (1990) show that considerable increases in erosion can be expected as a result of changes in future climate and land use.

By world standards rates of erosion in Britain are probably low, although little reliable data exists for most countries. In Britain, locally high rates of erosion on loamy, silty and sandy soils under intensive arable cultivation cause concern. Caesium-137 measurements suggest that certain fields have been subject to surprisingly high rates of erosion for the last thirty years. Most attention has been focused on the issue of flooding by runoff from fields of winter cereals and there have been local attempts to deal with the problem by engineering approaches and land use change (e.g. Robinson and Blackman, 1990; Boardman and Evans, 1991) but no coherent policy exists. Government response to the threat of soil erosion, flooding and associated pollution problems has been disappointing. Reform of the EC Common Agricultural Policy, with less incentive for farmers to grow arable crops or graze sheep on unsuitable land, is generally agreed to be necessary for environmental and economic reasons including the control of erosion.

REFERENCES

ADAS 1985. Soil erosion by wind. Ministry of Agriculture Fisheries and Food Leaflet 891.

ARDEN-CLARKE, C. and EVANS, R. In press. Soil erosion and conservation in the UK. In: Pimental, D. (ed.).World soil erosion and conservation. Cambridge University Press.

AUSTIN, R. and BROWN, D. 1982. Solids contamination resulting from drainage works in an upland catchment, and its removal by flotation. Journal of the Institution of Water Engineers and Scientists 36:281-288.

BOARDMAN, J. 1984. Erosion on the South Downs. Soil and Water 12:19-21.

BOARDMAN, J. 1986. The context of soil erosion. SEESOIL 3:2-13.

BOARDMAN, J. 1988. Severe erosion on agricultural land in East Sussex, UK, October 1987. Soil Technology 1:333-348.

BOARDMAN, J. 1990a. Soil erosion on the South Downs: a review. In: Boardman, J., Foster, I.D.L. and Dearing, J.A. (eds). Soil erosion on agricultural land. John Wiley, Chichester. pp.87-105.

BOARDMAN, J. 1990b. Soil erosion in Britain: costs, attitudes and policies. Social Audit Paper No. 1. Education Network for Environment and Development. University of Sussex, Brighton.

BOARDMAN, J. 1992. The sensitivity of Downland arable land to erosion by water. In: Thomas, D.S.G. and Allison, R.J. (eds). Environmental sensitivity. John Wiley, Chichester.

BOARDMAN, J. and EVANS, R. 1991. Flooding at Steepdown. Report for Adur District Council.

BOARDMAN, J., EVANS, R., FAVIS-MORTLOCK, D.T. and HARRIS, T.M. 1990. Climate change and soil erosion in England and Wales. Land Degradation and Rehabilitation 2:95-106.

BOARDMAN, J. and HAZELDEN, J. 1986. Examples of erosion on brickearth soils in east Kent. Soil Use and Management 2:105-108.

BULLOCK, P. 1987. Soil erosion in the UK - an appraisal. Journal Royal Agricultural Society of England 148:144-157.

BURT, T. and LABADZ, J. 1990. Blanket peat erosion in the Southern Pennines. Geography Review 3(4):31-35.

BUTCHER, D.P. 1992. The Southern Pennines: peat erosion and reservoir sedimentation. In: Boardman, J. (ed.). Post Congress Tour Guide. European Society for Soil Conservation, Silsoe.

COLBORNE, G.J.N. and STAINES, S.J. 1985. Soil erosion in south Somerset. Journal of Agricultural Science, Cambridge. 104:107-112.

CROSSON, P. 1984. New perspectives on soil conservation policy. Journal of Soil and Water Conservation 39:222-225.

EDWARDS, A.M.C. 1986. Land use and reservoir gathering grounds. In: Solbe, J.F. de L.G. (ed.). Effects of land use on fresh waters. Ellis Horwood, Chichester. p.534-537.

EVANS, R. 1971. The need for soil conservation. Area 3. pp.20-23.

EVANS, R. 1981. Assessments of soil erosion and peat wastage for parts of East Anglia, England. A field visit. In: Morgan, R.P.C. (ed.). Soil conservation: problems and prospects. John Wiley, Chichester. pp.521-530.

EVANS, R. 1988. Water erosion in England and Wales 1982-1984. Soil Survey and Land Research Centre, Silsoe.

EVANS, R. 1989. Soil erosion - the nature of the problem. Paper given to Symposium, Scottish Geographical Society.

EVANS, R. 1990a. Water erosion in British farmers' fields - some causes, impacts, predictions. Progress in Physical Geography 14:199-219.

EVANS, R. 1990b. Soils at risk of accelerated erosion in England and Wales. Soil Use and Management 6:125-131.

EVANS, R. 1990c. Erosion studies in the Dark Peak. Northern England Soils Discussion Group Proceedings 24:39-61.

EVANS, R. and CATT, J.A. 1987. Causes of crop patterns in eastern England. Journal of Soil Science 38:309-324.

EVANS, R. and COOK, S. 1986. Soil erosion in Britain. SEESOIL 3:28-58.

FROST, C.A. and SPEIRS, R.B. 1984. Water erosion of soils in south-east Scotland - a case study. Research and Development in Agriculture 1:145-152.

FULLEN, M.A. 1985. Compaction, hydrological processes and soil erosion on loamy sands in east Shropshire, England. Soil and Tillage Research 6:17-29.

FULLEN, M.A. and REED, A.H. 1986. Rainfall, runoff and erosion on bare arable soils in east Shropshire, England. Earth Surface Processes and Landforms 11:413-425.

HEPWORTH, P. 1987. Soil erosion by wind in the Vale of York. BA Dissertation, Brighton Polytechnic.

MACKNEY, D., HODGSON, J.M., HOLLIS, J.M. and STAINES, S.J. 1983. Legend for the 1:250,000 soil map of England and Wales. Soil Survey of England and Wales, Harpenden.

MORGAN, R.P.C. 1980. Soil erosion and conservation in Britain. Progress in Physical Geography 4:24-47.

MORGAN, R.P.C. 1985a. Assessment of soil erosion risk in England and Wales. Soil Use and Management 1:127-130.

MORGAN, R.P.C. 1985b. Soil erosion measurement and soil conservation research in cultivated areas of the UK. The Geographical Journal 151:11-20.

MORGAN, R.P.C. 1985c. Soil degradation and erosion as a result of agricultural practice. In: Richards, K.S., Arnett, R.R. and Ellis, S. (eds). Geomorphology and soils. George Allen & Unwin. pp.379-395.

MORGAN, R.P.C. 1987. Sensitivity of European soils to ultimate physical degradation. In: Barth, H. and L'Hermite, P. (eds). Scientific basis for soil protection in the European Community. Proceedings of Commission for the European Communities Symposium October 1986. Berlin. Elsevier, London.

QUINE, T.A. and WALLING, D.E. 1991. Rates of erosion on arable fields in Britain: quantitative data from Caesium-137 measurements. Soil Use and Management 7:169-176.

REED, A.H. 1979. Accelerated erosion of arable soils in the United Kingdom by rainfall and runoff. Outlook on Agriculture 10:41-48.

ROBINSON, D.A. and BLACKMAN, J.D. 1990. Some costs and consequences of soil erosion and flooding around Brighton and Hove, autumn 1987. In: Boardman, J., Foster, I.D.L. and Dearing, J.A. (eds). Soil erosion on agricultural land. John Wiley, Chichester. pp.369-382.

ROBINSON, D.A. and BOARDMAN, J. 1988. Cultivation practice, sowing season and soil erosion on the South Downs, England: a preliminary study. Journal of Agricultural Science, Cambridge. 110:169-177.

SPEIRS, R.B. and FROST, C.A. 1985. The increasing incidence of accelerated soil water erosion on arable land in the east of Scotland. Research and Development in Agriculture 2:161-167.

STAMMERS, R. and BOARDMAN, J. 1984. Soil erosion and flooding on downland areas. The Surveyor 164:8-11.

STRETTON, C. 1984. Water supply and forestry - a conflict of interests: Cray Reservoir, a case study. Journal of the Institution of Water Engineers and Scientists 38:232-330.

TRIMBLE, S.W. 1983. A sediment budget for Coon Creek Basin in the Driftless area, Wisconsin, 1853-1977. American Journal of Science 283:545-474.

WALLING, D.E. 1983. The sediment delivery problem. Journal of Hydrology 65:129-144.

WALLING, D.E. and QUINE, T.A. 1991. Use of ^{137}Cs measurements to investigate soil erosion on arable fields in the UK: potential applications and limitations. Journal of Soil Science 42:147-165.

WALLING, D.E. and WEBB, B.W. 1981. Water quality. In: Lewin, J. (ed.). British rivers. Allen & Unwin. pp.126-169.

WILSON, S.J. and COOKE, R.U. 1980. Wind erosion. In: Kirkby, M.J. and Morgan, R.P.C. (eds). Soil erosion. John Wiley, Chichester. pp.217-249.

Chapter 2

ASSESSMENT OF WATER EROSION IN FARMERS' FIELDS IN THE UK

R. Evans[1] and J. Boardman[2]
[1]Department of Geography, University of Cambridge,
Downing Place, Cambridge, CB2 3EN, UK
[2]School of Geography, University of Oxford,
Mansfield Road, Oxford, OX1 3TB, UK

Summary

An alternative to techniques based on the Universal Soil Loss Equation or similar plot-based models is needed in the assessment of erosion by water. There is a need to know the actual extent, frequency and rates of erosion. Measurements of sediment volumes eroded and deposited from rills can be easily and quickly made in the field, but are the results repeatable? Evidence from various British sources suggests they are, and that the results are sensible. Data from the National Erosion Monitoring Project are compared with those obtained by other workers. Comparisons are also made of estimates by different workers of amounts eroded at the same sites. A simple technique for assessing rill erosion, described here, has widespread applicability. Wash or splash erosion are not considered because sediment volumes eroded in Britain are largely derived from rills, so any underestimates of amounts transported will be slight.

1. INTRODUCTION

The Universal Soil Loss Equation (Wischmeier and Smith, 1978), or similar techniques (Elwell and Stocking, 1982), have been used widely to predict mean rates of erosion. They are derived from plot-based measurements. How these predictions relate to erosion in farmers' fields, which often have complex topography, has hardly been explored. Here, erosion is defined as the transfer of soil particles across a distance of more than a few metres. Soil may be redistributed within the field such that the topsoil which has been subject to erosion or deposition is markedly changed, either in texture or depth. There is a need to gather 'real' data on erosion extent, frequency and rate over time. Stocking (1987), commenting on the problems of measuring land degradation, made a plea for the development of simple field measurements of erosion.

Field measurements of cross-sectional areas of rills and gullies, channel lengths, areas covered by deposits and depths of deposition are made easily and quickly. Kaiser (1940-1978) made visual estimates of amounts eroded. Rill volumes were measured in eastern Europe (Holy, 1980; Zachar, 1982). Evans began measuring erosion in English arable fields in the 1970s (Evans and Morgan, 1974; Evans and Nortcliff, 1978); Boardman (1983, 1990), on the South Downs of Sussex, and Frost and Speirs (1984; Speirs and Frost, 1985), in southern and eastern Scotland, using similar methods. Reed (1979) and Evans (1980)

13

mapped eroded fields and related their distribution to soils, morphology and rainfall, and showed how the numbers of eroded fields varied from year to year (Evans and Cook, 1986).

From these *ad hoc* approaches the National Monitoring Scheme was developed, which was concerned with erosion extent, frequency and rates in 17 localities in England and Wales for a five-year period in the 1980s (Evans, 1988, 1990a). Eroded fields were identified on 1:10,000 scale aerial photographs taken in spring or early summer and checked in the field in August and September. Land at risk of erosion was also checked.

Although field measurements of rills are easily and quickly made, doubts exist about their reliability and repeatability. Sufficient data from a variety of sources has now accrued in Britain for some hypotheses to be tested. Thus, does the extent of erosion in areas with similar weather, soils, cropping and morphology vary in a similar way over time? Does the frequency of erosion vary in a similar manner? Do extent and frequency vary with soil type? Do the amounts measured in the field vary with soil type, and for similar conditions, do rates vary in a similar manner from year to year? And do measurements of eroded soil vary between workers?

2. COMPARISON OF DATA ON THE EXTENT OF EROSION

Data from the National Monitoring Scheme for nearby areas in Shropshire and Staffordshire (Figure 1) and Norfolk (Figure 2) showed that the extent of erosion did vary in a similar manner from year to year.

The aerial photograph transects in Shropshire and Staffordshire were only about 6km apart in places. Similar soils occurred in both localities, mainly the sandy Bridgnorth and Newport 1 associations and the coarse loamy Bromsgrove association (Evans, 1988; SSEW, 1983). Of the 197 eroded fields in Shropshire over a five year period, 48.7% were in the Bridgnorth association, 18.8% in the Newport 1, 6.6% in the Bromsgrove and 25.9% in other associations. In Staffordshire comparative figures for 205 fields were respectively 48.8%, 12.2%, 2.4% and 36.6%. Crops were also similar in the two transects. Of all eroded fields, 29.9% in Shropshire (28.3% in Staffordshire) were drilled to sugar beet, 21.3% (28.3%) to winter cereals, 18.3% (12.2%) to potatoes, 10.7% (19.5%) to spring cereals and 19.8% (11.7%) to other crops.

From 1967-1976, Reed (1979) mapped eroded fields in three parishes in Shropshire. Parts of these parishes were covered in the 1980s by the Monitoring Scheme transect. The area of land affected by erosion between 1967 and 1976 averaged about 2-3% per year, compared with 9% per year in the 1980s. Reed found that winter cereals were infrequently grown, whereas they were an important crop by the 1980s. In the Claverly catchment (which is very similar in area to the parish of Claverly, monitored by Reed), arable land increased at the expense of pasture between 1974 and 1988 from a ratio of 1.52 to 2.15 (Fullen and Mitchell, 1992). This may explain the apparent increase in area affected by erosion.

Figure 1. Variation in extent of erosion in Shropshire and Staffordshire.

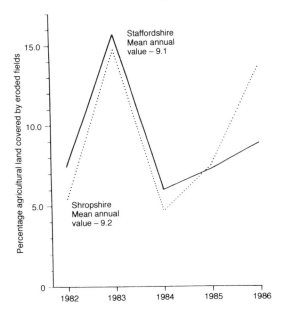

Figure 2. Variation in extent of erosion in Norfolk.

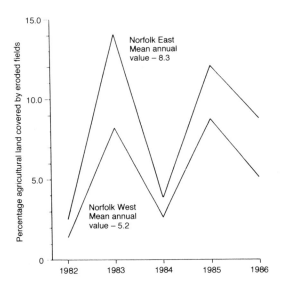

In Norfolk (Figure 2) the two transects were about 15km apart at their closest. On the eastern transect, 38.1% of all eroded fields occurred on sugar beet, compared with 42.7% on the western transect. Winter cereals accounted for 33.9% and 28.2% of all eroded fields, for the east and west transects respectively. Spring cereals accounted for 6.8% and 6.4%, and various other crops accounted for 21.2% and 22.7% respectively. Sandy soils occurred more widely in the east.

Evans (1988, unpublished) and Boardman (1990) have collected information on eroded fields on the South Downs near Lewes, Sussex. An aerial photograph transect comprised part (approx 50%) of the area monitored by Boardman. Erosion was predominantly in winter cereals on silty soils on chalk. Evans noted the number of fields which eroded; Boardman the number of eroded sites (catchments) within the fields. Rills did not occur throughout a field, but often in isolated sites such as in a minor valley floor. The ratios between eroded fields and sites for 1983-1986 were 2.43, 1.17, 2.50 and 4.08 respectively. Most fields/sites eroded in 1983, and fewest in 1984.

The results of the National Monitoring Scheme shows most erosion was on easily worked sandy soils, where a greater range of crops were grown, as in Shropshire. There was much less erosion on clayey soils, generally containing >35% clay (<2μm) content. For example, on average, only 2.0% of agricultural land on the chalky boulder clay of eastern England eroded over the 5 year period. Here, 64.6% of all eroded fields were in winter cereals, with a further 7.7% in autumn sown oil seed rape.

3. COMPARISON OF DATA ON THE FREQUENCY OF EROSION

Over the five year period the number of fields which eroded twice or more in Shropshire and Staffordshire was very similar (Table 1). In Nottinghamshire, eroded fields were mostly (89.6%) on the sandy Cuckney association (SSEW, 1983). Over the five years, erosion was in fields of sugar beet (35.5%), potatoes (21.3%), spring cereals (17.5%), winter cereals (14.7%) and other crops (10.9%). These proportions were different in Shropshire and Staffordshire. However, the frequencies at which fields eroded were similar.

Table 1. Frequency of erosion within five years in three localities with sandy soils.

	All fields	% eroded twice	% eroded 3 times	% eroded 4 times
Shropshire	145	28.3	9.0	1.4
Staffordshire	147	27.2	8.8	5.4
Nottinghamshire	172	29.6	7.0	--

Boardman mapped the distribution of eroded fields on the South Downs from 1983-1988 and noted how often they eroded (Boardman, 1990). The aerial photograph transect can be plotted on this map and the frequencies of eroded fields on the transect was similar to those on adjacent Downland (Table 2).

Table 2. Frequency of erosion on the South Downs, Sussex, 1982-1987.

	All fields	Frequency (%) eroded				
		Twice	3 times	4 times	5 times	6 times
Air photo transect	54	27.8	13.0	5.6	3.7	--
Adjacent downland	56	16.1	8.9	7.1	1.8	--

4. COMPARISON OF DATA ON AMOUNTS ERODED

Much of the data described here was collected for the National Monitoring Scheme. However, a number of projects were also set up to assess whether volumes measured in the field by different workers were comparable. Measurements of cross-sectional areas of rills were made at at least three points (top, middle and bottom) along the rills, or at mid-points of rills along a slope, or usually a combination of these. Depths of deposits were measured at at least three points, on a grid, or along length and breadth axes.

4.1 Comparison of data from the National Monitoring Scheme with that from other sources

Where landscapes and weather were similar, amounts eroded varied similarly from year to year. The mean amount eroded per hectare per year varied similarly in Shropshire and Staffordshire in three out of the five years (Figure 3).

In Norfolk, amounts eroded were generally greater on the eastern transect where more sandy soils occurred. Volumes eroded varied similarly on both transects from year to year, except in 1985, when on the western transect more fields in vegetables eroded in summer (Figure 4).

Figure 3. Variation in amounts eroded in Shropshire and Staffordshire.

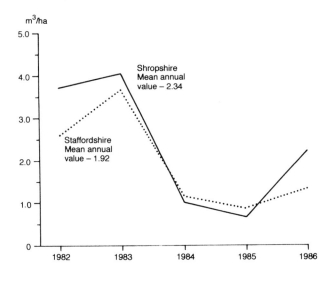

Figure 4. Variation in amounts eroded in Norfolk.

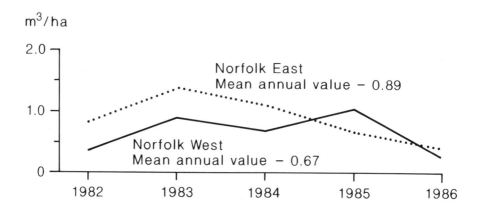

Boardman (1990) quoted median amounts eroded from contributing areas in fields on the South Downs. Evans' median rates (unpublished) are for fields. Ratios for the two sets of data are 3.9, 2.9, 6.1 and 2.0 respectively for the years 1983-1986. For 1988 on the South Downs, Boardman (1988) quoted median erosion rates of 3.3 m^3 ha^{-1} on a field basis and 5.5 m^3 ha^{-1} for areas contributing runoff to erosion, a ratio of 1.7.

Comparing data from the different localities in the National Monitoring Scheme shows mean annual amounts eroded per field over the 5 year period for different soil types as:
- coarse silty - 3.64 m^3 ha^{-1} (Somerset);
- sandy - 2.70 (maximum) - 1.48 (minimum) (Isle of Wight and Nottinghamshire);
- clayey - 0.38 (eastern England).

The values for other localities are mostly around 1.0 m^3 ha^{-1}.

Comparing data from the Monitoring Scheme with that derived from plots is fraught with difficulties. Often plots are of insufficient size for rills to form. For example, Morgan (1985) in the 1970s measured splash rates of about 0.5m^3 ha^{-1} yr^{-1} on clay soils. This compares with a mean for all rilled fields on the monitored transect 18km away of 0.38m^3 ha^{-1} yr^{-1}. Most erosion (76.9%) on the transect was in valley floors, 16.9% on slopes and valley floors, and only 6.2% just on slopes. Estimated rates for valley floors and slopes gave mean rates of 20.0-57.3m^3 ha^{-1} yr^{-1} over the period 1982-1984 and 0.32-1.81m^3 ha^{-1} yr^{-1} for contributing areas. For the four fields in which erosion was restricted to slopes, the mean rate was 0.17m^3 ha^{-1} yr^{-1}. The difficulties of comparing rates from fields and plots are related to scale factors and to the fact that plots may not assess erosion in that part of the landscape where it occurs most frequently.

Morgan's (1985) erosion rates measured on plots on sandy soils were similar to those noted by Fullen and Reed (1986). Rills occurred on all plots. Morgan's annual values over a seven year period ranged from 10.0 - 45.0t ha^{-1}. Those of Fullen and Reed ranged from

41.8-46.9t ha^{-1} between June 1982 and June 1984. Using a bulk density of 1.24g cm^{-3}, Fullen and Reed's rates convert to 29.9-37.8m^3 ha^{-1}. Their plots were always kept bare of vegetation so amounts eroded would be higher than those recorded on a cropped plot. The nearest equivalent to these plots in the Monitoring Scheme is given by the data for areas within a field which contribute runoff which caused rilling. The values for Shropshire were 10.9, 16.2 and 6.0m^3 ha^{-1} for 1982-84 respectively. The increase in erosion of sandy plots compared to clayey ones was greater than it was for fields on sand compared to those on clay, but the increases were of a similar order in size.

In Somerset from November 1982 to June 1983 data were gathered from eroded fields (Colborne and Staines, 1985) located within and adjacent to a monitored aerial photograph transect. The mean rate for 58 fields was 2.04m^3 ha^{-1} compared with 4.59m^3 ha^{-1} for 37 eroded fields on the transect.

Over the five year period of the Monitoring Scheme in Somerset, three people were involved at different times in collecting the data. The amounts eroded in this locality were nearly always the greatest, except one year (1985) when erosion was more severe in nine other localities, and in 1984 and 1986 when higher mean rates were recorded in Kent and Gwent respectively.

In those fields identified on Monitoring Scheme transects where deposits could be identified, comparisons were made between volumes eroded and deposited (Evans, 1990a). Amounts of soil deposited within fields related well to soil texture. Where soils were sandy, amounts deposited on average equalled amounts eroded. Where soils were of medium or clayey textures amounts deposited were <20% of those that eroded, that is, >80% of the soil eroded from rills had been transported out of the field. For light loams with high sand contents deposition accounted for 90% of the erosion, whereas for light loams with high silt contents deposits accounted for only 50% of the erosion.

4.2 Assessment of data on volumes eroded collected from other sources

Near Quiddenham, in Norfolk, eastern England, in March 1985, a large gully appeared on sandy soils drilled to a winter cereal. A complete suite of deposits, from gravels through to organic rich silts existed within the field. Thus, estimates of amounts eroded and deposited would allow a fair test of measuring erosion quickly and easily in the field.

Of the 320m^3 eroded, 304m^3 (95.0%) was found as deposits. The deposits comprised 95.6% sand and gravel, 4.2% silt and 0.2% organic-rich deposits. Similar sandy soils in the nearby Breckland (Corbett, 1973) contained 2-3% clay (<2μm), 2-3% fine silt (2-20μm) and 3% coarse silt (20-50μm) in the top metre. The topsoil contained 2-3% organic matter, with 1-2% in the lower horizons. Thus, in-situ soil contained 7-11% fine textured material (<50μm), of which about half was retained within the field.

In eastern Scotland in the winter of 1985/1986 there was widespread erosion in fields under winter cereals. Many fields were visited jointly. One worker made field measurements of erosion and another estimated amounts eroded in some of these fields from colour transparencies taken at ground level. The results compared well (Watson and Evans, 1991). The mean ratio between the two estimates of erosion for 11 fields was 1.02 (Table 3).

19

Table 3. Soil volumes (m ^{3}) eroded in winter 1985-1986, in 11 fields, as determined by measuring rills and gullies in autumn 1986 and subsequent visual estimates from ground level colour slides (Table from Watson and Evans 1991).

	Measurements	Colour slides	Ratio
	1.3	1.5	0.87
	4.0	6.0	0.67
	7.0	6.3	1.11
	9.0	9.1	0.99
	9.5	10.5	0.90
	19.1	18.0	1.06
	20.0	18.4	1.09
	22.3	10.5	2.12*
	77.0	90.0	0.86
	146.3	180.0	0.81
	187.2	144.0	1.30
Total	502.7	493.8	1.02(mean)

*This slide gave a foreshortened effect, so rill length was underestimated on the slide (35m vs. 65m measured). Spearman rank correlation coefficient, $r_s = 0.96$, $P<0.01$

Near Beaminster in Dorset in the winter of 1985/1986 there was severe erosion in a field under winter cereal. Two people measured rill cross-sectional areas along a transect midway along the most eroded part of the slope and assessed the areas and depth of deposits on slope concavities and in the valley floor (Table 4). Worker 2 estimated rill lengths twice that of Worker 1. Worker 1 considered that estimates made of rill cross-sectional areas along the transect applied only to the most eroded part of the slope, and above that rills were small or consisted of traces (Colborne and Staines, 1985). The area of the deposits was often difficult to assess because of the complex distribution of eroded material.

A bare soil field near Newmarket, Suffolk, was selected in autumn 1989 to test a rainfall simulator. Runoff generated by the simulator down tractor wheelings cut rills. Three teams of three mostly inexperienced people each assessed the amounts eroded and deposited (Table 5).

Table 4. Estimates of volumes eroded and deposited in a field near Beaminster, Dorset.

	Volumes (m ^{3}) moved	
	In rills	In deposits
Worker 1	96.57	276.72
Worker 2	212.85	106.00
Mean	154.71	191.36
Ratio	1.60-0.73	1.80-0.69

Table 5. Estimates of volumes eroded and deposited in two tractor wheelings in a field near Newmarket, Suffolk.

| | Amounts (m^3) moved | | | |
| | In rills | | In deposits | |
	Rill 1	Rill 2	Deposit 1	Deposit 2
Team 1	0.12	1.32	0.11	0.39
Team 2	0.06	0.56	0.04	0.58
Team 3	0.50	1.90	0.03	0.50
Mean	0.23	1.26	0.06	0.49
Ratio*	3.83-0.46	2.25-0.68	2.00-0.54	1.26-0.84

*Ratio of individual values to mean value

5. DISCUSSION

Where transects of the National Monitoring Scheme were close-by the variation in extent and frequency of erosion was similar over the five year period. So were the values of extent and frequency for similar soils. Silty soils eroded most, clays least, but sandy soils on which a wider range of crops were grown eroded most frequently. Loss of eroded sediments from fields also related well to soil texture, more soil being lost where soils were finer textured.

Problems arise when relating measured volumes of eroded soil to actual amounts. Complete suites of deposits equated well with volumes eroded. Comparisons between data from erosion plots and the Monitoring Scheme showed that relativities were of the correct order. Sandy soils eroded more than clayey ones, and values were of the same order of magnitude. Estimates made by different workers of amounts eroded were often similar in value, especially where workers were experienced in assessing erosion.

Marked differences in estimates of eroded volumes were often because estimates of rill lengths were dissimilar. Care must be taken when estimating rill lengths in the field. All the larger rills and gullies must be identified in a field, as they account for a disproportionate amount of eroded soil (Watson and Evans, 1991). At Beaminster in Dorset, 41.2% of all rills had cross-sectional areas $<4cm^2$ and accounted for only 4.8% of all the erosion. 69.6% of all rills had cross-sectional areas $<28cm^2$, and they accounted for 23.4% of all the soil eroded.

Rill erosion was assessed in all the projects noted above. It could be argued that splash and sheet erosion should also have been taken into account. Field observations and experimental data showed that in a temperate European climate splash and wash moved only a fraction of the amount of soil transported through rills (Evans, 1990a), probably $<0.3m^3$ ha^{-1}. Where rills do not occur, gross rates of wash and splash erosion, as measured by soil movement across a line, may be high. Net rates, however, will be low as it is only on upslope parts of convexities that soil transported downslope will exceed the incoming supply.

Unlike plot studies, measurements taken in the field show clearly (i) the importance of morphology in controlling the location of erosion; (ii) that rates of erosion measured at a site cannot be extrapolated to the landscape as a whole; and (iii) that less frequent events of large magnitude are important in governing mean rates of erosion.

Erosion was closely linked with the soils and land use of the locality. Thus, where soils were heavier textured and/or erosion was mostly in fields drilled to winter cereals, rills occurred predominantly in valley floors. On the Chalky Boulder Clay of eastern England, 76.9% of all erosion was in valley floors, in Gwent 70.3% and in Devon 57.1% over five years. On sandy soils rilling was confined only to valley floors in 25.9% of all instances of erosion in Shropshire, 25.4% in Staffordshire, 22.0% in the Isle of Wight and 14.8% in Nottinghamshire.

Large parts of the landscape were not susceptible to erosion because they were covered in crop or grass when erosive rains fell or because their slopes were too gentle or lacked a convexity (Evans, 1990a). Extrapolating rates of erosion from vulnerable sites to the wider landscape will overestimate erosion in the landscape by orders of magnitude (Table 6).

Table 6. Mean amounts eroded within fields, fields and transects in Shropshire in 1983 and 1984.

	Amounts eroded m^3 ha^{-1}	
	1983	1984
Area within the field directly affected by erosion	69.41	40.99
Contributing area within field	16.22	3.57
Field	4.06	1.41
Arable transect	0.63	0.05

Mean rates of erosion were governed by the low frequency high magnitude event (Evans, 1988, 1990a), and were often two to three times greater than median rates. Most fields eroded at rates $<1.0 m^3$ ha^{-1}. At these low rates rills did not impede the farmer's activities nor did deposits bury much crop. When amounts eroded were high the farmer became aware of erosion and wanted to combat it, but this happened infrequently. Less than 5% of all erosion events in a year were of such severity. So erosion is allowed to continue, but its long term impacts have not been negligible over large areas of England and Wales where up to 250mm of topsoil has been stripped away (Evans, 1990b), with a consequent loss of agricultural productivity.

6. CONCLUSIONS

The measurement of rill erosion in the field is a quick, easy and widely applicable technique. Its accuracy is generally within a range twice to half of the mean value estimated.

Many fields need to be monitored so that the results are representative. Monitoring only a small area may seriously over- or underestimate the impacts of erosion (Evans, 1992).

How results are reported is also important. Often they are calculated for areas affected by erosion or for the areas contributing runoff. Often large parts of fields and landscapes are not affected by erosion, so that extrapolation of such results will overestimate erosion. Also, different workers may interpret differently the areas contributing runoff to erosion from maps, whereas estimates of the areas confined within field boundaries are much less liable to vary.

Field measurement can only account for what is seen to have been transported or deposited. Rills, however small, are easily identified, as are their zones of deposition. However, estimates of amounts of soil moved only centimetres rather than metres by splash and wash are difficult to assess in the field. In terms of net loss of soil from a slope or valley floor, however, such processes in a temperate climate are probably negligible.

Emphasis should be given to measuring the distance particles are transported by splash, wash and rills in an erosion event.

REFERENCES

BOARDMAN, J. 1983. Soil erosion at Albourne, West Sussex, England. Applied Geography 3:317-329.

BOARDMAN, J. 1988. Severe erosion on agricultural land in East Sussex, UK, October 1987. Soil Technology 1:333-348.

BOARDMAN, J. 1990. Soil erosion on the South Downs: a review. In: Boardman, J., Foster, I.D.L, and Dearing, J.A. (eds). Soil erosion on agricultural land. John Wiley, Chichester. pp.87-105.

COLBORNE, G.J.N. and STAINES, S.J. 1985. Soil erosion in south Somerset. Journal of Agricultural Science 104:107-112.

CORBETT, W.M. 1973. Breckland Forest soils. Special Survey No. 7. Soil Survey, Harpenden.

ELWELL, H.A. and STOCKING, M.A. 1982. Developing a simple yet practical method of soil loss estimation. Tropical Agriculture 59:43-48.

EVANS, R. 1980. Characteristics of water-eroded fields in lowland England. In: De Boodt, M. and Gabriels, D. (eds). Assessment of erosion. John Wiley, Chichester. pp.77-87.

EVANS, R. 1988. Water erosion in England and Wales 1982-1984. Report for Soil Survey and Land Research Centre, Silsoe.

EVANS, R. 1990a. Water erosion in British farmers' fields. Progress in Physical Geography 14:199-219.

EVANS, R. 1990b. Soil erosion: its impact on the English and Welsh landscape since woodland clearance. In: Boardman, J., Foster, I.D.L. and Dearing, J.A. (eds). Soil erosion on agricultural land. John Wiley, Chichester. pp. 231-254.

EVANS. R. 1992. Erosion at Dalicott Farm, Shropshire - extent, frequency and rates. In: Boardman, J. (ed.). Post-Congress Tour Guide. First International ESSC Congress, European Society for Soil Conservation, Silsoe. pp.52-55.

EVANS, R. and COOK, S. 1986. Soil erosion in Britain. SEESOIL 328-59.

EVANS, R. and MORGAN, R.P.C. 1974. Water erosion of arable land. Area 6:221-225.

EVANS, R. and NORTCLIFF, S. 1978. Soil erosion in north Norfolk. Journal of Agricultural Science 90:185-192.

FROST, C.A. and SPIERS, R.B. 1984. Water erosion of soils in south-east Scotland - a case study. Research and Development in Agriculture 1:145-152.

FULLEN, M.A. and MITCHELL, D.J. 1992. Soil erosion studies in the West Midlands. In: Boardman, J. (ed.). Post-Congress Tour Guide, First International ESSC Congress, European Society for Soil Conservation, Silsoe. pp.32-47.

FULLEN, M.A. and REED, A.H. 1986. Rainfall, runoff and erosion on bare arable soils in east Shropshire, England. Earth Surface Processes and Landforms 11:413-425.

HOLY, M. 1980. Erosion and environment. Pergamon Press, Oxford.

KAISER, V. 1940-1978. Annual erosion survey of Whitman County. USDA-SCS, Spokane, WA 9921.

MORGAN, R.P.C. 1985. Soil erosion measurement and soil conservation research in cultivated areas of the UK. Geographical Journal 151:11-20.

REED, A.H. 1979. Accelerated erosion of arable soils in the United Kingdom by rainfall and runoff. Outlook on Agriculture 10:41-48.

SPEIRS, R.B. and FROST, C.A. 1985. The increasing incidence of accelerated soil water erosion on arable land in the east of Scotland. Research and Development in Agriculture 2:161-167.

SSEW 1983. Soil map of England and Wales. Soil Survey of England and Wales, Harpenden.

STOCKING, M. 1987. Measuring land degradation. In: Blaikie, P. and Brookfield, H. Land degradation and society. Methuen, London. pp.49-63.

WATSON, A. and EVANS, R. 1991. A comparison of estimates of soil erosion made in the field and from photographs. Soil & Tillage Research 19:17-27.

WISCHMEIER, W.H. and SMITH, D.D. 1978. Predicting rainfall erosion losses - a guide to conservation planning. Handbook No. 537. United States Department of Agriculture, Washington.

ZACHAR, D. 1982. Soil erosion. Elsevier Scientific Publishing Company, Amsterdam.

Chapter 3

SOIL EROSION MEASUREMENTS ON LOESS SOILS

A. Kerényi
Kossuth University, 4010 Debrecen, POB 9, Hungary

Summary

The two hectares of terraces in the area of Tokaj Big-Hill (north eastern Hungary) were selected for a detailed quantitative survey of small erosion features on loess soils. From the field data the total sediment loss from the loess erosion features was calculated as 476m[3], which represents a 24mm depth of loess assuming uniform removal of soil. Average annual soil loss amounts to 1mm. In the laboratory, measurements showed that solution erosion is four to five times more intense underground than on the surface. Further investigations were carried out to clarify the effects of various treatments on the dissolving power of water. The results proved that on soil covered loess, establishment of terraces requires special care to be taken of the soil water balance.

1. PREVIOUS RESEARCH WORK

The specialist literature emphasises the mechanical character of pipeflow (Hudson, 1981; Kügler et al., 1980). Morgan (1979) outlines the dissolution process during subsurface flow: 'Essential plant nutrients, particularly those added in fertilisers, can be removed by this process, thereby impoverishing the soil and reducing its resistance to erosion.' Roose (1970) published data about soil water flow, which only contributes about 1% of the total material eroded from a hillside. According to his experiments, Hahn (1977) says that the majority of depressional forms in loess areas cannot be explained by decalcification alone. He has established that the size of the features mentioned above exceeds the amount of material that could be removed by decalcification.

Jakucs (1977) stresses the polymineralic character of loess, and attributes an important role to the selective dissolution of the minerals that build up the loess.

2. RESEARCH TASKS

Experiments have been designed to identify which and how many ions are transferred into solution by water seeping through or running off the surface of the loess. Also, the question arises as to how does dissolution initiate conditions for mechanical pipeflow, and to what extent does loess degrade under rainfall of different drop energies? What was the rate of material loss over 25 years on loess-based terraces covering an area of two hectares?

3. THE RESEARCH AREA

The Tokaj Nagy-hegy (Big-Hill) is located in NE Hungary, at the southern end of the Tokaj mountains, slightly isolated from the main range which is of volcanic origin. This andesite cone of 516m altitude (Figure 1) is covered, except for the summit, by loess. In the Rákóczi valley terraces were built on the loess in 1959-60, which were then planted with vines, with the intention of applying mechanised cultivation. The terraces were horizontal or inward sloping, with gentle depressions on the bench itself (Figure 2a). Excess surface water convergence resulted in local flush-outs. Several years later ducts of pipeflow appeared, which grew so large over the years that the terraces were dissected, and mechanised cultivation became impossible (Figure 2b).

Figure 1. Location map of the area of investigation.

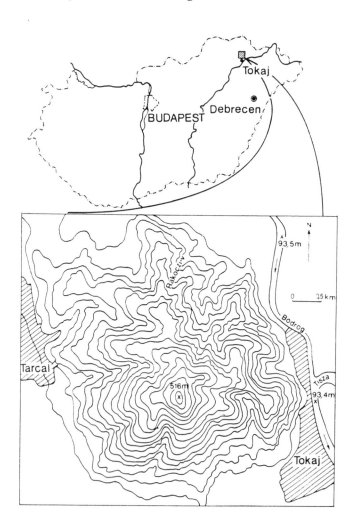

Figure 2. Bench terrace morphology.

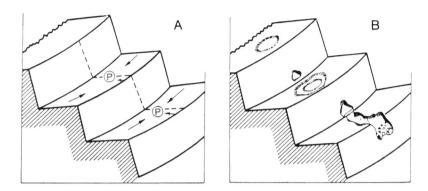

4. METHODS AND MATERIALS

In addition to field observations, the loess erosion features have been surveyed in the Rákóczi valley in the northern part of the hill. This survey allowed us to estimate the volume of loess lost during the development of the erosional forms. The amount lost was estimated by establishing which geometrical solids (such as a cylinder, cone or prism) represented the given morphological features. The dimensions of the geometrical solids were measured to the nearest centimetre, their volumes were calculated and these data added up. As the geometrical solids only approximated the shape of the erosional features, 55% of the calculated volumes were larger than the measured ones, and 45% were smaller. Thus, the differences cancelled each other out fairly well. Given the volumes of the individual features, further calculation showed that the data obtained for *total* loss of loess may deviate by ±5% from the real value.

Field measurements were supplemented by laboratory experiments. First, 56 × 36 × 10cm loess monoliths were used to study rainfall erosivity by rainfall simulation (Kerényi, 1984). Control measurements were carried out on monoliths of a brown forest soil (Table 1). Another series of experiments studied the effect of water solution on undisturbed samples of identical size. The ion contents of both surface runoff and water percolating through the monoliths and Ca^{2+}, Mg^{2+}, Na^+ and K^+ concentrations were determined.

The sampling places are denoted in Figure 3, and the data from the basic examination of the monoliths are summarised in Table 1. The grain size composition of the loess is typical of the material, however the lime content is lower than for loess generally. This fact suggests that the present material has been eluviated to various degrees. On the terraces, it was the artificially created excess water that caused eluviation (on the inverse sloping terraces), whereas the wall of the deep-cut track (Figure 3b) is continuously exposed to the immediate dissolving effect of precipitation. The differences in humus content are due to cultivation. On the terraces, cultivation requires the regular use of manure, which accounts for the prolonged increase in organic matter content. Even then it does not reach 1%.

Table 1. Data from basic examination of the loess monoliths L_1, L_2 and L_3 (Figure 3) and brown forest soil (BS).

Monolith	Mechanical composition							
	0.2-0.1	0.1-0.05	0.05 - 0.02	0.02 - 0.01	0.01 - 0.005	0.005 - 0.002	0.002 - 0.001	0.001>
L_1	0.4	13.7	48.6	15.0	4.7	4.0	2.3	11.3
L_2	3.8	12.0	47.8	14.1	5.6	4.3	4.6	7.8
L_3	2.8	8.4	48.9	18.6	6.2	3.2	3.9	8.0
BS	6.7	4.6	26.8	15.3	9.0	4.9	7.0	25.7

Monolith	Humus	pH prior to the experiment	pH at the end of the experiment	$CaCO_3$ (%)
L_1	0.95	7.83	7.61	4.5
L_2	0.55	8.12	8.12	2.5
L_3	0.38	7.55	7.41	1.1
BS	1.60	6.8	6.8	0.0

Figure 3. Location of sampled monoliths.

5. RESULTS

5.1 Field measurements

The first results of the erosion feature survey are contained in Table 2. Depressions are the most frequently occurring feature, and contribute a great deal to the amount of loess removed. They are oval landforms of great horizontal extension, being 8m wide and 10-20m long. They are shallow in depth, and result from the collapse of the loess structure. Loess wells are also common. They are mostly several metres deep and continue downslope in a pipe. They are of vertical cylindrical form on the terraces and semi-cylindrical form on the terrace margins, with pipes surfacing on the terrace riser.

Sinks are funnel-shaped features with convex walls. They extend in long, often cavernous, pipes. One of these pipes has also been investigated. The throughflowing rainwater formed a channel 20cm × 20cm in cross section at the bottom of the pipe. Due to repeated wetting on the bottom and walls of the small channel, the $CaCO_3$ content was heavily reduced (to 0.9 and 1.4%, respectively) while on the side walls and ceiling average $CaCO_3$ values of the Tokaj loess (5.4 and 5.0% respectively) were recorded.

A penetrometer was used to measure the mechanical resistance of the loess in the small channel and on the walls. It was found that with reduced $CaCO_3$ content, the mechanical resistance of the loess in the small channel is half that on the walls.

When pipe growth reaches a certain limit (which depends on the thickness of the loess layer overlying the pipe and the diameter of the pipe), the overlying loess mass cannot support its own weight and the pipe collapses. These collapsed pipes develop further into gullies, as surface runoff easily erodes loess fragmented after the collapse.

Table 2. Frequency of loess erosion forms and their contribution to the amount of loess removed (Kerényi and Kocsisné-Hodosi, 1990).

Form	Frequency (%)	Contribution to the amount of loess removed	
		m^{-3}	%
Depression	27.1	299.8	62.9
Loess well on terrace margin	25.0	22.2	4.7
Loess well on terrace flat	4.1	3.6	0.7
Swallow hole with pipe	20.8	58.0	12.2
Cauldron-like collapse	6.2	21.7	4.6
Collapsed pipe	16.8	71.1	14.9

Cauldron-like collapses are among the least common loess erosion features. They differ from depressions, with smaller, horizontal dimensions and greater depth, which makes them resemble cauldrons. The preconditions to their development are locally increased leaching of loess and, consequently, larger scale collapse of the structure itself.

From the data measured in the field the total material loss from the loess erosion features over the two hectare area has been calculated at 476m[3], which represents a 24mm depth of loess, assuming a uniform removal of material over the surface. Since it is known that the features have developed over a 25-year time span (since the terracing was carried out in the area), average annual material loss amounts to 1mm.

In the vineyards, however, it is not the average value that poses a problem, but the fact that the surface is covered with depressions and the risk of pipe collapse is common which makes viticulture difficult.

5.2 Laboratory investigation of loess erosion

Rainfall simulation was used to study the erosive impact of surface water on a loess and a brown forest soil (Table 3). The data reflect the following characteristics in the mechanical erosion of loess. Relatively few grains are dislocated by rainsplash. The material loss by splash under medium-intensity rainfall (505 $J.m^{-2}h^{-1}$) is less than half of that for the brown forest soil. This can be explained by the grain size composition and the relatively high water holding capacity of the loess. The infiltration of water into the porous loess means there is less splash-water than for the brown forest soil. With low raindrop energy (23 $J.m^{-2}h^{-1}$) the ratio is even higher for the brown forest soil (Table 3). In absolute terms, however, the figures for splash erosion are equally low for both soils.

Table 3. Splash erosion (Es), sediment concentration (qc) and sheet wash (Sw) during medium and low-intensity simulated rainfall (slope angle 10°, intensity 30mm.h^{-1}, duration 30 min).

Soils	Raindrop energy: 505 $J.m^{-2}h^{-1}$			Raindrop energy: 23 $J.m^{-2}h^{-1}$		
	Es(g)	qc(g dm^{-3})	Sw(g)	Es(g)	qc(g dm^{-3})	Sw(g)
Brown forest soil	77.91	13.26	30.02	2.95	4.01	6.25
Loess (L$_1$)	31.03	37.73	62.26	0.52	28.63	41.38

In contrast, under the higher raindrop energy, water flowing over the surface as a sheet removes twice as many loess particles than for the forest soils. Sheet erosion under low intensity rainfall on the loess is seven times that for the brown forest soil. The differences are even greater for sediment concentration: a unit volume of runoff contains many more loess particles than the runoff from the brown forest soil under identical conditions.

During the investigations the concentrations of four ions (Ca^{2+}, Mg^{2+}, Na^+ and K^+) were studied, in surface runoff and in the water infiltrating the different loess monoliths (Tables 4 and 5). After the first experiment the surface of the monolith was covered with a metal plate and left for five days. The monolith was enclosed on all sides except between the covering plate and the loess surface where there was a gap of 1cm. During the five days the CO_2 produced by the living micro-organisms accumulated in the monolith. The increased CO_2 concentration resulted in a nearly one and a half times increase in the solvent effect of the water (Table 5). Considering that the quantity of dissolved ions decreased after the first measurement in every case, the relative increase is even more significant.

The L$_1$ loess monolith was covered by a brown forest soil layer in the next stage of the experiment. The water seeping through this soil and on through the loess carried nearly five times as many Ca^{2+} ions into solution than in the experiments without soil cover. This markedly increased solvent effect is primarily related to the changes of three correlated factors: 1, Biological activity is higher in the brown forest soil, thus, there is more CO_2 in the soil air than in the loess; 2, The humic acid content is higher for the soil; and 3, As a consequence, the pH of the soil solution is more than 1 unit lower than that for the loess.

Table 4. Elements dissolved in water running off the surface of the loess monolith (mg/l). Intensity of simulated rain: 20mm.h⁻¹. T = 20°C.

Time passed since the beginning of runoff(min)	Treatment	Ca^{2+}	Mg^{2+}	K^+	Na^+
a. Loess monolith from the cultivated terrace (L_1)					
30		14.82	3.81	6.04	7.33
60		9.81	3.11	3.30	5.24
90		9.08	2.94	2.20	5.33
30	Superposition of soil	13.41	4.65	10.50	8.37
210		7.98	2.70	4.75	3.96
b. Loess monolith from the bank of the terrace (L_2)					
30		14.48	3.66	0.001	0.295
90		12.72	3.45	0.059	0.558
180		13.46	2.61	0.068	0.328
c. Monolith from the wall of the deep-cut track (L_3)					
30		56.29	9.42	6.50	27.03
90		9.55	1.31	1.09	5.33
180		11.41	2.27	1.36	5.81

As a further step in the experiments, the soil and the loess were mechanically compacted in a wet state by applying a pressure of 30kP dm⁻². After compaction, a slight increase in concentration was observed in the seepage water as compared to the former interference. This was because the duration of seepage was increased by the narrowing of the pores, and a longer period was available for dissolution.

In the concluding experiments the soil was removed, whereas the loess was left intact. The rate of dissolution initially decreased by nearly 30%, then, in the later experiments, to a lesser degree. However, it did not return to the original low value, which may be related to the prolonged contact (4 days) of the loess with the soil. The biological activity of the brown forest soil is stronger than that for the loess The soil having contact with the loess may have resulted in the enrichment of micro-organisms in the surface layer of the loess. In addition, the soil solution decreased the original pH of the loess, and provided better conditions for dissolution. The correctness of our assumption is supported by the fact that at the end of the series of experiments the pH of the loess was 7.61 compared with 7.83 initially. There was a clear decrease in the pH (0.14) of the monolith from the wall of the deep-cut track (Figure 3b, L_3). However, this decrease is related to the fact that a great deal of salt was transferred from the monolith to the solution, thus, eluviation was highest in this sample.

The concentration of magnesium increases in the surface runoff water when brown forest soil caps the loess, as does calcium ion concentration. Covering the loess for 5 days results in a small increase in concentration, whereas the superposition of soil onto the loess results in a nearly six-fold one. After compaction and the removal of the soil the changes measured followed the same trend as for calcium.

Table 5. Elements dissolved in the water seeping through the loess monoliths. T = 20°C.

Time passed since the beginning of the experiment	Treatment	Ca^{2+}	Mg^{2+}	K^+	Na^+
a. Loess monolith from the cultivated terrace (L_1)					
12 hours	Sprinkling for 90 min, then soaking	42.94	4.21	19.04	5.16
5 days	Coverage for 5 days	62.63	5.48	18.41	1.71
30 min	Superposition of soil	208.40	32.27	28.98	22.52
60 min		142.30	18.92	25.03	6.16
210 min	Sprinkling	88.94	8.90	20.50	2.07
2 days	Compaction inundation	104.10	10.52	33.28	2.00
4 days	Removal of soil	80.85	8.46	23.71	1.87
66 days	Slow drying for two months	75.61	6.54	24.61	5.42
68 days		63.84	4.84	22.68	2.98
b. Loess monolith from the bank of the terrace (L_2)					
3 days	Inundation	72.65	34.37	0.001	3.91
4 days	Inundation	69.77	29.70	0.001	3.45
5 days	Inundation	57.01	22.73	0.001	3.06
60 days	After 2 months of slow drying	48.54	21.74	0.110	3.55
61 days		39.00	16.94	0.390	2.94
c. Monolith from the wall of the deep-cut track (L_3)					
120 min	Sprinkling	3339	466.7	27.03	661.1
150 min	Sprinkling	2663	424.5	18.27	583.5
180 min	Sprinkling	2141	377.9	14.54	489.6

On the L_3 monolith the solvent effect of seepage water increases nearly identically as for the water running off the surface, as in the case of calcium.

The changes in the concentration of potassium differ from those of calcium and magnesium. After the superposition of the brown forest soil onto the loess, concentrations increased nearly fivefold in the runoff water, whereas that of calcium and magnesium hardly increased at all. In the seepage water the opposite was observed.

The transport of potassium into solution was least affected by the various interferences, and the amplitude of the changes was also the smallest. This is the only element where the increase in concentration was greatest as a result of soil compaction. In the experiments performed on the monoliths from three different places, the greatest differences were measured for this element (Tables 4 and 5).

The significant potassium concentration measured on the tilled loess and its 'irregular' alterations clearly show the anthropogenic origins of this potassium. Vines require large amounts of potassium, which is provided by artificial fertilisers usually in the form of KCl.

High potassium concentration was measured in the water seeping through the monolith from the deep-cut track. In this case the anthropogenic effect can be ruled out. The amount of potassium increases considerably in the wall of the deep-cut track due to the saturation and precipitation of soil solution at the edges. Large amounts of potassium may naturally get into the water running down the deep-cut track.

Plate 1. Scanning microscope image of loess sample, treated in part with carbonic acid.

Placing the soil onto the loess monolith exerted the greatest effect on the concentration of sodium dissolved in seepage water (13-fold increase) compared with the other elements. However, this effect lasted the shortest time. In the water seeping through the monolith from the deep-cut track, extremely high concentrations were measured.

Unlike the case of potassium, mechanical compaction proved to be ineffective (Table 5) with sodium. On the other hand, after an interval of 66 days - while the moisture content of the loess slowly decreased from the state of capillary saturation, the Na^+ concentration increased. After 68 days it greatly diminished.

The dissolution processes taking place in the loess loosen the structure of the material. A few drops of carbonic acid were applied to a sample of loess. In a few seconds one could see under the microscope that pores, a fraction of a millimetre large, developed in the loess, and the loess structure collapsed in places. When more carbonic acid was added, the pores widened. With a scanning microscope, pictures were taken of a sample of loess, of which only one part was treated with carbonic acid, while the other was left unchanged (Plate 1). In places only thin partitions were left between pores of millimetres in size. It is the dissolution processes that play a decisive role in subsurface flow, whereas in the ducts of several millimetres diameter the mechanical processes became predominant.

6. CONCLUSIONS

In loess areas it is not satisfactory to express soil loss in terms of mm year^{-1} or t ha^{-1}year^{-1}. However low the loss of material on the surface, degradational processes may take place under the surface, which may adversely influence plant cultivation later. In view of the processes, the water balance of the soil is the determining factor, which in turn is affected by precipitation, generation of runoff and evaporation. The CO_2 content of the air in the soil is also important.

Loss of soil due to subsurface forms of erosion can only be determined with great difficulty. The ducts of pipeflow may, after a time, attain the size at which they are unable to bear the weight of the overlying loess any longer and they collapse. In this way gullies, several metres in width and depth may develop, which hinder soil tillage. Loss of material can be easily estimated in this phase. However, the economic damage is much greater than one would imagine from the loss of soil per unit area. Vegetation is completely destroyed where gullies exist, and depressions hinder the movement of tractors in the field.

The intensity of loess degradation can be realistically judged if we determine the loss of soil on the surface, measure the $CaCO_3$ content of the loess, and the amount of Ca^{2+} and Mg^{2+} in the water seeping through the loess. Finally, loess structure is examined under the microscope. From the size and frequency of the pores one can make conclusions on the stability of the loess structure and the danger of its collapse. A further task of investigation may be the numerical expression of loess structural stability, on the basis of the number and size of ducts developed.

REFERENCES

HAHN, G.Y. 1977. Lithology and genesis of the loesses in Hungary: their geomorphological and chronological subdivision (in Hungarian), - Földrajzi Értesító 26:1-28.

HUDSON, N. 1981. Soil conservation. Batsford, London.

JAKUCS, L. 1977. Morphogenetics of Karst Regions. Variants of Karst evolution 263. Akadémiai Kiadó, Budapest.

KERÉNYI, A. 1984. Method of laboratory experiment for the investigation of soil erosion (in Hungarian), - Földrajzi Értesító 33:266-267.

KERÉNYI, A. and KOCSISNÉ HODOSI, E. 1990. Quantitative investigation of erosion forms and processes on loess in a vineyard (in Hungarian), - Földrajzi Értesító 39:29-54.

KÜGLER, H., SCHWAB, M. and BILLWITZ, K. 1980. Algemeine Geologie, Geomorphologie und Bodengeographie. 216 - VEG Hermann Haack Geographisch - Kartographische Anstalt, Gotha-Leipzig.

MORGAN, R.P.C. 1979. Soil erosion. Longman Group Limited, London.

ROOSE, E.J. 1979. Importance relative de l'érosion, du drainage oblique et vertical dans la pédogenèse actuelle d'un sol ferrallitique de moyenne Côte d'Ivoire. Cah. ORSTOM, sér. Pédol. 8:469-82

Chapter 4

SOIL DEGRADATION INDUCED BY OIL-SHALE MINING

L. Reintam and A. Leedu

Estonian University of Agriculture, Viljandi Road, Eerika,
EE2400 Tartu, Estonia

Summary

Underground oil-shale mining occupies an area of nearly 200 km ^{2} in north eastern Estonia (Toomik, 1991). Even in the early stages of mining, widespread subsidence is characteristic of these areas, which have expanded rapidly during the last decade. This expansion is connected with changes in technology and the fact that old excavations have become flooded. Subsidence results in the formation of artificial hollows and hillocks, several metres high, giving rise to abrupt changes in the relief of the area. Waterlogging in the depressions and droughtiness on the hillocks makes agrotechnology, agricultural management and land use difficult.

1. MATERIALS AND METHODS

To clarify the character, range and rate of soil deformation and crop yield reduction induced by the oil-shale mining, investigations were carried out in north eastern Estonia (59°17' - 59°22'N, 44°51' - 45°02'E) on fields adjacent to the villages of Vörnu, Sompa, Sininömme and Edise, in 1989-91. The Ordovician limestone bedrock lies at a depth of 60-110cm, although at Sompa it is nearer the surface. Calcic Luvisols on calcareous yellow-grey till prevail here, but Calcic Cambisols and Rendzic Leptosols can also be found.

Soil profiles in the waterlogged hollows and on the dry hillocks were studied and sampled. Their general description and identification were related to the concepts of FAO (1977; 1985). Analytical techniques were carried out in the laboratories of the Department of Soil Science and Agrochemistry of the Estonian University of Agriculture, using methods described in various handbooks (Klute, 1986; Sokolov, 1975). The first data obtained were presented at regional meetings at Jöhvi, north eastern Estonia (Reintam and Leedu, 1990) and in Tallinn (Reintam, 1991).

The moisture content by weight of each 10cm soil layer was determined twice a month from late April to late September, with four replications at Sininömme in 1989-1990. The averaged data were compared with the mean values of hydrological constants (permanent wilting point, discontinuous capillary moisture capacity and field moisture capacity) calculated from nomograms on the basis of bulk density and specific surface area of each horizon (Kitse, 1978). Thus the categories of available moisture for plants were found and chronoisopleats (isolines of different moisture categories) were designed. In 1991 the yields of fully matured winter rye and summer barley were measured in five replications of one square metre. The results were calculated at 14% moisture content, at the significance level of 95%.

2. RESULTS AND DISCUSSION

2.1 Surface deterioration

The deformation of the land surface and topsoil is widespread where continuous oil-shale mining by means of a combine has taken place. Subsided depressions can be several metres deep, and run parallel to unmined oil-shale pillars which support the excavations. Across these depressions and parallel to combine rows other subsidence has taken place in the form of undulating neorelief. These features distort the land surface and complicate land management activities. Cracks and cleavages more than 1m deep and up to 30cm wide were often found at the bottom and on the slopes of these hollows. These cracks often formed overnight. Although tillage has led to a slight levelling of relief over time, the more severe deterioration of relief seems to be permanent.

On the basis of subsidence at different times, due to the anthropogenous (agrotechnical) erosion, some changes in the structure of the topsoil may occur. Deepening of a humus horizon in the depressions is accompanied by a decrease in the depth of the topsoil. Both subsoil stones and bedrock slabs may be ploughed up, as well as there being an increase in stoniness on the remnant hillocks. It is possible that the bedrock limestone could be prone to cracking in its natural state, but the large, deep and fresh cleavages and Ordovician strata inclined towards the depressions suggest the effect of modern mining-induced fractures. Hence variable depths of soil occur, as the cracks and cleavages become filled with weathered material to various degrees.

The subsidence often correlates with soil type on different glacial and/or Holocene deposits. Furrow-like discontinuities in the bedrock at Sininõmme have been filled with sandy-silty Holocene material, underlain with calcareous till. As the limestone ceilings of the excavations are relatively thin, extensive subsidence has caused great deformation of relief.

The stagnation of surface leakage water and waterlogging in depressions caused by the subsidence are concurrent with water deficiency and aridification on the remnant hillocks. The process of modern gleyization is the direct result of this water stagnation. Numerous small ferric and manganese nodules, as well as fresh, friable microconcretions show an alteration of the reduction-oxidation processes, whereas greyish ferrous spots and efflorescences on aggregates and peds have formed in argillic horizons influenced by seasonal over-wetting.

Such gleyic features appear quickly, sometimes straight after the occurrence of subsidence in winter. Also the stagnation of leakage water has been followed by subsidence of the oil-shale face. Subsidence and gleyic degradation were found in the spring after autumnal cultivation and sowing of rye. In later years this was seen as the plants germinated after a normal sowing season. On these occasions it was possible to observe truncated machinery tracks on the edge of hillock, which then continued several metres below in the subsided hollow. Cracks and cleavages usually mark this boundary. Fresh hollows became overgrown with weeds, and stagnant water and ice damaged any crops there.

2.2 Moisture regime and crop yields

In 1989 the total supply of available moisture under clover-grass swards was the same for the hollows and hillocks (Table 1). However, the amount was doubled in mid-June for potatoes grown in a subsidence hollow. Even in July-August the moisture was 20-30% higher there than on the remnant hillock. In spite of a relatively dry season, the spring stagnation of leakage water in the hollow resulted in waterlogging there until the end of May and mid-June at 40cm and 20-30cm depth, respectively (Figure 1). At any time, for the remnant hillock, only 8-10mm of productive (available) water was measured at 30-50cm depth.

Table 1. Supplies of productive moisture in the half-metre Luvisol profile (mm).

Time of investigation	First area				Second area			
	Potato (1989)		Barley (1990)		Clover-grasses (1989)		Clover-grasses (1990)	
	Hollow	Hillock	Hollow	Hillock	Hollow	Hillock	Hollow	Hillock
Late Apr.	162	94	107	84	105	105	109	108
Early May	131	75	91	84	107	95	85	70
Late May	128	65	110	95	71	71	86	84
Early June	119	63	70	65	88	84	67	51
Late June	78	67	90	34	79	88	32	52
Early July	107	79	53	29	71	78	35	61
Late July	104	76	130	90	72	87	129	117
Early Aug.	112	89	90	88	114	114	91	103
Late Aug.	104	80	140	94	98	109	114	105
Early Sept.	110	92	122	94	125	118	128	115
Late Sept.	106	96	122	108	99	108	119	127

Seasonal differences in moisture content were greater in 1990; up to 2.5-3 times on hillocks and up to 4 times in hollows (Table 1). The variance of the replicates did not exceed 4-5%. As in the previous year, the stagnation of leakage water in spring and autumn under the clover-grass swards was characteristic (Figure 1). The soil dried through the profile and a summer decrease in moisture below the permanent wilting point had taken place close to the remnant hillocks, and in subsidence hollows. The mining-induced subsidence lead to highly variable moisture relationships due to changes both in circulation and accumulation of moisture in the deformed solum and/or bedrock stratum. The alternating waterlogging and drying in the hollows has been accompanied by the continuous drying out on the hillocks.

Unlike in 1989, leakage water was absent in 1990 under barley following potatoes. The differences in the soil moisture status between the hollows and hillocks are still evident. Such highly heterogeneous moisture relationships are typical in areas of mining subsidence in north eastern Estonia. This leads to a variation in crop yields of up to 2-108 times (Table 2). Variable crop conditions and the heterogeneity of their yields tend to be due to the stagnant water in the depression, but also to the deficiency in water and the stoniness of the soil on the hillocks. This is especially true for the barley at Vörnu. The yield differences between hillock and depression are statistically significant only in the case of rye at Vörnu and in a late variety of barley at Sininömme. In other cases, there is a clear tendency for lower yields of early barley varieties in the hollows, but due to high internal variances this statistical significance is limited. The yield of rye in the hollows comprised only 60% of that on the hillock, mainly due to the stagnant water, and the presence of ice and weeds. By 1989, the potatoes completely perished in the hollows at Sininömme. The average yield of barley was only 16% of the yield on the hillock, which was the highest of all the areas investigated. Only a quarter of the hollow areas were able to reproduce a yield equal to the seeds sown.

Table 2. Crop yields (g.m^{-2} at 14% moisture content).

Location and crop		Remnant hillock			Subsidence hollow		
		Yield	Yield variability	1,000 grain mass	Yield	Yield variability	1,000 grain mass
Vörnu:	rye	393 ± 77	316...470	28.8	325 ± 57	178...292	27.3
	barley	281 ± 128	153...409	34.1	272 ± 44	228...316	32.6
Edise:	barley	307 ± 61	246...368	37.1	254 ± 95	159...349	38.8
Sompa:	barley	409 ± 29	380...438	41.3	379 ± 141	238...520	39.6
Sininömme:	barley	420 ± 122	298...542	38.4	67 ± 62	5...129	27.9

2.3 Changes in soil characteristics

There were no essential differences in the quantity and quality of humus, acidity nor physico-chemical features for either the Luvisol or Cambisol profiles in the year when subsidence occurred, nor during the following year. There was only a slight tendency towards an increase in fulvicity and solubility of humus in the hollows. These changes seemed to increase as variance of humus characteristics increased (Table 3). A decrease in the activity of humic acids seems to be influenced by surface gleyization, although a slight increase in the rate of humification against a background of increasing solubility of total humus has also taken place. This is a sign of the early stages of hydromorphic humus accumulation, leading to the process of ferrolysis (Brinkman, 1979). An increase in fulvic complexes has been previously associated with the formation and migration of Ca-fulvates. The C:N ratio of the humus is quite acceptable, but its tendency to increase in hollows indicates a weakening of the nitrogenous bridges in the molecules of humus acids (Flaig, 1971) under the influence of seasonal stagnant water.

Such a decrease in the quality of humus results in a rapid transformation of the non-siliceous sequioxides relationship (Table 4) by way of ferrolytic activities (Brinkman, 1979). An accumulation of dithionite- and oxalate-extractable iron appears to be characteristic of seasonal over-wetting, probably associated with changes in a magnetic situation (Ghabru et al., 1990). Seasonal drying out of argillic horizons on the remnant hillocks tends to promote the formation of crystalline compounds of non-siliceous iron and a decrease in iron activity (Schwertmann, 1964). At the same time, the reoxydation of ferrous neoformations into ferric ones in the composition of nodules and concretions may also cause a decrease in iron activity, even in the horizons which are seasonally gleyed. Only a slight tendency towards the activization of amorphous aluminium can be observed where gleyization has taken place.

Figure 1. Soil moisture at Sininõmme, 1989 - 1990.

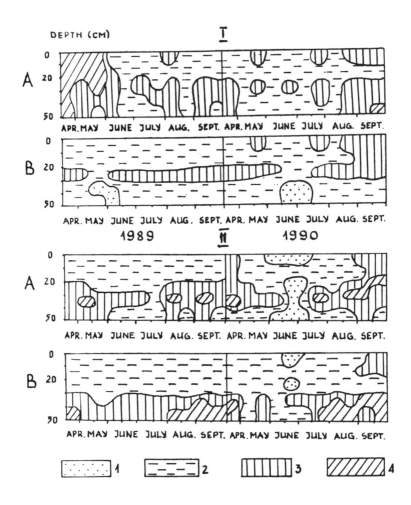

I - First area with potato (1989) and barley (1990) II - Second area with field grasses and red clover

A - Subsidence hollow   B - Remnant hillock.

Moisture status:

1. Below permanent wilting point (water unavailable to plants). **2**. Wilting point to discontinuous capillary moisture (water marginally available to plants). **3**. Discontinuous capillary moisture to field moisture (water available to plants). **4**. Over field moisture capacity (available to plants, but poor aeration is characteristic of these conditions).

There were no changes in acidity or physico-chemical features, whereas the neutrality, high BEC and base saturation had been maintained everywhere. It is possible that mobilization of aluminium is poor. A textural differentiation of soil profiles (Siuta, 1966) is characteristic of synchronous argillization in the calcic contacts with bedrock, lessivage and ferrolysis, whereas the processes above are progressing in mine-induced subsidence depressions (Table 4). This leads to an increase in specific surface area and to a progressive development of water stagnation.

Table 3. Some characteristics of humus relationships in the A-horizon (depth in cm) (Sininömme Experimental Area).

Characteristics	Remnant hillock		Subsidence hollow	
	A 0-15	A 15-30	A 0-15	A 15-30
Total C (%)	1.68-1.95	1.40-1.97	2.01-2.06	1.78-1.89
Total N (%)	0.18-0.21	0.17-0.20	0.19-0.20	0.17-0.18
C : N	9.3	8.2-9.9	10.3-10.6	9.9-11.1
Soluble humus (% of total C)	58-69	58-61	52-86	58-73
Degree of humification (% of total C)	16-20	18-24	18-21	20-24
Fulvic acids (% of total C)	33-47	32-35	28-58	37-42
Humic acids : Fulvic acids	0.3-0.6	0.5-0.8	0.3-0.7	0.5-0.6
Active humic acids (% of humic acids)	26-81	29-69	29-37	9-32
1st fraction : 2nd fraction	0.7-1.2	0.7-1.9	0.4-0.9	0.1-0.6

As the result of modern gleyization and ferrolysis in the depressions, the former Cambisols (A-Bm-D), Luvisols (A-El-Bmt-D) and even Rendzic Leptosols (A-D) have been transformed into the respective gleyic formations over the years. The presence of ferric- and ferrous-mottled ELg and Bmtg is typical of the gleyed profiles in these subsidence hollows.

3. CONCLUSION

Underground mining of oil-shale deposits in north eastern Estonia results in an enormous deformation of both the relief and the soil, due to rapid subsidence. The stagnation of leakage water in these depressions is accompanied by a drying out of the soil on the remnant hillocks. This results in highly unstable, spatially complicated moisture relationships. Mining-induced waterlogging and aridification alternate over short distances and are accompanied by complications in agrotechnology and variations in crop yields of up to 108 times.

Textural differentiation tends to allow the stagnant water to change the humus and sesquioxide relationships and this leads to modern gleyization in the soil profiles. An increase in humus´solubility, fulvicity and gleyic degradation and ferrolysis are the first indications of the mining-induced changes to soil genesis and productive capacity. Automorphic Luvisols, Cambisols and even Rendzic Leptosols on former till plains have been transformed into gleyic soils in the subsidence hollows. Soil heterogeneity has increased rapidly. Soil moisture status changed by at least 2-3 grades from moderately moist to droughty on the hillocks and to waterlogged in the hollows. It is important that these processes are taken into account when carrying out soil and land evaluation.

Table 4. Some soil characteristics in Luvisol horizons (on hillock and in neighbouring hollow).

Characteristics	Sininõmme - 1			Sininõmme - 2		
	A	EL/EL Bmg	Bmt/Bmtg	A	ELBm/EL	Bm/Btg
Dithionite-soluble Fe$_2$O$_3$ (%) by Coffin	0.7 / 1.3	0.7 / 1.1	1.8 / 0.9	1.6 / 1.9	1.2 / 1.7	1.4 / 3.0
Oxalate-soluble (%) by Tamm: Fe$_2$O$_3$	0.4 / 0.5	0.4 / 0.5	0.6 / 0.4	0.6 / 0.8	0.5 / 0.7	0.5 / 0.8
Al$_2$O$_3$	0.4 / 0.5	0.4 / 0.5	0.4 / 0.4	0.4 / 0.5	0.5 / 0.4	0.5 / 0.6
SiO$_2$	0.03 / 0.2	0.4 / 0.2	0.2 / 0.2	0.3 / 0.3	0.4 / 0.2	0.4 / 0.3
Fe-activity by Schwertmann (%)	60 / 42	60 / 47	32 / 43	42 / 60	43 / 39	35 / 25
Specific surface area (m^2.g^{-1}) by Puri-Murari	57 / 86	31 / 78	85 / 82	71 / 80	58 / 62	46 / 140
Clay, %	8.4 / 11.6	6.8 / 16.8	27.6 / 22.0	14.0 / 12.0	12.8 / 16.0	30.8 / 52.0
Sand. Coarse silt	3.6 / 1.8	3.9 / 2.2	1.7 / 1.9	1.9 / 1.7	2.0 / 1.8	1.4 / 0.5
Medium & fine silt:clay	3.3 / 2.7	3.5 / 2.1	1.8 / 2.1	2.9 / 3.2	2.7 / 2.3	1.8 / 1.4
Profile differentiation by Siuta	1.2 / 0.7	1.0 / 1.0	4.1 / 1.3	1.1 / 0.9	1.0 / 1.0	2.4 / 3.9

REFERENCES

BRINKMAN, R. 1979. Ferrolysis, a soil forming process in hydromorphic conditions. Wageningen.

FAO 1977. Guidelines for soil profile description. FAO, Rome.

FAO-UNESCO 1985. Soil map of the world. 1:500,000. Revised Legend. FAO, Rome.

FLAIG, W. 1971. Organic compounds in soil. Soil Science 111(1):19-33.

GHABRU, S.K., ARNAUD, R.J.St. and MERMUT, A.R. 1990. Association of DCB-extractable iron with minerals in coarse soil clays. Soil Science 149(2):112-120.

KITSE, E. 1978. Mullavesi, 142. Valgus, Tallinn.

KLUTE, A. (ed.). 1986. Methods of soil analysis. Part 1. Physical and mineralogical methods. 2nd ed. Madison, Wisconsin, USA.

REINTAM, L. 1991. Huumusseisundist kaevandamisega deformeerunud muldades. Inimene ja geograafiline keskkond (Man and geographical environment). Tallinn, 128-132. Institute of Geology of the Estonian Academy of Sciences.

REINTAM, L. and LEEDU, E. 1990. Muutustest kaevandamisega deformeerunud põllumuldades. Tootmine ja keskkond (Industry and environment). Tallinn - Kohtla-Järve, 61-64. Tallinn Botany Garden of the Estonian Academy of Sciences.

SCHWERTMANN, U. 1964. Differenzierung der Eisenoxide des Bodens durch Extraktion mit Ammoniumoxalat-Lösung. Zeitschrift für Pflanzenernährung, Düngung und Bodenkunde 105(3):194-201.

SIUTA, J. 1966. The relationship between the differentiation of mechanical composition, type and agricultural properties of soil. Part 1. Soil type and texture. Pamietnik Pulawski. PRACE IUNG 22:271-313.

SOKOLOV, A.V. 1975. (ed.). Agrochemical methods for soil investigations (In Russian), Nauka, Moscow.

ГOOMIK, A. 1991. Land deterioration due to underground mining and the ways to reduce it. Problems of contemporary ecology. Ecology and Energetics. Tartu, 175-176. Institute of Ecology and Marine Research of the Estonian Academy of Sciences.

Chapter 5

THE RESPONSE OF ABANDONED TERRACES TO SIMULATED RAIN

A. Cerda-Bolinches

Department of Geography, University of Valencia
22060, 46080 Valencia, Spain

Summary

Eight experiments with simulated rainfall were carried out on two groups of abandoned fields in the south east of Spain. The silty loam soils from the dry area of Petrer (north of Alicante Province) show a lower infiltration capacity (30%) and a higher erosion value than the sandy loam soils from the wet area of Genovés (south of Valencia Province). The soil cover and soil organic matter content have a positive relationship with the soil hydrology and erodibility.

Suggestions of soil conservation in both landscapes are made.

1. INTRODUCTION

Abandoned fields constitute a very important part of the landscape in the Spanish mountains (Lasanta, 1989; Rodriguez-Aizpeolea, 1990). The economic changes in Spain since the 1950s caused a depopulation of mountain areas and a reduction in the area of cultivated fields. The area of abandoned agricultural fields is around 20%, according to the agricultural census (official dates), but some researchers have demonstrated that this is an underestimate. In the Pyrenees, Lasanta (1988) showed that the abandoned area was 63% in 1957 and 97% in the 1980s. Near the area of this study, Rodriguez-Aizpeolea (1992) found that the abandoned area in the Vall d'Ebo (north of Alicante) was 58% at the end of the 1980s.

The abandonment of agricultural fields has important implications for the ecosystem and geomorphological processes (Garcia-Ruiz and Lasanta, 1990). From a geomorphological and biogeographical point of view, heterogeneity increases after abandonment (Ruiz-Flaño et al., 1991). When human activity is low, the regeneration is fast (Godron et al., 1981). Some surfaces are quickly covered by dense vegetation. The soil is colonised very quickly by herbs and later shrubs (Francis and Thornes, 1990; Garcia-Ruiz, 1991). In areas with matorral cover, the sediment yield is very low and reafforestation is not necessary (Francis and Thornes, 1990; Ruiz-Flaño et al., 1991). Moreover, the infiltration capacity of soils increases on sites which have been left fallow for a long time (Francis, 1986a).

However, in other cases, the surfaces have a slow rate of revegetation and are severely eroded by wash, rills and sometimes gullies (Lopez and Gutierrez, 1982; Romero et al., 1988). In some places terrace walls are destroyed by several processes and large gullies remove the sediment that has been stored on hillslopes for many centuries (Lopez and Torcal, 1986).

The regeneration of vegetation increases the stability of soil (Thornes, 1985; Ruiz-Flaño, 1992), but during the first years after abandonment there may be some problems related to the slow generation of soil cover (Rodriguez et al., in press). The process of vegetation regeneration is neither linear nor uniform, and is faster immediately after abandonment (Francis, 1986b).

The reduction in human influence can favour the degeneration of terraces and other retaining walls so that soil losses increase. New human activities, especially grazing and forest fires (Garcia-Ruiz, 1991) do not improve this situation (Garcia-Ruiz et al., 1988).

The evolution of soil surfaces after land abandonment is controlled by climate, lithology and soil characteristics. The distribution and amount of rainfall are important factors affecting revegetation under Mediterranean conditions (Mooney and Kummerow, 1981), together with human activities.

One objective of the paper is to present information on the hydrologic and erosive response of abandoned fields, including abandoned terraces, in south east Spain. Two locations were selected: one on a limestone area, with high precipitation, and another on an area of marls under very dry climatic conditions. Eight experiments were used to define both environments. This methodology has been used by some researchers (Scoging and Thornes, 1980; Bryan, 1981; Yair and Lavee, 1985; Imeson and Verstraten, 1986; Imeson et al., 1992).

2. THE STUDY AREAS

2.1 The landscape

In south east Spain different kinds of abandoned terraces can be found. The geomorphology, topography and lithology have a great influence on field morphology. In both study areas terraces were built in the valley bottoms. The fields are flat, the last crop was olives and the fields were abandoned in 1973 (Genovés) and in 1977 (Petrer).

In the Genovés area, on homogeneous Betic limestone (IGME, 1981), the terraces are situated in a small watershed oriented W-E. The climate is characterised by a dry summer and a wet autumn and spring. The annual rainfall at the nearest meteorological station (Xàtiva) is 688mm. October is the wettest month, usually with 20% of total annual rainfall.

The slopes are characterised by a vegetation cover of about 90% on north facing slopes and 60% on south facing ones. The dominant plants are *Thymus vulgaris, Erica multiflora, Chamaerops humilis, Rosmarinus officinalis, Quercus coccifera, Pinus halepensis, Globularia alypum* and *Sedum sediforme*.

In the abandoned fields, the vegetation is characterised by *Cistus albidus, Ulex parviflorus, Anthyllis cystisoides, Cistus monpeliensis* and *Olea europaea*.

The Petrer area is situated in the south of the Prebetic Mountains. The watershed is oriented E-W over marls (IGME, 1978) and the landscape is dominated by badlands (Calvo and Harvey, 1989). Slope morphology is characterised by rilling and crusting. Soil erodibility is very high (Harvey and Calvo, 1991).

South facing slopes are bare and the only plant is *Moricandia arvensis*. On north facing slopes the vegetation is dominated by *Brachipodium retusum, Salsola ginestoides, Centaurea scaliosa, Dorycnium pentaphilum, Bupleurum subfructicosum* and *Polygala calcarea*.

In the abandoned fields, the vegetation is richer with plants like *Phragmites australis, Daphne gnidium, Orobanche, Inula viscosa, Dactiles glomerata, Foeniculum vulgare, Cistus albidus, Hyparrihenia hista*, and *Coris ginestoides*.

The climate is characterised by a low annual precipitation (290mm), and a very irregular annual and seasonal distribution. Autumn and spring are the wetter seasons and the long dry summer has a duration of four or five months.

In both areas the terraces were built perpendicular to the thalweg. At Petrer, especially, the sedimentation of material eroded from the slopes has led to an increase in the elevation of the original field surface. In Genovés some drainage channels have been found along the terraces. These lateral drains reduce the runoff from upslope of the fields. The walls give protection from runoff.

At Petrer the walls are simple (30cm wide), with field surfaces higher than the terraces (Fig. 2). Today, the terraces are dissected by large gullies and in the future more walls will collapse, increasing soil losses. At Genovés the walls are double, with two faces (more than one metre wide) which are higher than the field surface. At present, the field infrastructure, such as lateral drains or walls, is not being destroyed and appears to be stable.

2.2 Soils and surfaces

In Table 1, some important characteristics of the soil cover (lichens, moss, plants and rock fragments) are shown. The sample plots are representative of the general landscape of the area.

The marls area is characterised by a low surface cover. Lichens, moss and stones are few and plants are only important in M3 and M4. At Genovés, only L1 has a low cover, and is representative of trampled areas such as paths. The other three plots (L2, L3 and L4) have a higher surface cover.

The soil characteristics of the two areas (Table 2) are very different. At Petrer, the soils are characterised by a high calcium carbonate content (75.25%) (using Bernard calcimetry) and a low organic matter content (0.54%) (Walkey-Black method). The texture is silty loam, but the percentage of sand and silt is very different for the bare (21.65 and 23.65% of sand in M1 and M2) and vegetated soils (48.28 and 51.32% in M3 and M4), because the sedimentation of material eroded from the badlands slopes is determined by the vegetation patterns.

Figure 1. Wall and field morphology in Genovés and Petrer (Cross sections)

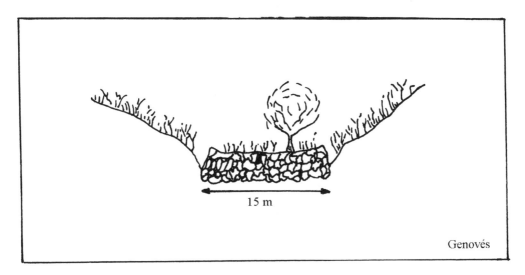

15 m

Genovés

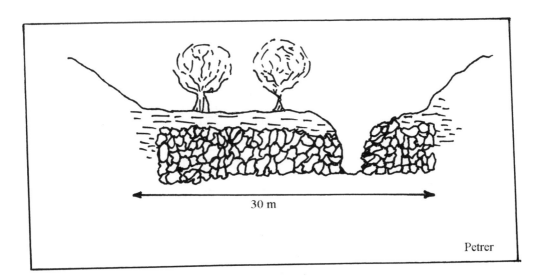

30 m

Petrer

Table 1. Characteristics of soil cover (%).

PETRER-MARLS	M1	M2	M3	M4	AVERAGE
Lichens (%)	0.00	0.00	4.00	0.00	1.00
Moss (%)	0.00	0.00	5.00	20.00	6.25
Plants (%)	3.00	10.00	85.00	75.00	43.25
Rock fragment (%)	2.00	2.00	3.00	8.00	3.75
Total cover (%)	5.00	12.00	97.00	103.00	54.25

GENOVES-LIMESTONES	L1	L2	L3	L4	AVERAGE
Lichens (%)	15.00	60.00	10.00	0.00	21.25
Moss (%)	0.00	5.00	70.00	90.00	41.25
Plants (%)	18.00	55.00	98.00	100.00	67.75
Rock fragment (%)	12.00	13.00	10.00	10.00	11.25
Total cover (%)	45.00	133.00	188.00	200.00	141.50

Table 2. Soil characteristics.

PETRER-MARLS	M1	M2	M3	M4	AVERAGE
Organic matter (%)	0.21	0.34	0.58	1.02	0.54
Texture (%)					
Sand	21.65	23.65	48.28	51.32	36.23
Silt	59.48	57.43	36.45	35.46	47.21
Clay	18.87	18.92	15.27	13.22	16.57
Calcium carbonate (%)	75.43	74.00	77.70	73.98	75.28

GENOVES-LIMESTONES	L1	L2	L3	L4	AVERAGE
Organic matter (%)	1.23	2.89	4.97	5.43	3.63
Texture (%)					
Sand	61.23	59.53	50.10	50.53	55.35
Silt	15.25	12.65	17.94	15.09	15.24
Clay	23.52	27.82	31.96	34.38	29.42
Calcium carbonate (%)	7.32	5.43	4.09	3.98	5.21

At Genovés soils have a low calcium carbonate content (5.21), and the organic matter ranges from 1.23 to 5.43% (3.62% on average). The soil texture is clay loam, and the variability between plots is smaller than in the marls area. At Petrer, the soil colour is 2.5 YR 8/2 (white) (Munsell chart) on the surface and at depth (4-6cm). At Genovés the soil colour is 7.5 YR 4/4 (brown/dark brown) on the surface and 7.5 YR 5/4 (strong brown) at depth.

Soil samples were taken before the experiments at the surface (0-2cm) and at 4-6cm depth in order to determine the antecedent soil moisture content. As the experiments were carried out in summer (1990), the volume of water in the soil was very low (approximately 1%). The soil water content was measured gravimetrically.

The antecedent soil moisture content was very low in all of the plots (Table 3). At the surface, it was around 1% in both cases and was 3.12% in Petrer and 2.15% in Genovés at depth. There are no significant differences between areas and among plots because the experiments were carried out at the end of July, when all the soils were very dry.

Table 3. Water soil content (%).

PETRER-MARLS	M1	M2	M3	M4	AVERAGE
(0-2)	1.01	1.16	1.33	1.23	1.18
(4-6)	2.32	2.12	4.83	3.21	3.12
Average	1.67	1.64	3.08	2.22	2.15

GENOVES-LIMESTONES	L1	L2	L3	L4	AVERAGE
(0-2)	0.89	0.99	0.62	1.30	0.95
(4-6)	2.27	2.45	2.34	1.54	2.15
Average	1.58	1.72	1.48	1.42	1.55

3. METHODS

Eight experiments were carried out using a sprinkler rainfall simulator (Calvo et al., 1988). The simulator operates on plots of $0.25m^{-2}$ with a rainfall intensity of $55mm.h^{-1}$. The duration of the experiments was 30 minutes and deionised water was used. The discharge of plots was measured at 0.3-1 minute intervals and sediment samples were collected three, four or five times during the experiment. At least one sample was taken at the beginning, end, and in the middle of runoff generation from the plots.

The rainfall simulator is similar to others used in semi-arid environments (Scoging and Thornes, 1979; Imeson, 1977; Wilcox et al., 1986, Wood, 1987, Alexander and Calvo, 1990; Sala and Calvo, 1990).

4. RESULTS AND DISCUSSION

4.1 Hydrological Response

The hydrological response (Table 4) of the Petrer plots is characterised by fast ponding and rapid runoff generation. For the Genovés plot, ponding and surface runoff start later, but in both cases ponding is positively related to total cover. Plant interception and soil organic matter content seem to determine the soil saturation.

Surface cracks closed first in the plots where they were smallest (L1, L2, M3 and M4 have cracks with less than 1mm). In M1 and M2 the cracks were wider than 5mm and had a polygonal structure. The differences between the Petrer plots is that on bare soils (M1 and

M2) the cracks closed after runoff starts, after 6.42 and 7.33mm of rain. On the vegetated plots (M3 and M4), the cracks close after only 3.66 and 3.03mm of rainfall. This occurred before runoff started.

Table 4. Soil surface changes and hydrological response (nr) no runoff, (nc) no cracks.

PETRER-MARLS	M1	M2	M3	M4	AVERAGE
Surface changes (min)					
Ponding	1.00	1.10	2.10	3.32	1.88
Surface runoff	1.30	2.00	8.00	6.20	4.38
Spout	2.05	2.30	12.00	7.00	5.84
Crack closed	7.00	8.00	4.00	3.30	5.58
End runoff	2.21	3.00	2.30	1.56	2.27
Runoff (mm/h)					
Average	39.64	36.51	14.52	11.00	25.42
Coefficient	0.70	0.61	0.16	0.15	0.40
Infiltration (mm/h)	5.24	10.82	35.80	37.96	22.46

GENOVES-LIMESTONES	L1	L2	L3	L4	AVERAGE
Surface changes (min)					
Ponding	0.50	3.10	4.50	nr	2.70
Surface runoff	1.10	4.20	14.00	nr	6.43
Spout	1.30	5.10	17.00	nr	7.80
Crack closed	5.00	4.67	nc	nr	4.84
End runoff	1.30	1.00	1.03	nr	1.11
Runoff (mm/h)					
Average	19.70	7.30	3.23	0.00	7.56
Coefficient	0.34	0.11	0.03	0.00	0.12
Infiltration (mm/h)	27.07	44.31	50.77	>55	31.04

The cracks in Genovés are smaller (1mm) and closed after only five minutes of rainfall. In L1 the cracks were closed after the runoff started but this occurred before runoff began on L2. In any case, the time taken for the cracks to close modified the hydrological response (Table 4).

The time between the cessation of rainfall and the end of the runoff is twice as long at Petrer (1.11 minutes) than it is at Genovés (2.27 minutes). On the terraces the volume of water ponding on the plot surface has a positive control over this parameter.

The volume of runoff is very high in Petrer, especially for bare surfaces with less than 10% of plant cover. On the bare marl plots the average runoff ranges from 39.64 to 36.51mm h^{-1}, and the infiltration capacity is very low (5.24 and 10.82mm h^{-1}). On the other hand, on the bare limestone plots, the infiltration capacity is more than three times higher than in Petrer, and the average amount of runoff is only half.

The vegetated surfaces at Petrer had an infiltration capacity of approximately 37mm.h^{-1}, more than 12mm.h^{-1} lower than that at Genovés. In the limestone area the average runoff is about one third of that in the marl area.

The runoff coefficient of plots show that the bare surfaces in Petrer produce a lot of runoff (0.70 and 0.61), followed by the trampled bare surface in Genovés (0.34). In Petrer, the vegetated plots had low runoff coefficients (0.15 and 0.16), and this is also true for the poorly vegetated surfaces of Genovés (0.11). The lowest runoff coefficients were from the densely vegetated surfaces of Genovés (0.03 and 0).

If the marl and limestone areas are compared, there are some clear differences: the infiltration capacity is lower (30%) on the marls, and the average runoff and runoff coefficient are more than three times higher in Petrer than in Genovés (Table 4).

4.2 Erosional Response

Erosion rates indicate more clearly the differences between the Genovés and Petrer soils. Among the eight plots, two had a high sediment concentration (7.27gr.l^{-1} on average), three had moderate ones (2.2gr.l^{-1}), and the rest were low (0.57gr.l^{-1}). On the marls area the sediment concentration of the runoff is 6.7 times higher than in the limestone area. Every soil in Genovés has a low sediment concentration except for the areas trampled by people and animals.

In every experiment the sediment concentrations of the runoff decrease with time. This is clearer in the degraded soils, especially on the marls. Some authors find an increase in the sediment concentration over time on badlands slopes (Gerits, 1991).

The erosion rates are very high for bare soils of Petrer (314.32 and 211.97gr.m^{-2}·h^{-1}), low for the vegetated soils of Petrer and the trampled areas of Genovés (22.83gr.m^{-2}·h^{-1} on average), and very low for the vegetated soils on limestones (2.3gr.m^{-2}·h^{-1}) in Genovés.

The interrill erosion is nearly 20 times higher for the terraces on the marls than for those on limestones.

4.3 Terrace Morphologies

The hydrological and erosional responses of the surfaces affect the evolution of the terraces. The areas with poor soils, low infiltration capacities and high erosion rates develop flat surfaces (terraces) dissected by large gullies in less than 15 years. At Genovés the low erosion rates and runoff coefficients explain the longevity of the walls and channels after nearly twenty years of abandonment.

The implications of these results suggest that improving the terrace infrastructure reduces erosion. However it is much more important to improve the infiltrability of soils to reduce the soil losses. In areas with low annual rainfall on marl soils we advise vegetation regeneration or cultivation of abandoned fields.

On the other hand, for the terraces on limestone, with high precipitation, it is not necessary to encourage vegetation regeneration, because the matorral colonize the abandoned soils and reduce the sediment and water losses. Possibly, the human impacts like reafforestation or new cultivation produce an increase in erosion.

5. CONCLUSIONS

The evolution of abandoned fields in south east Spain shows differences depending on the parent material of the terraces (marls or limestones). The regeneration of vegetation is faster and the total cover is higher in the limestone area, while in the marl area the total cover is less than 50%. The different evolution of vegetation explains why the soil organic matter content is almost seven times higher in Genovés.

These differences, together with the characteristics of the lithology (more calcium carbonate and silt in Petrer) explain the contrast between the hydrological and erosional response of the two sites.

In the marls area (Petrer), the infiltration capacity is 30% lower than in the limestone area (Genovés) and the average runoff and runoff coefficients are more than three times greater in Petrer.

The erosional response is similar to the hydrological one. The sediment concentration of runoff is almost seven times higher in Petrer than in Genovés, and, as a consequence, the erosion is twenty times higher. In every case, the sediment concentration decreases from the first to the last sample, and these changes are greater in Petrer than in Genovés.

Plant cover, and especially the organic matter content of soils have a positive relationship with the infiltration capacity. On the other hand, erosion and silt content (in Petrer) have a positive relationship with the runoff coefficient.

In general, the total cover of soil controls the hydrology, and even more clearly the erosional response. The terraces with fast regeneration of vegetation and a high plant cover (Genovés) have low water and sediment losses. When the vegetation is sparse (Petrer) the erosion and runoff is high and the morphology is dominated by gullies.

There is a clear relationship between vegetation regeneration and soil infiltration capacity, which can explain the different evolution of both kinds of terraces. The results of eight experiments with simulated rain are related to the morphology and the geomorphological processes dominant on the current terraces.

ACKNOWLEDGEMENTS

I would like to thank Mario Payá and his family, Jorge Payá, Enrique Terol and Joanma Calatayud for their help in field work. I am also grateful to Adolfo Calvo and the anonymous referees whose corrections and suggestions have greatly improved this paper. The CICYT-PNICYDT, project NAT89-1072-C06-04, has sponsored one of the field studies.

REFERENCES

ALEXANDER, R.W. and CALVO, A. 1990. The influence of lichens on slope processes in some Spanish Badlands. In: Thornes, J.B. (ed.). Vegetation and erosion: processes and environments. Wiley, Chichester. pp.385-98.

BRYAN, R.B. 1981. Soil erosion under simulated rainfall in the fields and laboratory: variability of erosion under controlled conditions. IAHS Publication 133:391-402.

CALVO, A., GISBERT, B., PALAU, E. and ROMERO, M. 1988. Un simulador de lluvia de fácil construción. In: Gallart, F. & Sala, M. (eds). Métodos y técnics para la medición de procesos geomorfológicos. pp.6-16.

CALVO, A.M. and HARVEY, A.M. 1989. Distribution of badlands in south east Spain: Implications of climatic change. In: Imeson, A.C. & De Groot, R.S. (eds). Landscape-ecological impact of climatic change.

FRANCIS, C.F. 1986a. Soil erosion on fallow fields: an example from Murcia. Papeles de Geografia Fisica 11:21-8.

FRANCIS, C.F. 1986b. Variaciones sucesionales de la vegetación en campos abandonados de la provincia de Murcia, España. Ecologia 4:35-47.

FRANCIS, C.F. and THORNES, J.B. 1990. Matorral: Erosion and reclamation. In: Albaladejo, J., Stocking, M.A. & Diaz, E. (eds). Degradacion y regeneración del suelo en condiciones ambientales mediterráneas. Consejo Superior de Investigaciones Cientifics, Murcia. pp.86-115.

GARCIA-RUIZ, J.M. 1991. Consecuencias ambientales del abandono agrícola. In: Rubio, J.L. & Calvo, A. (eds). Seminario. Procesos de desertificación en condiciones ambientales mediterráneas. UIMP.

GARCIA-RUIZ, J.M. and LASANTA, T. 1990. Land use changes in the Spanish Pyrenees. Mountain Research and Development 10(3):267-79.

GARCIA-RUIZ, J.M., LASANTA, T. and SOBRON, I. 1988. Problemas de evolución geomorfológica de campos abandonados: El Valle del Jibera (Sistema Ibérico) Zubia 6:99-114.

GERITS, J.J.P. 1991. Physico-chemical thresholds for sediment detachment, transport and deposition. PhD Thesis, Universiteit van Amsterdam.

GODRON, M., GUILLERM, J.L., POISSONET, P., THIAULT, M. and TRABOUD, L. 1981. Dynamics and management of vegetation. In: De Castri, F., Goodall, D.W. & Specht, R.L. (eds). Mediterranean-Type Shrublands. Elsevier, Amsterdam.

HARVEY, A.M. and CALVO, A. 1991. Process interactions and rill development on badlands and gully slopes. Z. Geomorph. N.F. Suppl. Bd. 83:175-94.

IGME 1978. Mapa geológico de España (E:1/50.000), Petrer. Madrid.

IGME 1981. Mapa geológico de España (E:1/50.000), Jativa. Madrid.

IMESON, A.C. 1977. A simple field-portable rainfall simulator for difficult terrain. Earth Surface Processes 2:431-36.

IMESON, A.C. and VERSTRATEN, J.M. 1986. Erosion and sediment generation in semi-arid and Mediterranean environments: the response of soils to wetting by rainfall. Journal of Water Resources 5(1):388-417.

IMESON, A.C., VERSTRATEN, J.M., VAN MULLIGEN, E.J. and SEVINK, J. 1992. The effect of fire and water repellency on infiltration and runoff under Mediterranean type forest. Catena 19:345-361.

LASANTA, T. 1988. The process of desertification of cultivated areas in the Central Pyrenees. Pirineos 132:15-36.

LASANTA, T. 1989. Evolución reciente de la agricultura de Montaña. El Pirineo aragonés. Geoforma Ediciones, Logroño.

LOPEZ-BERMUDEZ, F. and GUTIERREZ-ESCUDERO, J.D. 1982. Estimación de la erosión y aterramiento de embalse en la cuenca hidrográfica del río Segura. Cuadernos de Investigación Geográfica. VII:3-18.

LOPEZ-BERMUDEZ, F. and TORCAL-SAINZ, L. 1986. Procesos de erosión en tunel (Piping en cuencas sedimentaria de Murcia (España)). Estudio preliminar mediante difracción de rayos x y microscopio electrónico de barrido. Papeles de Geografía Física 11:7-20.

MOONEY, H.A. and KUMMEROW, J. 1981. Phenological development of plants in Mediterranean climate regions. In: De Castri, F., Goodall, D.W. & Specht, R.L. (eds). Mediterranean-Type Shrublands. Elsevier, Amsterdam.

RODRIGUEZ-AIZPEOLEA, J. 1990. Evolució i situació actual als bancals abandonats en el parc natural de Montgó. Aguaits 5:29-54.

RODRIGUEZ-AIZPEOLEA, J. 1992. Distribució espacial y evolucióde l'agricultura a la Vall d'Ebo. Actes III Congrés d'Estudis de la Marina Alta, Instituto Juan Gil-Albert, Alicante.

RODRIGUEZ-AIZPEOLEA, J., PEREZ-BADIA, R. and CERDA-BOLINCHES, A. In Press. Colonización vegetal y producción de escorrentía en bancales abandonados: Val de Gallinera, Alacant. Cuaternario y Geomorfología.

ROMERO-DIAZ, M.A., MARTINEZ-FERNANDEZ, J., FRANCIS. C.F., LOPEZ-BERMUDEZ, F. and FISHER. G.C. 1988. Variability of overland flow erosion rates in a semi-arid Mediterranean environment under matorral cover (Murcia, Spain). Catena Supplement 13:1-11.

RUIZ-FLAÑO, P. 1992. La evolución geomorfológica de campos abandonados en áreas de Montaña. El ejemplo de valle de Aisa. Pirineo Aragones. PhD Thesis, Universidad de Zaragoza.

RUIZ-FLAÑO, P., LASANTA-MARTINEZ, T. and GARCIA-RUIZ, J.M. 1991. The diversity of sediment yield from abandoned fields of the Central Spanish Pyrenees. Sediment and stream water quality in a changing environment: trends and explanation. Proceedings of the Vienna Symposium, August 1991. IAHS Publication 203.

SALA, M. and CALVO, A. 1990. Response of four different mediterranean vegetation types to runoff and erosion. In: Thornes, J.B. (ed.). Vegetation and erosion: process and environments. Wiley, Chichester. pp.347-62.

SCOGING, H.M. and THORNES, J.B. 1979. Infiltration characteristics in a semi-arid environment. The hydrology of areas of low precipitation. Proceedings of the Canberra Symposium. IAHS Publication 128.

SCOGING, H.M. and THORNES, J.B. 1980. Infiltration characteristics in a semi-arid environment. IAHS Publication 128. p.159-168.

THORNES, J.B. 1985. The ecology of erosion. Geography 70(3):222-36.

WILCOX, B.P., WOOD, M.K., TROMBLE, J.T. and WARD, T.J. 1986. A hand-portable single nozzle rainfall simulator designed for use on steep slopes. Journal of Range Management 39(4):331-5.

WOOD, M.K. 1987. Plot numbers required to determine infiltration rates and sediment production on rangelands in south central New Mexico. Journal of Range Management 40(3):259-63.

YAIR, A. and LAVEE, H. 1985. Runoff generation in arid and semi-arid zones. In: Anderson, M.G. and Burt, T.P. (eds). Hydrological forecasting. pp.183-220.

Chapter 6

SOIL PROPERTIES AND GULLY EROSION IN THE GUADALAJARA PROVINCE, CENTRAL SPAIN

J.L. Ternan[1], A.G. Williams[1] and M. Gonzalez del Tanago[2]
[1]Department of Geographical Sciences, University of Plymouth,
Plymouth PL4 8AA, UK
[2]Escuela Tecnica Superior de Ingenieros de Montes,
Universidad Politecnica de Madrid, 28040 Madrid, Spain

Summary

Substantial land degradation has occurred in the west of Guadalajara province, north east of Madrid, on highly erodible soils. This paper presents a preliminary assessment of the problem, with an investigation of the principal soil characteristics influencing erosion and the likely land use impacts. Representative sites under different land uses and parent geology were identified for soil sampling on the basis of a simplified terrain analysis procedure. Laboratory rainfall simulation tests found that cultivated soils developed on Pliocene Rañas were the most readily dispersed, with no aggregates surviving after 10mm of simulated rain. Gullies occur on areas of Miocene conglomerates, and where recent bench terracing of hillslopes prior to afforestation has taken place, erosion rates appear to be high. Caesium 137 analyses found that 40cm of deposition had occurred at sites in the gully floor in the past 27 years, which indicates that erosion is active, although an intergully site under undisturbed *Cistus Rosmarinus* scrub showed little evidence of erosion in the recent past. Analysis of soil moisture characteristic curves for each site showed that the Miocene conglomerate soils had a substantial lack of transmission pores and hence overland flow may occur even during moderate intensity storms.

1. INTRODUCTION

Accelerated soil erosion is a particular problem in the basin of the Rio Jarama, in the Guadalajara province north east of Madrid. Although there is evidence of soil erosion on the older basement rocks of the catchment, the most severe erosion occurs on a variety of unconsolidated Miocene and Pliocene alluvial sediments. These show considerable lateral and vertical variations in composition and range from calcareous marls to siliceous conglomerates and breccias (IGME, 1980). In the Puebla de Valles-Puebla de Belena area (Figure 1) the landscape is characterised by a series of wide planation terraces and benches with steeply incised valleys of the Rio Jarama and its tributaries. Gully erosion is extensive on the valley sides and, despite a generally good vegetation cover, gives rise to a badland landscape on the margins of the benches (Plate 1).

Since the 1950s extensive soil conservation measures including check dams, have been undertaken to reduce potential problems of reservoir sedimentation. Other measures, principally conifer afforestation (*Pinus nigra, P. pinaster*), have been carried out, with the aim of reducing soil loss rather than to crop the timber commercially. This afforestation programme has continued to the present day and in more recent years has been in conjunction with bench terracing of the steeper hillslopes (Plate 1). Agriculture in the area is dominated by extensive cereal cultivation on the flat or sloping planation benches, although nomadic sheep rearing continues on poorer grasslands and areas of *Cistus* scrub. Areas of mixed oak forest (*Quercus ilex, Q. faginea, Q. pyrenaica*) survive on some of the poorer, steeper slopes.

Plate 1. Gullied hillslopes on margins of planation terraces. *Pinus* afforestation with bench terracing adopted as part of the soil conservation strategy.

According to Muñoz et al. (1989) much of the western region of Guadalajara province receives between 600-800mm.yr^{-1}. At higher altitudes precipitation may exceed 800mm. At Pantano El Vado (1000m) precipitation averaged 815mm between 1941-1970, with 107mm of rain in November. Maximum daily rainfall intensity averaged 20mm day^{-1} with a maximum of 65mm day^{-1} recorded at El Vado in November 1976. The summers are very dry, with potential evapotranspiration exceeding precipitation from May to September.

This paper presents the results of a reconnaissance study of soil properties in the area with a view to assessing the likely impact of different land uses on soil erosion.

Figure 1. Location of the sampling sites in west Guadalajara province, central Spain. Sites are listed in Table 1.

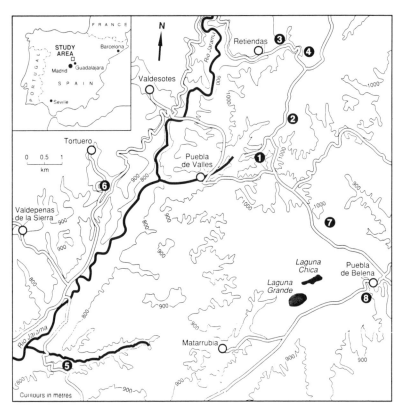

2. METHODS

Soil sampling sites were selected on the basis of a modified terrain analysis procedure similar to that devised by Williams (1981). A random sample of ten 1:18,000 scale air photos was selected from coverage of the 100km ^{2} area. Land systems, based on geology, were identified and further subdivided into land units according to land use and vegetation cover. Eight sites (Table 1) were selected at the land unit scale and where practical, both surface and sub-surface horizons were sampled. The main focus was on the Miocene conglomerates and Pliocene Rañas soils and sediments (Figure 2). Within each land unit, sampling sites were chosen on gentle slopes (<5°) close to drainage divides. At this reconnaissance stage no attempt was made to evaluate soil variability in relation to slope or topographic controls.

Figure 2. Location of study sites in relation to geology. Geology based on the 1:200,000 Mapa Geologico de España (Segovia - 38).

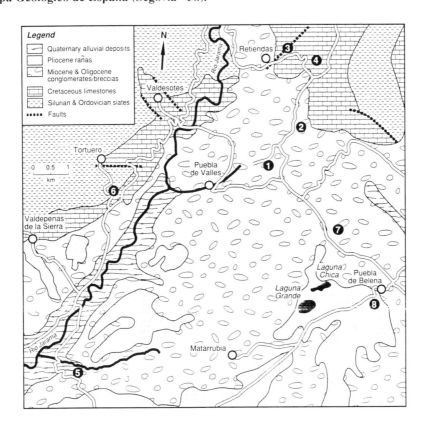

3. SOIL CHARACTERISTICS

In general the dominant soil type is the brown earth which extends over the planation terrace areas and gently sloping hillsides. On the Miocene and Pliocene alluvial sediments the soils range in colour from reddish yellow (5YR 6/6) to light yellowish brown (10YR 6/4), often within a distance of 100m. Such colour variation is more likely to result from pedogenic processes or differences in parent material than from topsoil removal by erosion. In general, the topsoil was around 30cm in depth, with a fine subangular blocky structure, and a silt loam texture (Table 2). Abundant fine fibrous roots were noted in the topsoil under semi-natural vegetation communities, while relatively fewer were observed in agricultural areas. The subsoil was similar to the topsoil in texture but was less well structured. No indurated layers were observed.

On the slate and limestone regions lithomorphic soils were found, rankers on slate and rendzinas on limestone. These were less variable in texture than soils on the Miocene and Pliocene sediments (Table 2).

Table 1. Summary of characteristics of principal sites identified from a simplified terrain analysis.

Site	LAND SYSTEM*	LAND UNITS	
	Parent Material	Vegetation Community	Management
1	Miocene Conglomerates (M2-4)	*Pinus pinaster* *P. nigra* *Cistus* understorey	Afforested
2	Miocene Conglomerates (M2-4)	*Pinus pinaster* *P.nigra* *Cistus* understorey	Afforested, bench terraced
5	Miocene Conglomerates (M2-4)	*Quercus ilex* *Q. faginea* *Q. pyrenaica* *Cistus*	Natural woodland
7	Miocene Conglomerates (M2-4)	Bare	Ploughed, arable
8	Pliocene-Rañas (PL)	Bare	Ploughed, arable
3	Cretaceous Limestone (C1-6)	*Quercus ilex* *Q. coccifera* *Cistus Rosmarinus*	Woodland and sheep grazing
4	Cretaceous Limestone (C1-6)	*Quercus ilex* *Cistus Rosmarinus*	Rough grazing
6	Silurian Slate	*Cistus* scrub	Rough grazing

* Based on the 1:200,000 Mapa Geologico de España (Segovia - 38). Site locations are shown in Figure 1.

Organic matter determined by weight loss on ignition at 350°C ranged from 2-10%. The values were relatively high despite summer temperatures in excess of 35°C, which encourages oxidation of the organic matter. However, in undisturbed situations the organic matter inputs may be high. The lowest values were found on cultivated Rañas soils (2.5%). Higher values were found under oak woodland (3.3%) and conifer plantation (3.1%). On the slate and limestone soils the values exceeded 5%. According to Greenland (1977) soils with less than 3.5% organic matter or 2% carbon are most vulnerable to erosion because of the lack of organic polymers or binding agents.

Water soluble salts were determined from saturated pastes and the results are presented in Table 3. The low sodium values recorded suggest that this is not a significant factor in soil dispersion, and values approximately 100 times greater have been recorded for soils in southern Spain (Imeson and Verstraten, 1985) and Morocco (Imeson et al., 1982). pH values were determined on the saturated extracts. Under mature conifer forest on the Miocene conglomerates (Site 1) the surface horizon is more acid (pH 4.8) and lower in calcium when compared with young conifers on the bench terraces (pH 5.8). Exposure of parent material may account for the higher values. The mature oak woodland at site 5 had the highest pH and calcium levels of the Miocene conglomerate sites (pH 6.6). Slate soils had moderate acidity (pH 5.8) and limestone soils were the most alkaline (pH 7.4). In terms of erosion susceptibility conifer afforestation has been shown to lead to reduction of mineral cation exchange capacity associated with leaching, and to structural degradation (Nys and Ranger, 1985).

Table 2. Soil particle size characteristics.

Particle size	Parent material		
	Miocene conglomerates (10 analyses)	Cretaceous Limestone (7 analyses)	Silurian Slate (6 analyses)
% >2.00	0.7 - 39.0	20.4 - 50.3	7.6 - 39.9
% Sand*	17.4 - 57.3	29.8 - 44.5	33.8 - 54.1
% Silt*	37.3 - 71.1	44.6 - 54.2	37.6 - 53.9
% Clay*	2.7 - 17.7	7.8 - 16.0	8.1 - 16.3

*Expressed as a % of the <2.00 mm fraction. Particle size determined by a combination of wet sieving procedures (>63μm fraction), and a Fritsch Scanning photo-sedimentograph for the silt and clay fractions.

3.1 Soil aggregate stability

Soil aggregate stability was assessed by the use of a laboratory rainfall simulator inside an environmental chamber. Twenty five aggregates of 4.0-5.6mm diameter were placed on a 2.8mm sieve and equilibrated for 24 hours at 50% relative humidity to ensure similar initial moisture status (Farres and Cousen, 1985).

Rainfall was simulated at approximately 90mm.hr^{-1} intensity, with a mean drop size of 580μm (SD 251μm) as determined by the filter paper technique adapted by Mason and Andrews (1960). The percentage of aggregates surviving at 0.5 minute time intervals was calculated.

Considerable variation in soil aggregate stability was observed for the surface horizons of soils overlying different parent geologies, and under different vegetation and land management (Figure 3). Limestone soils had the highest aggregate stability with <5% breaking down within 40mm of simulated rainfall. By contrast, aggregates from soils developed on Pliocene Rañas readily dispersed, and for site 8 no aggregates survived after 10mm of simulated rain. Similarly at site 7 (also under cereals) only 20% of the aggregates survived a 40mm rainstorm. Aggregates from Miocene-conglomerate soils were also readily dispersed where the land was terraced for conifers (site 2). Where slopes have been afforested without terracing (site 1) the aggregate stability was high, with only 5% breaking down in the 40mm simulation. Such stability would seem unusual in view of the low organic matter content and decalcified nature.

Table 3. Soil chemistry of the surface horizons determined from saturated extracts.

Site	Water soluble salts (saturated pastes) meq l^{-1}				
	pH	Na	K	Mg	Ca
1	4.8	1.6	0.3	0.1	0.2
2	5.8	2.1	0.5	0.2	1.0
3	7.4	1.5	0.3	0.1	1.5
4	7.4	1.8	0.1	0.2	1.5
5	6.6	2.3	0.2	0.2	1.3
6	5.8	2.1	0.1	0.1	0.7

Figure 3. Aggregate stability of the soils from sites listed in Table 1.

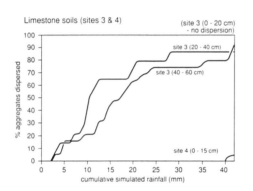

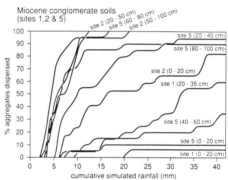

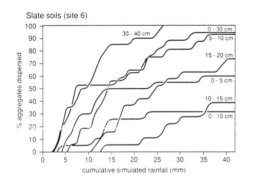

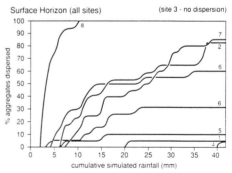

The soils also showed considerable variations in aggregate stability with depth (Figure 3). On Miocene conglomerates the surface soil horizons tended to be more resistant than sub-surface horizons although no clear pattern emerges with increasing depth. Limestone soils showed a clear contrast between resistant surface horizons and more dispersible sub-surface aggregates, possibly reflecting a higher organic matter content in the near surface horizons. Slate soils showed no clear pattern. As deeper soils on slates occur predominantly in hollows, much of the soil material may not have developed in situ.

4. SOIL WATER CHARACTERISTICS

The soil moisture characteristic is a useful measure of pore size distribution, and therefore an indirect measure of soil structure and hydraulic conductivity. Water retention characteristics were determined with a sand table using undisturbed 63cm^3 cores for retention at <50m bar and pressure plates (Soil Moisture Equipment Co.) for retention from 0.3 to 15 bars. The slope of the characteristic curve is proportional to the frequency distribution of pore sizes present, since pores are drained at tensions depending on their size. The greater the number, i.e. total volume of pores, greater than 50µm, the greater the ability of soils to transmit water through the profile. Thomasson (1978) classified soil structure in terms of the relationship between volume of pores greater than 50µm defined as the air capacity (C_a) and volume of pores less than 50µm known as available water capacity (C_w). A well structured soil has a substantial volume of both large and small pores, with C_a and C_w more than 15%, as indicated by a reasonably steep curve. Soils which have a poor structure possess limited pore volumes, with C_a and C_w being less than 10%, and this is indicated by a relatively flat soil moisture characteristic curve.

Miocene conglomerate soils under oak woodland had the steepest characteristic curves, and are therefore the best structured soils (Figure 4). A value of 14% for C_a indicates that the soil is well drained since there is a high volume of large pores able to transmit water rapidly. Similarly, there are large reserves of soil water available. These soils would be classified as having a good soil structure by Thomasson (1978). In contrast, conglomerate soils under *Pinus* forest and from bare areas where gullying occurred had relatively flat characteristic curves. The C_a value of less than 5% for each suggests low infiltration rates, with the result that during storms these areas would be prone to waterlogging and overland flow may ensue. Available water capacity of about 14% was reasonably high although the plants could be stressed in the arid summers. Soil structure would be classified as poor due to limited pore space for drainage.

The limestone soil under *Quercus ilex* and a slate profile under *Cistus* scrub had relatively flat characteristic curves, that were similar to those found on the conglomerate soils with *Pinus*. The C_a values of about 6% lead to low infiltration rates and the high silt plus clay fraction of the soils as shown in Table 2 may be responsible for this lack of large pores. Soils from these areas would be in the poor soil structure category.

In summary, there seem to be few differences between the soil water characteristics of the soils found over the various parent materials. The main variation is between soils under different vegetation and land use. Soils under oak woodland were the best structured; few infiltration problems would arise in the *Pinus* and *Cistus* scrub areas due to high interception rates, and the areas most at risk both in terms of overland flow and structural degradation are the bare areas.

Figure 4. Soil water characteristics of principal soils.

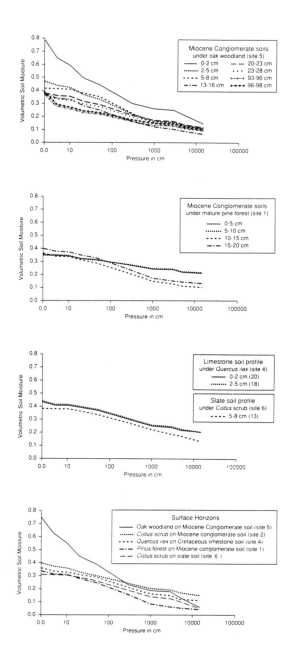

5. DISCUSSION

5.1 Parent material

The stability of soil aggregates depends on the magnitude of disruptive as opposed to stabilising forces (De Meester and Jungerius, 1978). In general the amount and type of clay mineral present is a major influence on aggregate stability (Morgan, 1986). Molina et al. (1986; 1990) have described the widespread occurrence of smectite clays in weathering mantles in central Spain. Clays such as smectite and illite readily form aggregates but their swelling and shrinking characteristics render them potentially unstable (Morgan, 1986). A reduction in the stabilising forces provided by living tissue and organic matter (Imeson and Jungerius, 1976) through adverse land management practices may therefore be critical.

Although further work is needed on the properties of the Rañas soils, their clay mineralogy, combined with low organic matter content may be significant factors. The soils developed on the Pliocene Rañas appear to be the most readily dispersed. When ploughed, these soils form large clods but the low aggregate stability suggests that they may readily disperse with the onset of rain. Due to the low relief of the planation terraces evidence of soil erosion tends to be confined to sloping sites where rills and soil deposition can be observed.

Field observation and aerial photograph analysis demonstrate that the most dramatic erosion occurs on the unconsolidated Miocene conglomerates with extensive gully development, particularly on convex slopes at the edges of the planation terraces and at the margins of incised ephemeral channels (Plate 1). Many gullies appear to be relict features and are now being colonised by scrub vegetation. The origin of these gullies is unknown but may be related to higher grazing pressures in the past. At other sites however gullying is very active with headscarp recession threatening road communications. Active gully erosion may be largely related to land management practices but spatial variations in the textural, mineralogical and chemical composition of the sediments may be significant. Considerable lateral variation in the sediment occurs due to changing sources of parent rock, varying distance from the source rock and variations in the original depositional environment (IGME, 1980).

Although piping does not appear to be a major factor in the development of gullies incised into coarse sediments, deep pipes similar to those recorded by Harvey (1982) in south-east Spain have been observed. These have developed in silt or clay rich horizons with little or no stone content, and collapse of these pipes has given rise to irregular gully floor profiles.

Due to the resistant nature of the parent rock, gully erosion does not occur on the slate or limestone areas. Nevertheless the extensive bare rock surfaces and thin skeletal soils of the slate region suggest that erosion by overland flow may be active. Deep soils only occur in valleys or depressions in the slate areas and analyses have demonstrated the low aggregate stability of these soils. In contrast, the limestone areas generally have a good soil cover and these soils show the greatest erosion resistance.

5.2 Role of land management

Although the areas of Miocene and Pliocene conglomerate have a high erosion hazard, evidence suggests that erosion rates may be strongly influenced by land management. Considerable amounts of sediment have been moved down gully systems, as evidenced by the accumulation of material on gully floors and behind check dams which were built from 1957 onwards. Many of these areas were terraced prior to afforestation. Caesium 137 analyses for an inter-gully site profile under natural *Cistus Rosmarinus* scrub show that the majority of caesium is held in the top 15cm of the soil (Figure 5). Although a full caesium inventory has to be carried out, the overall shape of the caesium profile suggests that this site has not been eroded to any great extent. Caesium analysis of a gully floor site located in an area of extensive recent terracing shows a distinct maximum at 40-45cm (Figure 5), which could represent the c. 1963 ground surface. The profile is typical of a site which is accreting fairly rapidly, and suggests a minimum of 40cm deposition in 27 years. Field investigations would suggest that much of this sediment is derived from the gully walls and the disturbed sediments comprising the terraces.

Figure 5. Caesium 137 profiles for inter-gully (*Cistus* scrub on Miocene conglomerates) and gully floor (*Pinus negra, P. pinaster* on bench terraced gully).

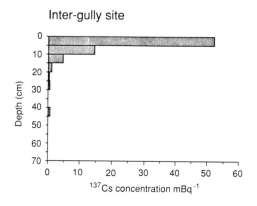

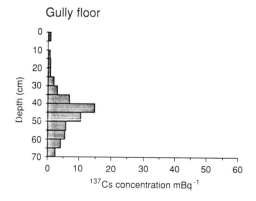

This site preparation for afforestation may be responsible for increased erosion rates for a number of reasons. Terracing disturbs the existing, often fairly dense *Cistus rosmarinus* scrub which provides a protective canopy to the ground surface. This ground surface appears to be stable as evidenced by the caesium profile. Areas under semi-natural vegetation are often characterised by a stone layer which appears to reduce erosion by armouring the surface. Although Poesen (1987) has demonstrated that small pebbles may be transported by interrill flow, the presence of an extensive lichen cover suggests that contemporary erosional losses are minimal (Alexander and Calvo, 1990). Terracing exposes the more erodible subsurface horizons, which have lower aggregate stability (Figure 3) and lower organic matter content. This site preparation also appears to affect the surface and sub-surface hydrology of the hillslopes. Where terraces intersect the gully walls there is evidence that lateral drainage from the terrace surfaces is leading to the development of lateral gullies along the line of the terrace.

Many of these effects appear to be important in areas where terracing has taken place since 1980. Areas under mature conifer forest (age >30 years) seem to be much more stable, suggesting that some kind of equilibrium may be re-established in the longer term.

Both past and present agricultural activity in the Humanes region may also have an important influence on erosion. Extensive Tertiary or Quaternary terrace surfaces are ideal for cereal cultivation. Crops are managed to the highest standards using pesticides, herbicides and growth retardants so that they are weed free and of uniform height. However this cereal production leaves large areas bare for some of the wetter winter months, and rills and small gullies were observed in cropped fields. Air photo analysis also shows that areas cultivated in the past and recolonised by semi-natural vegetation communities are now being brought into arable production. These areas are usually more steeply sloping than the terrace surfaces, and the erosion risk is likely to be high. The arable soils were found to be low in organic matter, and soil clods and aggregates readily dispersed under simulated rain. Air escape during wetting is likely to be affected by pore size distribution (Farres, 1980), and this may be altered by land management practices. Furthermore, the lack of large transmission pores in the soils suggest that overland flow on sloping sites is likely, and could lead to reactivation of semi-stable gullies, many of which have headscarps extending into these arable areas.

6. CONCLUSIONS

The Guadalajara area presents a complex mosaic of different materials, vegetation and land management practices. Soils developed on Miocene and Pliocene sediments are most susceptible to erosion due to their textural, chemical and water retention characteristics. Under oak woodland high organic matter and biological activity have created a good soil structure and the erosion risk is lessened. However where the soils have been disturbed particularly by bench terracing prior to afforestation erosion rates would appear to be very high locally. Expansion of cultivation in the area also presents an increased risk of reactivation of semi-stable gully systems.

ACKNOWLEDGEMENTS

Support for this project was under the Acciones Integradas Hispano-Britanicas Programme. Richard Hartley and Anne Kelly provided invaluable technical assistance, and Sheila Ternan assisted with field observations. Tim Absalom and Brian Rogers are thanked for cartographic work and Julie Sugden for typing of the manuscript. We are also grateful to Professor D. Walling, Exeter University for analyses and discussions relating to caesium 137.

REFERENCES

ALEXANDER, R. and CALVO, A. 1990. The influence of lichens on slope processes in some Spanish Badlands. In Thornes, J.B. (ed.). Vegetation and erosion: processes and environments. Wiley, Chichester.

DE MEESTER, T. and JUNGERIUS, P.D. 1978. The relationship between the soil erodibility factor K (Universal Soil Loss Equation), aggregate stability and micromorphological properties of soils in the Hornos area, S. Spain. Earth Surface Processes 3:379-391.

FARRES, P.J. 1980. Some observations on the stability of soil aggregates to raindrop impact. Catena 7:223-231.

FARRES, P.J. and COUSEN, S.M. 1985. An improved method of aggregate stability measurement. Earth Surface Processes and Landforms 10:321-329.

GREENLAND, D.J. 1977. Soil damage by cultivation. In Cooke, G.W., Pirrie, N.W. and Bell, G.D.H. (eds). Agricultural efficiency. Philosophical Transactions. Royal Society, London. 281:193-208.

HARVEY, A. 1982. The role of piping in the development of badlands and gully systems in south-east Spain. In Bryan, R.B. and Yair, E. (eds). Badland geomorphology and piping. Geobooks, Norwich.

IGME 1980. Mapa geologico de Espana E 1:200,000 Segovia (38) Segunda edicion, Instituto Geológico y Minero de España.

IMESON, A.C. and JUNGERIUS, P.D. 1976. Aggregate stability and colluviation in the Luxembourg Ardennes; an experimental and micromorphological study. Earth Surface Processes 1:259-271.

IMESON, A.C. and VERSTRATEN, J.M. 1985. The erodibility of highly calcareous soil material from southern Spain. Catena 12:291-306.

IMESON, A.C., KWAAD, F.J.P.M. and VERSTRATEN, J.M. 1982. The relationship of soil physical and chemical properties to the development of badlands in Morocco. In Bryan, R.B. and Yair, A. (eds). Badland geomorphology and piping. Geobooks, Norwich.

MASON, B.J. and ANDREWS, J.B. 1960. Drop size distributions from various types of rain. Quarterly Journal of the Royal Meteorological Society 86:346-353.

MOLINA, E., BLANCO, J.A., PELLITERO, E. and CANTANO, M. 1986. Weathering processes and morphological evolution of the Spanish Hercynian massif. In Gardner, V. International geomorphology 1986. Part 2.

MOLINA, E., CANTANO, M., VICENTE, M.A. and GARCIA RODRIGUEZ, P. 1990. Some aspects of paleoweathering in the Iberian Hercynian massif. Catena 17:333-346.

MORGAN, R.P.C. 1986. Soil erosion and conservation. Longmans Scientific and Technical, Harlow.

MUÑOZ, M.J., ALDEANUEVA, R.A. and ARNAIZ, J.M.R. 1989. El clima de la Provincia de Guadelajara. Paralelo 37° Revista de Estudios Geográficos 13:227-252.

NYS, C. and RANGER, J. 1985. Influence de l'espece sur le fonctionnement de l'écosystème forestier. Le cas de la substitution d'une essence resineuse à une essence feuillue. Sci. Sol. 21:203-216.

POESEN, J. 1987. Transport of rock fragments by rill flow - a field study. In Bryan, R.B. (ed.). Rill erosion - processes and significance. Catena Supplement 8:35-54.

THOMASSON, A.J. 1978. Towards an objective classification of soil structure. Journal of Soil Science 29:38-46.

WILLIAMS, D.F. 1981. Integrated land survey methods for the prediction of gully erosion. In: Townshend, J.R.G. (ed.). Terrain analysis and remote sensing. Allen and Unwin, London.

Chapter 7

A SIMPLE FIELD ERODIBILITY INDEX BASED ON SOME ENGLISH SOILS

A.W. Vickers
Silsoe College, Cranfield University, Silsoe,
Bedfordshire, MK45 4DT, UK

Summary

A one-month survey of seventeen soil series at twelve locations in England resulted in the creation of a unique database on soil erodibility. The database consisted of field measures of the following soil properties:

(i)   Soil surface shear strength
(ii) Bulk density of the top 19mm of the soil
(iii) The gravimetric moisture content of the soil at the time of sampling.

In addition, laboratory analysis of the various soils ability to withstand single water drop impacts was also performed. From this database an index was developed to describe soil erodibility. This index is designed for simple, quick use in the field. The index requires further work to validate and possibly link it to soil loss and hence erosion prediction.

1. INTRODUCTION

The assessment of soil erodibility has often involved the use of complex indices requiring the collection of vast quantities of data on the chemical, physical and biological status of the soil. Where simplicity has been sought, methods that bear little relation to erosive processes have been developed. Such indices work with varying degrees of success, but all require work in the laboratory and none can offer the field surveyor an instant and reliable assessment of soil erodibility. Bryan (1969) suggests that an ideal index of erodibility should be 'simple to measure, reliable in operation, and capable of universal application'. This survey was carried out with the hope of meeting some of these ideals.

Soil shear strength was measured using a Torvane shearometer. This device measures the top 5mm of the soil. It is very portable and consequently ideal for field work. The Torvane has been used by many workers in the past and has proved to be a reliable tool (Wall et al., 1988; Coote et al., 1988). Shear strength was used because it is a direct measure of a soil's ability to withstand shearing forces exerted on it. This means a soil with a high inherent shear strength is better able to withstand the shearing forces of raindrop impact and overland flow than a soil with a lower shear strength. Consequently a soil with a high shear strength is relatively less erodible under the same conditions of land use than a soil with a lower shear strength. Clearly shear strength is influenced by other soil properties such as bulk density and moisture content, so these properties were also measured. Hence different density/moisture regimes could be compared. This also allowed the influence of soil compaction and capping and their effect on erodibility to be studied.

Shear strength, as an indication of a soil's resistance to erosion has been shown to be a useful method by many workers. Rauws and Govers (1988) demonstrated that shear strength was an effective measure of soil erodibility with respect to erosion by flowing water. Cruse and Larson (1977) showed a close correlation between soil detachment caused by raindrop impact and soil shear strength. Almost all of this work has involved measuring shear strength at or near saturation. This is sensible as a soil is in its most erodible state in such conditions. The purpose of this study, however, was to produce a system which would allow a field worker to take a measurement of erodibility in the conditions found in the field at the time of survey. This is because it is unfeasible to saturate a soil prior to measurement if a number of sites are to be measured. It was hoped that by measuring shear strength at a variety of moisture contents and bulk densities the influence of differences in moisture content and bulk density could be taken into account in the final index. The second argument in favour of using shear strength as a survey criterion is its ease of measurement.

2. METHODS

The survey took place during May and June 1989 and was based around seventeen series located at ten sites through England.

The survey sites were all selected from areas used by Evans (1980), during a five year monitoring programme for the Soil Survey of England and Wales. The series to be surveyed were also selected from the series monitored by Evans and covered a range of textures (Table 1).

The samples included determinations of bulk density, oven dry moisture content and organic matter. At each sample point fifteen shear strength tests were performed with a Torvane within a 50cm radius of the central sampling point, at the natural moisture content. Bulk density was determined using a core of 52mm diameter. Oven dry moisture content was determined from the bulk density sample. A sample of soil was bagged to allow determination of organic matter content using the Walkley-Black method, and the resistance to single drop impact. The variation in the number of samples for each series is due to the number of locations at which the series is sampled. The purpose of the sampling was to test the soil in its most erodible state, which is usually a freshly prepared, unvegetated seedbed. All samples were taken from as close to this as possible. Samples were also taken where obvious variations in bulk density occurred, such as in wheelings, on crusted soil and in seed presses.

To examine the influence of soil moisture content on shear strength in an uncapped, uncompacted sandy loam soil, a column of Cottenham series was held at different moisture contents on a sand table. Fifteen shear strength determinations were performed at the moisture contents of 13.4%, 15.3%, 16%, 18.4% and 21.1%.

Table 1. Soil series, number of samples taken and their respective sampling locations.

Series	Location	Number of samples
Wickham	Beds, East and West Sussex, Dorset	10
Denchworth	Beds, East and West Sussex	8
Aylsham	Norfolk	6
Bridgenorth	Shrops, Staffs, Notts, Cumbria	10
Wickmere	Norfolk	4
Conway	Staffs.	2
Compton	Notts	4
Wick	Norfolk, Shrops, Staffs, Notts, Cumbria	11
Redlodge	Norfolk, Staffs	8
Newport	Norfolk, Shrops, Staffs, Notts, Cumbria	12
Dale	East and West Sussex, Dorset	10
Bromyard	Hereford, Shrops, Staffs	6
Eardiston	Hereford, Shrops, Staffs	10
Cuckney	Shrops, Staffs, Notts, Cumbria	10
Claverley	Shrops, Staffs, Cumbria	10
Clifton	Shrops, Staffs, Cumbria	10
Quorndon	Shrops, Staffs, Cumbria	9

3. RESULTS

The means for the results obtained for shear strength, bulk density and resistance to single drop impact are shown in Table 2. The results for moisture content are shown on Figure 1. The results for the shear strength/moisture content tests for an uncapped, uncompacted sandy loam soil appear in Table 3, along with the values for uncapped/uncompacted light textured soils taken in the field. The missing values are due to technical failure.

4. DISCUSSION

The heavier textured soils in the survey have a much higher shear strength than the light textured soils under the same conditions of land management. This is not very surprising, and is due almost entirely to the greater cohesion of the heavy soils due to a higher clay content. The values of shear strength for the soils in an uncapped/uncompacted state are, generally, four to ten times greater for the heavy soils than for the light soils. Even for soils in a capped or compacted state, there is still a difference of between 1.5 and 5 times that for the light soils.

The high coefficients of variance for some series are due to a number of factors. First, the moisture content at the various sample sites for each series varied. Second, all these soils were uncapped and uncompacted, hence small differences in bulk density would lead to small differences in the cohesion of the soil. This, when added to the differences in moisture content, could create relatively large differences in shear strength. The third reason for the differences is that unavoidably, the seedbed conditions for each sample area were not identical. The seedbeds all varied in age with differences in root density. This could easily create differences between replications. However, at each sample site the replications for

Table 2. Results for capped/compacted and uncapped/uncompacted shear strength. The number of drops required to decompose a 4mm aggregate and the capped/compacted, uncapped/uncompacted bulk density. Where appropriate the coefficients of variance are given.

Series	U.C. Shear strength (kg cm^{-2})	C/V	C. Shear strength (kg cm^{-2})	C/V	Drop No.	C/V	U.C. Bulk density (g cm^{-3})	C. Bulk density (g cm^{-3})
Denchworth	34.00	15.80	34.00	8.60	500	0.0	1.09	1.32
Wickham	15.00	18.00			483	8.0	1.04	
Dale	12.30	19.40			500	0.0	1.02	
Conway	22.30	14.00			500	0.0	1.19	
Compton	10.30	9.00			500	0.0	0.91	
Aylsham			7.16	7.20	48	12.6		1.26
Redlodge	2.40	24.00	8.30	8.90	52	40.0	1.10	1.44
Wickmere	9.10	7.00			498	0.0	1.02	
Newport	2.70	39.00	8.00	13.20	77	69.0	1.10	1.32
Wick	3.40	19.50	8.80	18.90	131	71.0	1.11	1.25
Bromyard	5.40	21.00	6.50	14.90	92	94.0	1.11	1.33
Eardiston	3.40	40.10	10.00	25.30	27	100.0	1.09	1.50
Clifton	2.50	32.60	6.30	11.70	61	77.0	1.12	1.44
Bridgnorth	2.40	24.19	7.60	19.73	33	30.0	1.02	
Quorndon	2.50	26.20	6.60	17.80	102	65.0	1.11	1.28
Claverley	2.90	29.80	7.50	11.60	67	45.0	1.12	1.27
Cuckney	1.90	16.10	8.20	16.20	60	100.0	1.21	1.31

Notes: U.C. = Uncapped   C. = Capped   C/V = Coefficient of variance

each individual sample were not significantly different at the 15% level. It is only when various samples from different sites are combined and described by one mean value that the differences become significant and the means less reliable.

Figure 1. Histogram showing moisture contents for all series sampled.

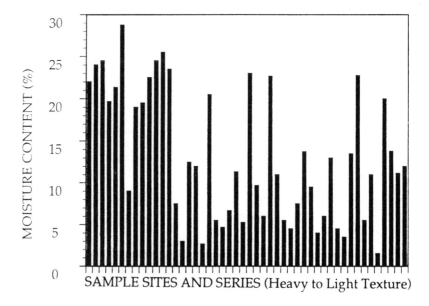

All the values for the capped/compacted soils with coefficients of variance greater than 15% still fall within the 25% level apart from the Eardiston series. This is due to the variety of cap thickness displayed at various sample sites by that soil.

Where a soil series has shear strength values for both a capped/compacted state and an uncapped/uncompacted state, the following pattern is seen. The capped/compacted soil has a mean shear strength value of 2.5 to 3 times that of the uncapped/uncompacted mean value. Only the light to medium textured soils are prone to surface capping and easy compaction by agricultural machinery. The heavier textured soils show signs of compaction but this does not seem to manifest itself in terms of increased shear strength.

There is a clear drop off in shear strength that occurs at the texture threshold between heavy to moderate soil. Capping and compaction on the light soils brings the shear strength value up to an equivalent value of the least cohesive heavy textured soils.

Table 3. Shear strength and moisture content values over a range of suctions for an uncapped sandy loam soil.

Suction (cms)	Shear strength ($kg\ cm^{-2}\times10^{-2}$)	Moisture content (%)	c/v.
1	1.7		23.0
2	4.8		7.5
4	5.8		10.9
5	6.1	21.1	6.3
7	6.5	18.4	10.0
10	7.2	16	7.9
12	7.9	15.3	8.6
15	8.5		7.3
20	9.0	13.4	13.8

Notes:
c/v. = Coefficient of variance

The results of the bulk density tests indicate that a dividing line between the capped, compacted values of bulk density and the uncapped, uncompacted values seems to exist. This dividing value is around 1.24g cm^{-3}. This seems to support the work of Fullen (1985), who recorded an average field bulk density of 1.24g cm^{-3} on capped or compacted arable soils in East Shropshire. The average value for soils in a similar state in this study is slightly higher, at 1.33g cm^{-3}. No uncapped/uncompacted soil had a bulk density greater than 1.22g cm^{-3}, and no capped/compacted soil had a bulk density below 1.24g cm^{-3}.

The results of gravimetric antecedent moisture content analysis show little pattern other than heavier textured soils having a higher oven dry moisture content than lighter textured soils. When the lighter soils dry out their shear strength seems to increase up to a point where cohesion between the soil particles is lost and shear strength drops away to a residual level. This is modelled by the laboratory work.

The results of the organic matter determination revealed no pattern. The results for resistance to single drop impacts were so variable no reliable pattern could be established.

5. THE INDEX OF ERODIBILITY

From the data collected in this survey the flow diagram shown in Figure 2 was developed. It must be stressed that if a soil is shown by this index to have a high degree of erodibility it does not mean that soil will definitely erode. There are many other factors involved in the erosion process. As yet, this index only describes the data set from which it was developed, so validation is required before it can be used to classify soil erodibility.

Figure 2. Flow diagram explaining the index of erodibility developed from the benchmark database collected in the survey. Categories are degrees of erodibility.

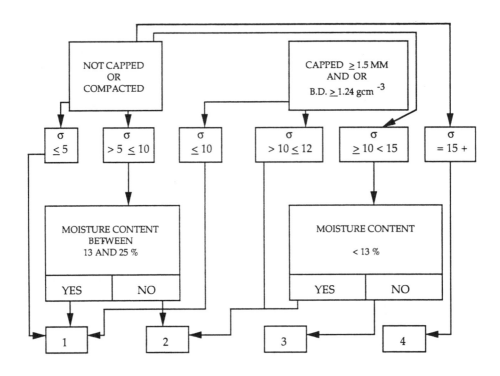

KEY

B.D. = BULK DENSITY

σ = SHEAR STRENGTH Kg cm^{-2} ($\times 10^{-2}$)

1 - 4 = DEGREES OF ERODIBILITY WHERE 1 IS
MOST ERODIBLE AND 4 IS LEAST ERODIBLE

6. CONCLUSIONS AND FINAL RECOMMENDATIONS

The similarity of soil shear strength values for soils of similar texture, under the same conditions of land use was identified. There was a uniform increase in shear strength displayed by the light textured soils when capped or compacted to similar bulk densities. The relationship between moisture content and shear strength for uncapped, uncompacted light textured soils reveals a critical point where shear strength rapidly diminishes with further drying to a residual low level.

There were large differences in shear strength between soils of different texture. These relationships were used to produce an index of erodibility. This index is a description of the benchmark database collected in the survey. The index requires validation, and provides a framework for future development and refinement.

Further work is also required to link the erodibility categories to possible soil loss under controlled conditions. This will give the categories some meaning in real terms and will allow a 'likelihood of erosion' to be added to the categories. As shear strength forms the basis of this database and hence the index, it may be possible to relate the index to critical shear velocity and, eventually, to basic erosion prediction.

The database only includes soils that are in their most erodible state, with no vegetation cover and in a seedbed condition. The database requires expansion to include soils in various states during the season. This would also need to include any seasonal changes in erodibility.

REFERENCES

BRYAN, R.B. 1969. The development, use and efficiency of indices of soil erodibility. Geoderma 2:5-26.

COOTE, D.R., MALCOLM-MCGOVERN, C.A., WALL, G.J., DICKINSON, W.T. and RUDRA, R.P. 1988. Seasonal variation of erodibility indices based on shear strength and aggregate stability in some Ontario soils. Canadian Journal of Soil Science 68:405-416.

CRUSE, R.M. and LARSON, W.E. 1977. Effect of soil shear strength on soil detachment due to raindrop impact. Soil Science Society of America Journal 41:777-781.

EVANS, R. 1980. Mechanics of water erosion and their spatial and temporal controls: an empirical viewpoint. In: Kirkby, M.J. and Morgan, R.P.C. (eds). Soil erosion. Wiley, Chichester. pp. 109-128.

FULLEN, M.A. 1985. Compaction, hydrological processes and soil erosion on loamy sands in east Shropshire, England. Soil and Tillage Research 6:17-29.

RAUWS, G. and GOVERS, G. 1988. Hydraulic and soil mechanical aspects of rill generation on agricultural soils. Journal of Soil Science 39:111-124.

WALL, G.J., DICKINSON, W.T., RUDRA, R.P. and COOTE, D.R. 1988. Seasonal soil erodibility variation in Southwestern Ontario. Canadian Journal of Soil Science 68:417-424.

Chapter 8

SOIL ERODIBILITY IN RELATION TO SOIL PHYSICAL PROPERTIES

P. Schjønning
Department of Soil Physics, Soil Tillage and Irrigation,
Danish Institute of Plant and Soil Science, Research Centre Foulum,
PO Box 23, DK-8830 Tjele, Denmark

Summary

In the early spring of 1991 soil susceptibility to erosion by water was tested for eleven Danish soils using a rainfall simulator. The shear strength, porosity and actual water content before water application were also measured. The measurements were applied to fields, which had been mouldboard ploughed and sown with a winter wheat crop in the autumn. Soils represented a range of Danish soils of different geological origin and soil texture.

Surface runoff and soil loss were highest for sandy soils, and declined with increasing clay content. Surface runoff also declined with an increase in soil porosity. No distinct influence of soil shear strength on soil loss could be detected.

Measurements of the same parameters in differently cultivated plots on a loamy sand and a sandy loam indicated the very pronounced influence of soil physical properties upon soil erodibility. In these plots, measurements of the surface roughness by means of an automated micro-relief meter estimated differences in the surface water storage capacity, which also influences the soil susceptibility to erosion by water.

1. INTRODUCTION

Early models describing soil erosion are rather empirical, using simple equations derived from plot studies (e.g. Wischmeier and Smith, 1978). In these models, parameter values are chosen from tables or graphs. However, in recent years research work tends to concentrate on developing more physically based descriptions of processes and their interactions (e.g. Nielsen et al., 1986). Thus it is very important to pin down the key parameters in the erosion process. Three soil properties are often highlighted as important in this connection: water infiltration capacity, surface roughness and structural stability.

For both simple, empirical and deterministic models, a simple method of integrating the effects of soil parameters on soil erodibility is needed. A true simulation of natural rainfall requires rather large, sophisticated equipment. However, soil spatial variability demands replicated measurements to give reliable estimates. A simple, easily transportable rainfall simulator can detect differences in soil erodibility, due to different soil management practices (Kamphorst, 1987). Estimates of soil erodibility as determined by the Kamphorst equipment can be related to soil physical characteristics.

2. MATERIALS AND METHODS

2.1 Erodibility of different soil types

Eleven winter wheat fields located in regions of different geological origin were selected. Topsoil and soil types are shown in Table 1.

Table 1. Soil types (Soil Taxonomy System) and top soil textures.

Location	Soil type	Texture (% w/w)				
		Org. matter	Clay	Silt	Fine Sand	Coarse sand
	μm:		<2	2-20	20-200	200-2000
Jyndevad	Orthic Haplohumod	2.3	3.6	4.3	3.8	86.0
Borris	Orthic Haplohumod	2.3	5.6	9.1	48.5	34.5
Foulum	Typic Hapludult	3.3	8.2	12.7	33.4	42.4
Askov	Typic Hapludalf	3.1	12.0	13.3	34.2	37.4
Årslev	Typic Agrudalf	2.0	12.3	12.8	37.7	35.2
Ødum	Typic Agrudalf	2.0	13.4	14.1	41.3	29.2
Tørring	Not classified*	2.4	13.8	12.6	36.6	34.6
Rønhave	Typic Agrudalf	2.6	14.2	17.0	41.7	24.5
Højbakkegård	Not classified˅	2.2	15.5	15.9	38.2	28.2
Langvad	Not classified˅	2.1	15.5	15.9	39.1	27.4
Aabenraa	Not classified	4.3	20.0	20.3	27.9	27.5

* same origin as Askov
˅ same origin as Årslev

In the early spring of 1991 a small rainfall simulator was used to estimate soil erodibility (Kamphorst, 1987). A test plot, measuring $0.0625m^2$ ($0.25m \times 0.25m$) was bounded by metal sheets, to a depth of 10cm within the plough layer. A rain shower of 18mm was applied over a period of 3 minutes. The average fall height of the drops was 400mm. A sample bottle placed in a small trench at the bottom of the plot was used to collect runoff and soil loss. Eight replicate measurements were carried out in each field.

Soil shear strength in the plough layer was estimated by 30 replicate measurements using a shear vane, as described by Serota and Jangle (1972). Fifteen samples ($300cm^3$) of undisturbed soil were taken from the 8-12cm layer, for determination of water content and bulk density. Porosity was calculated from bulk density and specific gravity.

2.2 Field plot studies

Comparative measurements of some soil physical parameters were conducted over two years in two field trials concerning soil erosion. The trials included six different soil management treatments, two of which were replicated. The trial design is described in detail by Sibbesen et al. (this volume). Shear strength was measured in the autumn using the procedure described above. Thirty replicates were carried out for each plot. Soil surface roughness was also estimated in the autumn. A level metal frame supported a motor driven vertical gauge rod, which measured the micro-relief within a 3cm × 3cm grid. In the spring, infiltration was measured with a double ring infiltrometer (Bower, 1986) with two replicates per plot. Ten samples of undisturbed soil were collected from depths of 0-5cm and 10-15cm. These samples were taken to the laboratory for the determination of bulk density and water content at a water potential of -100 hPa. Saturated hydraulic conductivity was also measured for these samples, using a steady state method described by Klute and Dirksen (1986) and by a technique described by Rasmussen (1976).

3. RESULTS AND DISCUSSION

3.1 Erodibility of different soil types

Mean results from the rainfall simulation and the soil physical measurements are given in Table 2. Surface runoff tends to decline with increasing content of clay (Figure 1a). This indicates a higher water infiltration capacity for aggregated soils, with higher porosity and pore continuity than for sandy soils. The fact that porosity is a major parameter determining runoff is also evident from Figure 2. Excluding the extremely coarse sandy soil (Jyndevad: 3.6% clay, 86% coarse sand), there is a clear trend of decreasing soil loss with an increasing clay content (Figure 1b). For the sandy soils, there appears to be a more pronounced influence of surface water flow than that experienced for the loamy soils with a higher infiltration. However, soil loss is lowest for soils with a high content of clay (Figure 1c). This indicates the higher resistance of soil particles to detachment.

There seems to be no clear relationship between the specific soil loss to the measured shear strength (Figure 3). This means that in this study shear strength is a poor estimate of soil detachability. However, several authors suggest soil shear strength *is* an effective measure of soil erodibility (Rauws and Govers, 1988; Brunori et al., 1989). The very poor relationship between detachability and soil strength in this study leads to some doubts as to the measuring technique used for the strength estimate. Based on a comparison of different methods, Brunori et al. (1989) suggest a pocket vane for soil erodibility studies. According to these authors, the shear vane used in this investigation correlates linearly with the pocket vane, but overestimates the strength because the shear surface cannot be controlled. However, such a linear deviation from the 'true' soil strength cannot explain the poor correlation of observed specific soil loss and soil shear strength.

Table 2. Results from field investigations of rainfall simulation. Standard error of the mean is given in brackets.

Location	Water runoff %	Soil loss g per surface	Soil loss g per 1 runoff	Soil porosity % v/v	Actual water content % v/v	Air filled porosity % v/v	Shear strength kPa
Jyndevad	58.5 (1.9)	27.5 (6.5)	41.0 (9.4)	41.1 (0.4)	16.3 (0.2)	24.8 (0.4)	6.9 (0.2)
Borris	83.0 (1.1)	55.5 (3.2)	59.4 (3.3)	40.0 (0.4)	26.7 (0.3)	13.3 (0.7)	27.0 (0.5)
Foulum	67.6 (3.7)	29.9 (3.5)	38.8 (3.3)	45.2 (0.7)	28.5 (0.4)	16.7 (0.8)	15.9 (0.6)
Askov	52.4 (4.2)	16.1 (2.9)	26.6 (3.3)	45.1 (1.0)	29.5 (0.4)	15.6 (1.3)	16.2 (0.6)
Årslev	73.6 (3.1)	24.9 (3.7)	29.6 (3.6)	43.1 (0.5)	25.7 (0.4)	17.4 (0.8)	17.8 (0.4)
Ødum	60.0 (4.5)	18.4 (2.7)	26.6 (2.6)	42.1 (0.6)	25.0 (0.5)	17.1 (1.0)	17.9 (0.6)
Tørring	56.0 (5.4)	12.8 (2.4)	19.2 (2.4)	46.8 (0.3)	24.1 (0.3)	22.7 (0.6)	19.4 (0.5)
Rønhave	53.8 (2.5)	10.3 (1.3)	17.0 (2.1)	42.6 (0.6)	27.8 (0.3)	14.8 (0.8)	54.2 (1.8)
Højbakkegård	38.9 (4.1)	6.7 (0.7)	15.5 (0.8)	43.4 (0.6)	24.4 (0.3)	19.0 (0.9)	20.3 (0.7)
Langvad	64.2 (3.7)	15.9 (3.2)	21.2 (3.6)	44.4 (0.5)	27.6 (0.4)	16.8 (0.4)	22.6 (0.7)
Aabenraa	9.3 (2.6)	1.2 (0.4)	15.0 (2.4)	53.8 (0.8)	34.2 (0.5)	19.7 (1.2)	16.6 (0.6)

Figure 1. Relationships between clay content and a) runoff (%), b) soil loss (g) and c) sediment concentration (g/l).

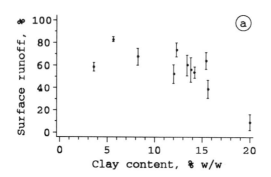

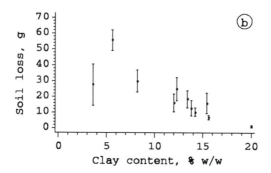

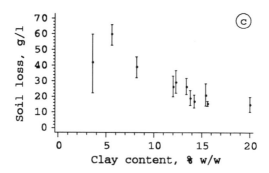

The water content of the soil at the time of measurement is probably a very important factor. Simple shear strength measurements cannot differentiate between the various components of strength and therefore will measure the apparent strength of soil, including the effective strength as well as the matric potential (a 'suction' holding particles together even in single grain sandy materials). As found by Watson and Laflen (1986), soil shear strength measured after rainfall gave a better prediction of interrill erosion than the strength measured before rain. In the present investigation shear strength was measured at around field capacity, before rainfall was simulated. The poor explanation of soil loss by shear strength measurements is probably due to the relatively high proportion of matric potential of the rather sandy soils in question. It is believed that in such soils especially, it is important to estimate shear strength at moisture contents close to saturation.

Figure 2. Relationship between surface runoff (%) and soil porosity (% v/v).

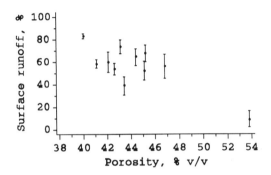

Figure 3. Relationship between sediment concentration (g/l) and shear strength (kPa).

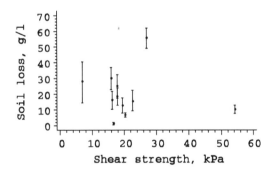

3.2 Field plot studies

As part of a comprehensive research programme on soil erosion (Sibbesen et al., this volume), some soil physical measurements were made in field trials of different soil management practices (Table 3). The measured parameters can be grouped into three categories:

(1) the field infiltration capacity, saturated hydraulic conductivity and related porosity, which are expressions of the soil's ability to assimilate rain and thereby eliminate water accumulation on the soil surface;

(2) the microrelief index, which is a measure of the soil surface roughness following the slope direction. This is related to the capacity of the surface to 'store' water and thereby prevent flow;

(3) the vane shear strength, which is a measure of soil structural stability or detachability of soil particles, which are of importance when water flow actually applies a shear stress upon the soil surface.

The generally better ability to assimilate water by the loamy soil at Ødum, compared to Foulum (Table 3) is probably the main reason for much lower runoff and soil loss from the Ødum plots compared to Foulum. The field measurements of infiltration at Foulum indicate a considerable decrease in the soil's ability to assimilate water, owing to the seedbed preparation. Also the laboratory measurements of saturated hydraulic conductivity show a great difference compared with the ploughed soil without further treatment. A decrease in soil porosity and the volume of coarse pores is probably one explanation of this effect. For the structurally more stable loamy soil at Ødum, the same operations in the field have not resulted in a clear difference in the soil's ability to assimilate water, as observed for the sandy soil. In other words, the two soil types respond quite differently to secondary tillage treatments. This observation is reflected in the large difference in surface runoff between the differently tilled plots at Foulum, while at Ødum runoff was about the same for the plots in question. It is noteworthy that in both soil types, a lower volume of both total and coarse pores is found in the plots which have been tilled across slope, compared with up/down tillage. This might be due to the trial design (no replication of treatments) but this was observed in both years and for both soil types. Therefore, perhaps it is due to a greater compaction effect when the secondary tillage is carried out perpendicular to the ploughing direction, unlike up/down slope treatments.

The index of surface roughness was calculated as the mean square error in a regression of 33 height measurements, 3cm apart in the direction of slope. The slope of the regression model represents an estimate of the slope of the soil surface. The tabulated values (Table 3) are the mean of 4 lines (4 regressions). For both soil types, but especially for the loamy soil, a larger index of roughness is observed in the across-slope tilled plots compared with the up-down tilled plots. The highest value was found at Foulum in the ploughed plot where the soil did not receive any further treatment. At Ødum this plot was harrowed by mistake in the year of measurement and therefore a rather low value is given. The highest soil loss from the field plots for both soil types was measured where soil had been tilled up and down the slope. This is probably due primarily to the difference in surface roughness (tillage orientation) quantitatively described here.

The vane shear strength is increased when the ploughed soil undergoes seedbed preparation (Table 3). As for the water assimilation parameters, the effect is most pronounced at Foulum. Soil strength is important as an erodibility index only when threshold values of water infiltration and surface storage capacity (roughness) have been exceeded. When this occurred, tillage orientation was observed to be of greatest significance to soil erosion occurrence. Therefore soil strength cannot explain the differences in erosion reported by Sibbesen et al. (this volume).

Table 3. Soil physical characteristics of the differently treated top soils. (Values are the means over two years; roughness index is based on only one year of study)

Location	Treatment	Water assimilation measures				Surface roughness measure	Soil strength measure
		Porosity. % v/v		Saturated hydraulic conductivity, mm/hour	Field infiltration capacity, mm/hour	Regression residual error, cm	Vane shear strength, kPa
		total	>30 μm				
Foulum (8% clay)							
	none	48.8	25.1	215	168	1.38	6.1
	harrowed, drilled "up-down" slope	46.9	22.0	78	27	0.63	12.9
	harrowed, drilled "across" slope	43.9	18.0	40	32	0.93	17.2
	SD	1.9	2.6	n.d.ʼ	n.d.ʼ	0.18	1.6
Ødum (13% clay)							
	none*	47.0	23.6	211	231	0.60	4.9
	harrowed, drilled "up-down" slope	46.9	23.6	222	289	0.55	6.6
	harrowed, drilled "across" slope	43.9	19.5	188	226	1.23	7.4
	SD	1.7	2.4	n.d.ʼ	n.d.ʼ	0.09	1.4

* harrowed by mistake in 1990

ʼ not determined (log-normal distribution)

4. CONCLUSIONS

Surface runoff and soil loss declined with an increasing content of clay in the soil. Higher clay content gave higher values of porosity, thus higher infiltration capacity and lower surface runoff. Soil loss could not be explained by shear strength estimates. The lack of correlation may be due to errors in the estimate of strength due to low soil-water content. Soil strength measurements for soil erosion studies should be made at moisture contents close to saturation.

Loamy soil generally had the best ability to assimilate water. On the loamy soil this soil property was only slightly modified by secondary tillage operations, whilst for the sandy soil, seedbed preparation induced a considerable decrease in infiltration. This observation and differences in surface roughness (tillage orientation) between treatments are thought to be the main factors determining surface runoff and soil loss.

REFERENCES

BOWER, H. 1986. Intake rate: cylinder infiltrometer. In: Klute, A. (ed.). Methods of soil analysis I. Physical and mineralogical methods. Agronomy Series No. 9. Madison, Wisconsin, USA.

BRUNORI, F., PENZO, M.C. and TORRI, D. 1989. Soil shear strength: its measurement and soil detachability. Catena 16:59-71.

KAMPHORST, A. 1987. A small rainfall simulator for the determination of soil erodibility. Netherlands Journal of Agricultural Sciences 35:407-415.

KLUTE, A. and DIRKSEN, C. 1986. Hydraulic conductivity and diffusivity: Laboratory methods. In. Klute, A. (ed.). Methods of soil analysis. I. Physical and mineralogical methods. Agronomy Series No. 9. Madison, Wisconsin, USA.

NIELSEN, S.A., STORM, B. and STYCZEN, M. 1986. Development of distributed soil erosion component for the SHE hydrological modelling system. International Conference on Water Quality Modelling in the Inland Natural Environment 10-13 June. Bournemouth, England.

RASMUSSEN, K.J. 1976. Soil compaction by traffic in spring. II. Soil physical measurements (in Danish, with English summary). Tidsskrift for Planteavl 80:835-856.

RAUWS, G. and GOVERS, G. 1988. Hydraulic and soil mechanical aspects of rill generation on agricultural soils. Journal of Soil Science 39:111-124.

SEROTA, S. and JANGLE, A. 1972. A direct-reading pocket shear vane. Civil Engineering ASCE 42:73-74.

SIBBESEN, E., HANSEN, A.C., NIELSEN, J.D. and HEIDEMANN, T. 1994. Runoff, erosion and phosphorus loss from various cropping systems in Denmark. In: Rickson, R.J. (ed.). Conserving Soil Resources: European Perspectives. CAB International.

WATSON, D.A. and LAFLEN, J.M. 1986. Soil strength, slope, and rainfall intensity effects on interrill erosion. Transactions of the American Society of Agricultural Engineers 29:98-102.

WISCHMEIER, W.H. and SMITH, D.D. 1978. Predicting rainfall erosion losses - a guide to conservation planning. US Department of Agriculture, Agriculture Handbook No. 537.

Chapter 9

RUNOFF, EROSION AND PHOSPHORUS LOSS FROM VARIOUS CROPPING SYSTEMS IN DENMARK

E. Sibbesen[1], A.C. Hansen[2], J.D. Nielsen[1] and T. Heidman[1]

[1]Danish Institute of Plant and Soil Science, Research Centre Foulum,
PO Box 23, DK-8830 Tjele, Denmark
[2]Hedeselskabet, Danish Land Development Service,
PO Box 110, DK-8800 Viborg, Denmark

Summary

Surface runoff, soil erosion and loss of phosphorus were measured from differently cultivated plots in Denmark during the autumn-winter periods of 1989/90 and 1990/91. Plots of 22.1 × 3.0m were laid out on slopes of approximately 10% at Foulum and Ødum, on soils containing 8 and 13% clay, respectively.

The following cultivation treatments were compared: 1) Permanent ryegrass pasture; 2) Spring barley followed by a ryegrass catch crop during winter; 3) Spring barley followed by a ploughed surface during winter; 4) Winter wheat drilled across the slope; 5) Winter wheat drilled up and down the slope; and 6) Fallow, ploughed in spring and harrowed from time to time to remove weeds.

The surface runoff and losses of soil and phosphorus ranged from 5-162 mm, 0.01-26 t ha^{-1} and 0.07-33kg ha^{-1}, respectively. Generally, the runoff and losses were greater at Foulum than at Ødum, and greatest in the second year because of a wetter autumn. No distinct thaw events occurred. The runoff and losses from the various treatments generally followed the order: Fallow = wheat-up-down > wheat-across >> barley-ploughed = barley-catch crop > pasture.

1. INTRODUCTION

Erosion by water seems to be an increasing problem on loams, sandy loams and sandy soils in northern Europe (Morgan, 1986; Colborne and Staines, 1985). According to Hansen (1989) erosion by water is not significant in Denmark from a farming point of view, but very few studies have been made (Hasholt et al., 1990), so our knowledge of the extent, frequency and rate of erosion by water and its relationship with soil, land use, land form and climate in Denmark is scarce. Furthermore, surface runoff and erosion can transport nutrients, especially phosphorus to aquatic environments, where they may cause damage.

Rainfall in Denmark is generally of low intensity. The USLE EI$_{30}$ index, in units of kJ.m^{-2}.mm.hr^{-1}, is 20-43 in Denmark (J.E. Olesen, pers. communic.) compared to 43-152 in Bavaria (Rogler and Schwertmann, 1981) and 40-1100 in the USA (Wischmeier and Smith, 1978). However, low intensity rain may cause ponding, surface runoff and erosion if infiltration capacity is reduced because the soil surface structure is degraded (Monnier and Boiffin, 1986) or the soil is frozen. Sharp frosts do not occur every year in Denmark.

About 60% of the total land area in Denmark is arable. In Germany, the Netherlands, Belgium, France, Ireland and the UK, arable land amounts to about 30% or less (Eurostat, 1988). To combat nitrogen leaching, Danish farmers have to keep at least 60% of their land covered with cereals, rape or other crops in winter. Under English conditions fields with winter cereals suffer more water erosion than ploughed ones (Colborne and Staines, 1985).

The aim of this work was to study the effects of different cropping systems on surface runoff, erosion by water and loss of phosphorus under Danish conditions.

2. METHODS

Two field experiments were started at Foulum and Ødum in Jutland, Denmark in autumn 1989. At both sites eight plots of 22.1 × 3.0m were laid out on slopes of 10.4% at Foulum, and 9.6% at Ødum. The soil at Foulum is a typic hapludult with 8.2% clay. At Ødum the soil is a typic agrudalf with 13.4% clay.

The plots were given the following six treatments:

1. Permanent ryegrass, cut four times per year;
2. Spring barley followed by a ryegrass catch crop during the winter, ploughed in the spring;
3. Spring barley, ploughed in the autumn;
4. Winter wheat drilled across the slope, ploughed in the autumn;
5. Winter wheat drilled up and down the slope, ploughed in the autumn;
6. Fallow, ploughed in the spring and harrowed from time to time to remove weeds.

Treatments 5 and 6 were replicated. All plots were framed by 10cm high boards. Paths of 0.5m wide between the plots allowed inspection of the plots without disturbance. Troughs at the lower end of each plot gathered surface runoff and eroded soil, and discharged them into 1300 litre tanks. The tanks were emptied from time to time and the amounts of water and suspended and precipitated sediments were determined. The water samples were taken with a depth integrating sampler (Nilsson, 1969). The surface soil and the samples of suspensions filtrates (0.45µm) and sediments were analysed for total P and various P fractions.

Precipitation was recorded every minute and temperature was measured at 100cm and 20cm above and 5cm and 10cm below the soil surface. Soil physical and hydraulic parameters were determined by Schjønning (this volume).

3. RESULTS AND DISCUSSION

Only the main results of the following two autumn/winter periods, 21 October 1989 - 9 April 1990 and 6 October 1990 - 6 March 1991 are reported. Little runoff and erosion occurred during the spring and summer. Both winters were relatively mild, with no distinct frost periods or thaw events.

Autumn 1990 was much wetter than that of 1989, resulting in considerably higher surface runoff (Figure 1). Widespread water erosion took place in Denmark during autumn 1990, especially in areas with newly sown winter wheat. The wet conditions probably favoured additional soil compaction during seedbed preparation, and early degradation of the surface soil structure. Both factors may have contributed to a lower than usual water infiltration capacity of the surface soil.

Figure 1. Surface runoff from differently cultivated plots at Foulum and Ødum, during autumn/winter 1989/90 and 1990/91.

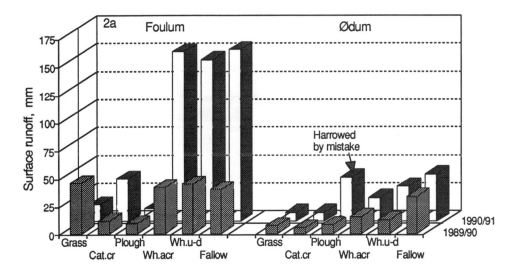

The surface runoff was generally much greater at Foulum than at Ødum. The difference is most likely to be due to a smaller infiltration capacity caused by a weaker soil structure at Foulum (Schjønning, this volume).

When comparing treatments, it should be noted that the wheat and fallow plots had a much greater surface runoff than both the ploughed plot and those under permanent grass and a catch crop (Figure 1). These effects probably arise from differences in storage capacity for surface water and water infiltration capacity. Seedbed preparation and drilling made the soil surface smooth compared to a ploughed surface, and thereby decreased the surface-water storage capacity.

The surface runoff detached and transported least soil under permanent grass and catch crop treatments, and most under wheat-cultivated-up-down slope (Figure 2). As expected, the mean sediment concentration generally increased with increasing surface runoff.

Figure 2. Mean sediment concentration from differently cultivated plots at Foulum and Ødum, during autumn/winter 1989/90 and 1990/91.

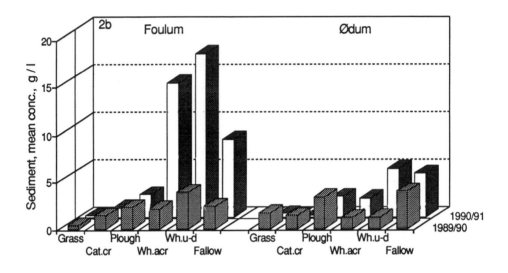

The resulting total sediment losses are shown in Figure 3. Our results agree with the observation of Colborne and Staines (1985) that fields in winter cereals are more prone to water erosion than ploughed fields. However, the expected effect of drill direction was not clearly illustrated here. Figure 4 shows that the mean concentration of dissolved P in the runoff was high in both places. Both soils are very P-fertile, with around 70mg $NaHCO_3$-extractable P/kg soil (according to Olsen et al., 1954). The concentration of dissolved P in the runoff was greatest for the permanent grass and catch crop treatments, which agrees with the findings of Uhlen (1989). According to Bromfield and Jones (1972) and Schreiber and McDowell (1985), half the phosphorus in plant litter can easily be washed out.

The mean concentration of particulate P (Figure 5) increased with increasing concentrations of transported sediment (Figure 2). Amounts of total P (Figure 6) also increased with increasing total sediment loss (Figure 3). About 90% or more of the total P-loss consisted of particulate P, except for the permanent grass plots and the catch-crop plot at Ødum.

In the only other Danish erosion plot study, much less surface runoff and losses of soil and phosphorus were recorded (Hasholt et al., 1990). However, three out of the four soils under that study were probably not susceptible to surface runoff and erosion, as two were coarse sands on 4% slopes and one was a loam soil on a 2% slope. Their fourth soil, a sandy loam on a 12% slope had a ploughed winter surface for all three years, which, according to our results, is not prone to surface runoff and erosion under Danish conditions.

Figure 3. Total sediment loss from differently cultivated plots at Foulum and Ødum, during autumn/winter 1989/90 and 1990/91.

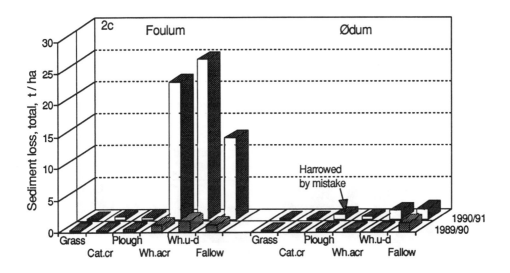

Figure 4. Mean concentrations of dissolved P (total loss of dissolved P divided by total runoff) from differently cultivated plots at Foulum and Ødum, during autumn/winter 1989/90 and 1990/91.

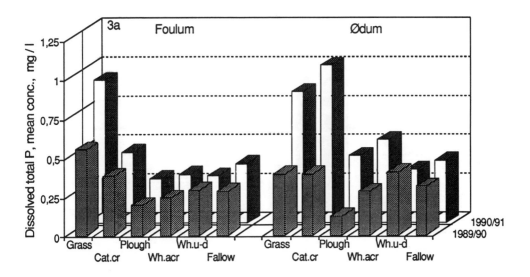

Figure 5. Particulate P (total; loss of particulate P divided by total runoff) from differently cultivated plots at Foulum and Ødum, during autumn/winter 1989/90 and 1990/91.

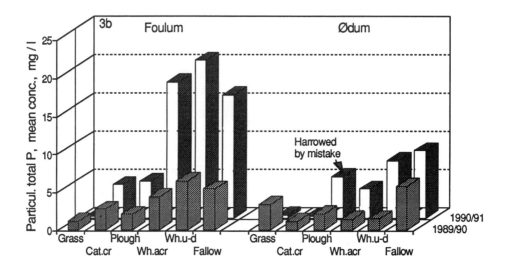

Figure 6. Total loss of P from differently cultivated plots at Foulum and Ødum, during autumn/winter 1989/90 and 1990/91.

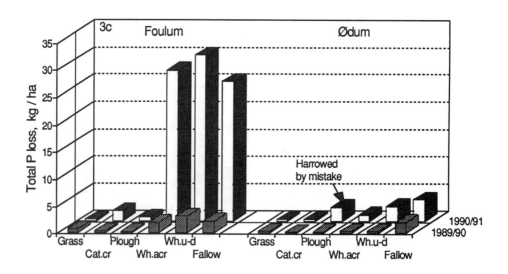

4. CONCLUSION

Under Danish conditions with no distinct thaw events, surface runoff, total soil loss and loss of phosphorus during the autumn-winter period are much affected by year, soil type and cropping system. A wet autumn seems to increase the losses compared to a dry autumn. Fields with winter cereals seem to be much more susceptible to surface runoff, erosion and P losses than ploughed land, pasture or fields with catch crops.

REFERENCES

BROMFIELD, S.M. and JONES, O.L. 1972. The initial leaching of hayed-off pasture plants in relation to the recycling of phosphorus. Australian Journal of Agricultural Research 23:811-824.

COLBORNE, G.J.N. and STAINES, S.J. 1985. Soil erosion in south Somerset. Journal of Agricultural Science 104:107-112.

EUROSTAT 1988. Agriculture. Statistical Yearbook, Theme 5, Series A. Luxembourg.

HANSEN, L. 1989. Soil tillage, soil structure and soil erosion in Denmark. Soil Technology Series 1.

HASHOLT, B., KUHLMAN, H., MADSEN, H.B., HANSEN, A.C. and PLATOU, S.W. 1990. Erosion og transport af fosfor til vandløb og søer. NPo-forskning fra Miljøstyrelsel Nr. C12. p.118 (In Danish).

MONNIER, G. and BOIFFIN, J. 1986. Effect of agricultural use of soils on water erosion: the case of cropping systems in Western Europe. In: Chisci, G. and Morgan, R.P.C. (eds). Soil erosion in the European Community. A.A. Balkema, Rotterdam & Boston. pp.17-32.

MORGAN, R.P.C. 1986. Soil degradation and soil erosion in the loamy belt of northern Europe. In: Chisci, G. and Morgan, R.P.C. (eds). Soil erosion in the European Community. A.A. Balkema, Rotterdam & Boston. pp.165-172.

NILSSON, B. 1969. Development of a depth-integrating water sampler. UNGI Report 2, Uppsala, Sweden.

OLSEN, S.R., COLE, C.V., WATANABE, F.S. and DEAN, L.A. 1954. Estimation of available phosphorus in soils by extraction with sodium bicarbonate. U.S. Dept. Agric. Circ. 939.

ROGLER, H. von and SCHWERTMANN, U. 1981. Erosivität der Niederschläge und Isoerodentkarte Bayerns. Zeitschrift für Kulturtechnik und Flurbereinigung 22:99-112.

SCHREIBER, J.D. and McDOWELL, L.L. 1985. Leaching of nitrogen, phosphorus and organic carbon from wheat straw residues: I. Rainfall intensity. Journal of Environmental Quality 14:251-256.

UHLEN, G. 1989. Nutrient leaching and surface runoff in field lysimeters on a cultivated soil. Nutrient balances 1974-81. Norwegian Journal of Agricultural Sciences 3:33-46.

WISCHMEIER, W.H. and SMITH, D.D. 1978. Predicting rainfall erosion losses - a guide to conservation planning. U.S. Department of Agriculture, Agriculture Handbook No. 537.

Chapter 10

NUTRIENT LOSSES AND CROP YIELDS IN THE WOBURN EROSION REFERENCE EXPERIMENT

J.A. Catt[1], J.N. Quinton[2], R.J. Rickson[2] and P. Styles[3]
[1]Soil Science Dept, AFRC Inst. of Arable Crops Research, Rothamsted
Experimental Station, Harpenden, Hertfordshire, AL5 2JQ, UK
[2]Silsoe College, Cranfield University, Silsoe, Bedfordshire, MK45 4DT, UK
[3]AFRC Institute of Arable Crops Research, Woburn Experimental Farm,
Husborne Crawley, Woburn, Bedfordshire, MK43 0XF, UK

Summary

The Woburn Erosion Reference Experiment, begun in 1988 by Silsoe College and Rothamsted Experimental Station, has two contrasting soil management treatments:
- cultivations across the slope, or up and down the slope;
- minimal tillage with crop residues incorporated, or standard cultivations with residues removed.

Results for the first three years of the experiment (1989/91) show that in generally dry periods the largest amounts of runoff, soil loss and nutrient loss were from plots cultivated up and down the slope. The smallest amounts were from plots cultivated across the slope. In the wetter winter of 1989/90, the largest amounts of runoff, soil loss and nutrient loss were from plots given standard cultivations and the smallest were from minimally cultivated plots. This was probably because cultivations for late sowing of winter wheat in wet conditions compacted the soil and decreased its infiltration capacity. Plots cultivated across the slope yielded 25% more potatoes, 9% more wheat and 7% more barley than those cultivated up and down the slope. Plots given standard cultivations yielded more than those given minimal cultivation. Nutrient concentration factors (transported sediment/original soil) were usually influenced by cultivation practice and direction as were the amounts of runoff, soil loss and nutrient loss.

1. INTRODUCTION

Experimental erosion plots were established in England and Spain in 1988 to provide data on runoff and soil losses for eventual validation of EUROSEM (Morgan et al, 1990). The English plots were laid out at Woburn Experimental Farm, Bedfordshire (National Grid Reference SP969358) in collaboration with Rothamsted Experimental Station. Rothamsted's interest in the Woburn experimental plots is to assess the long-term effects of erosion on soil fertility, crop production and nutrient losses, and to evaluate various cultivation techniques for minimising the detrimental effects of erosion.

This paper reviews results from spring 1989 to autumn 1991, during which period three crops have been grown - potatoes, sown in spring 1989 and harvested in summer 1989, winter wheat, sown in autumn 1989 and harvested in summer 1990, and winter barley, sown in autumn 1990 and harvested in summer 1991. Apart from the winter of 1989/90, the period was unusually dry; twenty four erosion events were recorded on one or more of the plots, but 17 of these occurred between 20 October 1989 and 25 February 1990. The atypical meteorological conditions limit the conclusions that can be drawn from the data at this early stage, but the results for such dry periods may be relevant to prediction of erosion in SE England in a future scenario of global climate change.

2. SITE CHARACTERISTICS

The 8 plots of the experiment are on a site (Great Hill III) subject to erosion since at least 1950 (Catt, 1992). The mean slope of the plots is 5° with a SW aspect, and their altitude ranges from 100 to 107m O.D. Each plot measures 24 × 36m (area 0.0864ha) and is bounded by grassed earth bunds, 1m wide and 30cm high, to prevent runoff and eroded soil from leaving the plot before they are measured in collectors at the lower end. The bunds can be crossed by farm machinery.

The soil of the site is mainly the Cottenham series, a brown sand (as defined by Avery, 1980, p.47) developed in the Cretaceous Lower Greensand or Woburn Sands (Catt et al., 1975). On lower slopes it gives way to the Lowlands series (previously Stackyard series), a colluvial brown earth (Avery 1980, p.47) formed in loamy superficial colluvial deposits derived from upslope by fieldwash (i.e. past soil erosion).

3. EXPERIMENTAL DESIGN AND METHODS

The eight plots are arranged in two blocks, each of which has four plots with different combinations of the following cultivation treatments:

(A) Direction of all cultivation and drilling
    (i) across the plots, parallel to the contour
    (ii) up and down the slope, perpendicular to the contour.

(B) Cultivation practice
    (i) standard cultivations (ploughing to 20cm and seedbed preparation techniques appropriate to the crop and weather conditions) after removal of crop residues
    (ii) minimum tillage (shallow tining) with retention of chopped crop residues.

All plots grow the same crop each year in the following rotation: potatoes, winter wheat, winter barley, sugar beet, winter wheat, winter barley. The two root crops are sown in the spring and harvested in late summer. The winter cereals are sown in autumn and harvested the following summer. This means that between the third and fourth crops in the rotation and at the end of each rotation the plots have no crop cover over the winter. However, this paper deals only with the period up to harvest of the first winter barley crop. The crop rotation, or others similar to it, are widely used by farmers working sandy soils in the Woburn area. The fertiliser and pesticide treatments for each crop are also typical of those recommended for the area. Crop yields are measured on representative areas within each plot and are quoted at 85% dry matter content for cereal grain, and as fresh weight for root crops.

The collector tanks at the foot of each plot are emptied periodically during or soon after erosion events, and samples of runoff and eroded sediment are taken for chemical analyses. Soil samples are taken from each plot at intervals of approximately one year. Total nitrogen in both the soil and eroded sediment samples was determined by Kjeldahl digestion and estimation as ammonium using a Technicon AA II continuous flow colorimetric analyser (Crooke & Simpson, 1971). In the runoff samples ammonium and nitrate were determined with an Alpkem rapid segmented flow colorimetric analyser; the nitrate after reduction to nitrite using a cadmium coil and diazotisation using sulphanilic acid and N-(1-napthyl) ethylenediamine (Litchfield, 1967). Total phosphorus and potassium were determined by inductively coupled plasma arc spectrometry. The soil and eroded sediment samples were pre-digested in aqua regia. For each erosion event the amounts of runoff were measured in litres, and losses of total soil, N, P and K in runoff and N, P and K in the eroded sediment were expressed in kg ha^{-1}. Runoff, soil loss and amounts of N, P and K were then summed for the periods between sowing and harvest of each crop.

Amounts of runoff, total soil and nutrient losses and nutrient concentration factors associated with each of the two cultivation directions and the two cultivation practices were compared by calculating mean values for the four plots given each of these four treatments. Because of the design of the experiment, the four values contributing to each mean for a cultivation direction include two pairs with contrasting cultivation practices. Likewise, the four contributing to the mean for each cultivation practice include two pairs each with contrasting cultivation directions. However, the means of four values give the best estimates of the effects of each cultivation direction or practice because the influences of the two other treatments cancel one another. Plots with identical treatments occur only in pairs, one plot from each block, and therefore provide means with larger standard errors than the means of four values. Also, as each pair has one cultivation direction and one cultivation practice, these means do not distinguish the effects of individual soil management treatments. Most differences between means were not statistically significant. Those that were significant are identified in the subsequent discussion.

4. RESULTS

4.1. Frequency of erosion events 1989-91

There was only one erosion event under potatoes in the early part of 1989. This resulted from a storm of 27mm rain on 7 July, which accounted for 18% of the total rainfall at Woburn from May to September 1989. In the following winter 17 events under winter wheat resulted from a total rainfall of 395mm from October 1989 to February 1990. Under winter barley there were 6 small events between 25 December 1990 and 8 January 1991, resulting from a total rainfall in these two months of 112mm.

4.2. Runoff and soil loss (Table 1)

The single erosion event of 7 July 1989 generated a mean of 1875 litres runoff on the four plots cultivated up and down the slope, but only 343 litres for plots cultivated across the slope. The deep furrows between potato ridges acted as effective channels to collect runoff and, where they were aligned up and down the slope, to conduct it down the plots and into the collectors. The 5.5-fold increase in runoff led to a 14.9-fold increase in soil loss. The differences in mean runoff and soil loss from the plots given standard and minimal cultivations were much less. Both were greater from the minimally cultivated plots.

Table 1. Mean values for runoff (1) and soil loss (kg ha^{-1}) for the four cultivation treatments under three crops (SED: standard error of the difference between means; not given for 1990/91 because of zero values for some plots).

	Across slope	Up and down slope	Standard, residues removed	Minimal, residues retained	SED
Potatoes 1989					
Runoff	343	1875	734	1484	611
Soil loss	65	967	487	545	429
Winter wheat 1989/90					
Runoff	18800	20999	22539	17261	3858
Soil loss	7228	9513	19154	6587	6460
Winter barley 1990/91					
Runoff	97	465	386	176	--
Soil loss	5	59	35	29	--

In the wet winter of 1989/90 there were large amounts of runoff and soil loss from all the plots. When these events occurred the wheat plants were still small and offered little protection to the soil. The different directions of cultivation resulted in much smaller differences in runoff (12% greater with up and down cultivations than with cultivations across the slope) and soil loss (32% greater with up and down cultivation) than occurred previously under potatoes. However, there was a stronger contrast between plots given standard and minimal cultivations. The losses were greater from plots given standard cultivations (31% more runoff and 54% more soil loss) than from those cultivated minimally.

The six small events of winter 1990/91 resulted in much less runoff and soil loss under winter barley than the single event under potatoes in July 1989. However, the increases in runoff and soil loss from plots cultivated up and down the slope relative to those cultivated across the slope (4.8-fold increase in runoff and 11.6-fold increase in soil loss) were similar to those under potatoes. As with winter wheat, standard cultivations led to greater runoff and soil losses than minimal cultivation.

4.3. Nutrient losses (Table 2)

The amounts of all three nutrients lost in solution in runoff were always less than those lost as solid particles in the transported sediment. Most of the nitrogen was probably lost as particulate organic matter, as was most of the phosphorus, with P adsorbed on sesquioxide and clay particles or as phosphatic sand minerals such as apatite and collophane, and most of the potassium as glauconite, muscovite, feldspars, illitic clays or other K-bearing minerals known to occur in soils on the Lower Greensand at Woburn (Catt et al., 1975). Losses of potassium both in solution and as solid particles were usually several times those of phosphorus or nitrogen. They were exceeded by losses of nitrogen on only a few plots in 1990/91.

97

Table 2. Mean nutrient losses (kg ha⁻¹) in runoff and transported soil sediment for the four cultivation treatments under three crops. (SED: standard error of the difference between means; not given for 1990 91 because of zero values for some plots).

	Across slope	Up and down slope	Standard, residues removed	Minimal, residues retained	SED
Potatoes 1989					
N in runoff	0.02	0-.09	0.04	0.07	0.066
P in runoff	0.01	0.04	0.02	0.03	0.031
K in runoff	0.07	0.36	0.13	0.30	0.264
N in sediment	0.03	2.59	0.79	1.84	0.918
P in sediment	0.12	3.71	0.95	2.88	2.220
K in sediment	0.35	7.49	3.43	4.41	3.301
Winter wheat 1989/90					
N in runoff	0.40	0.48	0.51	0.38	0.107
P in runoff	0.27	0.29	0.33	0.24	0.062
K in runoff	0.87	0.91	0.96	0.82	0.181
N in sediment	17.50	12.39	18.33	11.56	10.470
P in sediment	9.84	9.04	13.00	5.87	6.550
K in sediment	33.16	30.30	42.98	20.48	20.510
Winter barley 1990/91					
N in runoff	0.00	0.08	0.08	0.01	--
P in runoff	0.00	0.01	0.01	0.01	--
K in runoff	0.02	0.06	0.04	0.04	--
N in sediment	0.01	0.30	0.10	0.21	--
P in sediment	0.01	0.16	0.15	0.02	--
K in sediment	0.02	0.27	0.23	0.06	--

Under potatoes in 1989, the losses of all three nutrients in both runoff and transported sediment were largest from plots cultivated up and down the slope and least from plots cultivated across the slope. The minimally cultivated plots lost more than those given standard cultivations. This order of losses by treatment is the same as those for runoff and total soil loss under potatoes.

In contrast, the largest losses of both dissolved and particulate nutrients under winter wheat (1989/90) were from the plots given standard cultivations, and the least were from the minimally cultivated plots. For plots cultivated up and down or across the slope the losses were between these two extremes; amounts of all three nutrients in solution were greater where the cultivations were up and down the slope, but amounts in transported sediment were greater where the cultivations were made across the slope. This order of losses by treatment is the same as those for runoff and total soil loss under wheat.

Under winter barley (1990/91) the ranking of nutrient losses reverted almost to that under potatoes. The largest losses of all three nutrients in both runoff and transported sediment were from plots cultivated up and down the slope, and the least were from plots cultivated across the slope. However, under barley the losses of N in runoff and of P and K in particulate matter were greater from plots given standard cultivations than from plots cultivated minimally, whereas under potatoes the reverse was true.

The total nutrient losses (in solution and as solid particles) were very small compared with amounts applied in fertiliser each year (Table 3), especially under potatoes in 1989 and barley in 1990/91. Moreover, as only small percentages of the total nutrients lost were in solution, the amounts likely to have been derived from the immediately previous fertiliser applications are much smaller even than the percentages shown in Table 3. Even the relatively large amounts of soluble N lost in runoff during the wet winter of 1989/90 were equivalent to less than 1% of the N applied to the winter wheat. As no P or K was applied to the two cereal crops, the losses of these two nutrients cannot be expressed as percentages of the amounts applied and have therefore been omitted from Table 3.

Table 3. Nutrient losses from the four cultivation treatments as percentages of fertiliser applied to each crop.

	Across slope	Up and down slope	Standard, residues removed	Minimal, residues retained	Fertiliser applied (kg ha^{-1})
Potatoes 1989					
% N	0.02	1.15	0.35	0.81	234
% P	0.13	3.68	0.95	2.85	102
% K	0.14	2.63	1.19	1.57	299
Winter wheat 1989/90					
% N	9.17	6.60	9.66	6.11	195
Winter barley 1990/91					
% N	0.91	0.24	0.11	0.14	160

4.4. Soil nutrient composition and sediment/soil concentration factors (Table 4)

The mean amounts of total nutrients measured in the plot soils show no trends with plot position or time.

Because there were different numbers of erosion events in each year and some of the events were restricted to certain plots, there were different numbers of eroded sediment samples analysed from each plot and in each year. Mean concentrations of N, P and K were calculated for sediment eroded from each plot for the periods when each crop was growing. From these, mean concentration factors (sediment/soil) were calculated for each plot and for each crop period. Table 4 shows the values of these averaged over each cultivation direction and practice.

Table 4. Mean values for nutrient concentration factors (transported sediment/plot soil) for the four cultivation treatments under three crops (SED: standard error of the difference between means; not given for 1990/91 because of zero values for some plots).

	Across slope	Up and down slope	Standard, residues removed	Minimal, residues retained	SED
Potatoes 1989					
N	1.222	3.254	2.135	2.342	0.1716
P	0.887	2.162	1.303	1.746	0.0925
K	0.753	2.263	1.394	1.623	0.1218
Winter wheat 1989/90					
N	2.777	1.918	2.522	2.173	1.3273
P	2.118	1.453	1.820	1.751	0.9165
K	2.345	1.425	1.994	1.776	1.0496
Winter barley 1990/91	1				
N	0.373	1.398	1.224	0.546	--
P	0.253	0.893	0.750	0.397	--
K	0.227	0.917	0.786	0.358	--

Under potatoes (1989) the largest concentration factors for all three nutrients were for sediment from the plots cultivated up and down the slope, and the smallest were for sediment from plots cultivated across the slope. The differences between these cultivation directions were significant at <1% level for all three nutrients. Mean values for plots minimally cultivated were slightly greater for all nutrients than those for plots given standard cultivations. The order of nutrient concentration factors in relation to cultivations (up and down > minimal > standard > across) is the same as those for runoff, soil loss and amounts of all nutrients (both in runoff and transported sediment) under potatoes.

Under winter wheat (1989/90) the largest concentration factors for all three nutrients were for sediment from plots cultivated across the slope, and the smallest were for plots cultivated up and down the slope. The factors for plots given standard cultivations were slightly greater than those for plots given minimal cultivation. The order of concentration factors in relation to cultivations (across > standard > minimal > up and down) was different from the order for runoff, soil loss and amounts of nutrients in both runoff and transported sediment under wheat (standard > up and down > across > minimal).

Under winter barley (1990/91) all the mean concentration ratios for P and K and half of those for N were less than one (i.e. the sediment contained less nutrients than the original soil). As with potatoes, the largest ratios for all three nutrients were for plots cultivated up and down the slope and the smallest were for plots cultivated across the slope. However, unlike potatoes, the ratios for plots given standard cultivations were greater than those for plots cultivated minimally. The order of concentration factors in relation to cultivations (up and down > standard > minimal > across) was the same as the orders for runoff, soil loss and nutrient losses in runoff under barley.

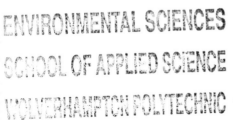

4.5. Crop yields (Table 5)

The yields of potatoes (1989) and winter wheat (1990) were smallest on plots cultivated up and down the slope and largest on plots cultivated across the slope (25% and 9% more, respectively). The plots with standard or minimal cultivation gave intermediate yields. In 1990/91 the yields of winter barley were smallest for plots given minimal cultivations and largest (33% more) for plots given standard cultivations. The yields of plots cultivated in different directions were then intermediate, though those cultivated across the slope gave yields 7% greater than those cultivated up and down the slope. Despite these changes in the order of yields, plots cultivated across the slope always out-yielded those cultivated up and down, and plots given standard cultivations always out-yielded those given minimal cultivation.

Table 5. Mean crop yields (t ha) for the four cultivation treatments (SED: standard error of the difference between means).

	Across slope	Up and down slope	Standard, residues removed	Minimal, residues retained	SED
Potatoes 1989	35.4	28.4	32.4	31.4	3.52
Winter wheat 1989/90	5.04	4.57	4.85	4.76	0.63
Winter barley 1990/91	7.50	7.03	8.29	6.23	0.35

5. DISCUSSION

The four different soil management treatments resulted in very large differences in runoff and soil loss, especially under potatoes in summer 1989 and winter barley in winter 1990/91. Under both of these crops the runoff and soil losses were approximately 5-15 times greater from plots cultivated up and down the slope than from plots cultivated across the slope. The associated losses of nutrients were approximately 3-86 times greater, but these can account for only a small part of the yield differences, because (a) the losses are very small percentages of the total nutrients applied as fertiliser to the potatoes and barley, (b) most of the nutrients lost were in organic matter or soil minerals and therefore not directly available to the crops, and (c) the nitrogen was applied to the barley crop as a spring top-dressing after the erosion had occurred in the preceding winter.

The yield differences for potatoes may be partly attributable to the different amounts of runoff, as this crop is sensitive to the amount of available water, especially on sandy soils like those at Woburn in dry summers (Catt et al., 1980). The potato ridges across the plots acted as small dams trapping the runoff, whereas those orientated up and down the slope conducted runoff down the plots to the collectors. The measured yield difference need not be entirely attributed to the single runoff event on 7 July 1989. The runoff difference on this date was equivalent to only 2mm rain. At other times rain which did not cause any runoff may have preferentially reached the potato roots where the ridges were orientated across the slope, because it would tend to infiltrate downslope from the furrows into the base of each ridge. This would not happen where the ridges were orientated up and down the slope because any water accumulating in the furrows would infiltrate the soil directly beneath the furrows.

Another factor likely to cause differences in soil water is the different aspects of ridges orientated across and up and down the slope. Those up and down the slope are orientated NE-SW, so neither side of the ridges is protected either from the prevailing (SW) wind at the site or from sunshine throughout the day. In contrast, the NE-facing slopes of ridges orientated across the slope (i.e. NW-SE) are in the lee of the prevailing wind and also receive little sun during the day, and so lose much less water by evaporation.

However, runoff differences cannot account for the differences in barley yields, as the runoff occurred entirely in the winter when the barley was growing very slowly and required little water. The exact cause of the surprisingly large (but not statistically significant) difference in yields of barley between plots given standard and minimal cultivations is therefore uncertain at the moment.

The smaller yield increases (7% and 9%) for cereals grown across the slope compared with those grown up and down the slope can perhaps be attributed to slower movement of the grain across the sieve within the combine harvester. As the harvester moves up the slope some of the grain may be lost among the straw and chaff carried over the sieve and ejected from the rear of the machine.

Compared with that for potatoes and barley, the different pattern of relative amounts of runoff, soil loss and nutrient loss from the four cultivation treatments under winter wheat in 1989/90 can probably be attributed to late planting of the crop in the wet soil conditions of that autumn. The strongest contrasts in runoff, total soil loss and losses of all three nutrients, both in solution and as solid particles, were between the plots given standard cultivations with the previous crop residues (potato haulm) removed and those given minimal cultivation with the haulm retained. The amounts of haulm were fairly small, so it is unlikely that this played much part in limiting runoff and losses from the minimally cultivated plots. The main factor was probably wheelings created by the cultivations on soil which was soft after harvesting potatoes and ploughing, and was also rather moist when the wheat was sown late in the autumn. The wheelings compacted the soil making it less permeable and creating more runoff than on plots given minimal cultivation.

Under wheat in 1989/90 the percentage differences in runoff and total soil loss between plots cultivated up and down the slope and those cultivated across the slope were much smaller than under the other two crops. The nutrient losses in solid particles were greater from plots cultivated across the slope, instead of greater from plots cultivated up and down the slope. This change in the pattern of nutrient losses was emphasised by the largest nutrient concentration factors under wheat occurring on plots cultivated across the slope and the smallest on plots cultivated up and down. It arose because in the frequent heavy rain of the 1989/90 winter, the wheeling channels orientated across the slope often acted as reservoirs, which overflowed at a few low points causing fewer but larger rills to develop compared with those on plots where the wheelings and other cultivation channels were orientated up and down the slope. The relative enrichment of nutrient-rich fine material in sediment from plots cultivated across the slope resulted from selective redeposition of nutrient-poor coarser sediment (sand and silt) within the plots. This happened because much of the runoff on plots cultivated across the slope had to travel a greater distance from the wheelings near the upper margins of these plots before reaching the sediment traps at the foot of each plot. Also crop rows aligned across the slope assisted redeposition of coarser sediment by decreasing the energy of the surface runoff more than rows aligned up and down the slope. Selective redeposition of coarse particles rather than preferential entertainment of fine particles on the across-cultivated plots under wheat is indicated by the fact that the change in order of particulate nutrient losses with cultivation direction (across > up and down under wheat, but up and down > across under potatoes and barley) did not occur with nutrient losses in solution (up and down > across under all three crops).

Despite the greater losses of total nutrients in 1989/90 from plots given standard cultivations than from those cultivated minimally and the greater losses from plots cultivated across the slope than from those cultivated up and down, the mean yields of wheat were least on plots where minimal cultivation was practised or the cultivations were up and down the slope. This supports the conclusion that the yield differences result from some factor other than the nutrient losses.

For all three crops, yields were slightly greater on plots given standard cultivations than on minimally cultivated plots. Although crop yields over a range of soil types in UK are not usually decreased by minimal cultivation compared with ploughing (Cannell, 1985), lighter soils can give slightly lower yields in the first few years of direct drilling because of soil surface compaction and poor germination (Ellis et al., 1982; Ball, 1988). In later years the aggregate stability and crop yields of direct drilled or minimally cultivated soils are improved because the organic matter content of the surface soil increases. This should also decrease runoff and soil losses, an effect that may already be evident at Woburn.

ACKNOWLEDGEMENT

We thank A.D. Todd for statistical help.

REFERENCES

AVERY, B.W. 1980. Soil classification for England and Wales (higher categories). Soil Survey Technical Monograph 14, Rothamsted Experimental Station, Harpenden.

BALL, B.C. 1988. Reduced tillage in Great Britain: practical and research experience. Proceedings of EEC Workshop on Energy Saving by Reduced Tillage, 10-12 June 1987, Göttingen, W. Germany.

CANNELL, R.Q. 1985. Reduced tillage in north-west Europe - a review. Soil and Tillage Research 5:129-177.

CATT, J.A. 1992. Soil erosion on the Lower Greensand at Woburn Experimental Farm, Bedfordshire - evidence, history and causes. In Bell, M. & Boardman, J. (eds). Past and present soil erosion. Oxbow Monograph 22.

CATT, J.A., KING. D.W. and WEIR, A.H. 1975. The soils of Woburn Experimental Farm. I. Great Hill, Road Piece and Butt Close. Report Rothamsted Experimental Station for 1974, Part 2:5-28.

CATT, J.A., WEIR, A.H., NORRISH, R.E., RAYNER, J.H., KING, D.W., HALL, D.G.M. and MURPHY, C.P. 1980. The soils of Woburn Experimental Farm. III. Stackyard. Report Rothamsted Experimental Station for 1979. Part 2:5-39.

CROOKE, W.M. and SIMPSON, W.E. 1971. Determination of ammonium in Kjeldahl digests of crops by an automated procedure. Journal of the Science of Food and Agriculture 22:9-10.

ELLIS, F., CHRISTIAN, D.G. and CANNELL, R.Q. 1982. Direct drilling, shallow tine cultivation and ploughing on a silt loam soil, 1974-1980. Soil and Tillage Research 2:115-130.

LITCHFIELD, M.H. 1967. The automatic analysis of nitrite and nitrate in blood. Analyst 92:132-136.

MORGAN, R.P.C., QUINTON, J.N. and RICKSON, R.J. 1990. Structure of the soil erosion prediction model for the European Community. In Proceedings of International Symposium on Water Erosion, Sedimentation and Resource Conservation. Central Soil and Water Conservation Research and Training Institute, Dehra Dun. pp.49-59.

Chapter 11

RUNOFF, SOIL EROSION AND PESTICIDE POLLUTION IN CORNWALL

T.R. Harrod
Soil Survey and Land Research Centre, Staplake Mount,
Starcross, Exeter, EX6 8PE, UK

Summary

The mild climate of west Cornwall has encouraged intensive farming of daffodils, cauliflowers, cabbage and early potatoes. All these crops are grown over winter and in ridges. Such land use carries inherent risks of soil erosion and turbid runoff, which are heightened by the soil damage caused by winter landwork. For nearly 30 years the pesticide aldrin was applied to several hundred daffodil fields, many of which still contain its degradation product, dieldrin. Runoff not only erodes soil from the land but preferentially removes, transports and sorts those soil fractions associated with pesticide residues. Such turbid discharges can pollute water even under low intensity rainfall of as little as $2\,mm.h^{-1}$ during the field capacity season.

1. INTRODUCTION

In climatically favoured parts of Cornwall field vegetables are cultivated. Winter cauliflower, cabbage and potatoes are grown in rotation with leys and cereal break crops. Double cropping is common in the more favoured sites. Daffodils have been a traditional crop and are still grown as part of these intensive systems. The mild winter climate means that there is much activity on the land. Use of the persistent pesticide aldrin has also been common for a number of years. Part of the work reported here resulted in the immediate withdrawal of aldrin from use in the UK.

In three catchments studied between 1989 and 1991, with a total area of about $140km^2$, about 2,700ha were in these intensive crops. Around 900ha were in potatoes, 800ha in winter cauliflower, 650ha in cabbage and 290ha in daffodils.

The land involved is largely on naturally well drained loamy soils of the Denbigh and Ludgvan series (Staines 1979; Findlay et al., 1984), which have been in intensive cultivation for more than a century. Heavy additions of composted seaweed, sea-sand and town waste (Staines, 1979) have maintained topsoil organic matter contents and, where large, have produced the man-made Ludgvan series. The landscapes are undulating, with slopes mainly in the range 0-11°. Steeper land is generally in pasture or semi-natural vegetation.

This paper describes investigations of movement of pesticide residues into watercourses in west Cornwall. It explores the relationship between soil degradation and cultivation of high-value crops. There, persistent pesticides have been applied to soils subject to ill-timed landwork and cultivation practices which can encourage runoff and soil erosion. An examination is made of how circumstances combine to affect water quality under commonly occurring and unexceptional weather conditions.

2. SOIL MANAGEMENT

Potatoes and daffodils are both planted in ridges. In Cornwall it is also common practice to plant cauliflower and cabbage on similar, though smaller ridges, to reduce wind damage. The usual practice is to cultivate and ridge up and down the slope. This is often encouraged by the shape of fields, though most farmers and growers consider that it makes planting and harvest machinery more efficient.

An important aspect of intensive cultivation of these high value crops is the amount of landwork involved in the field capacity period (the winter half of the year when rainfall exceeds evapotranspiration). In the Hayle-Penzance area the average duration of this wet season is over 200 days between September and April. During that time daffodils, winter cauliflower and cabbage are harvested and cultivations for potatoes, including de-stoning/de-clodding, planting and ridging, are all carried out. In addition, early potatoes are forced using plastic sheeting between January and April. The winter harvesting involves much foot traffic, which is especially intensive in daffodil fields. In addition, machinery is used to pack and remove the gathered flowers and brassica heads. Although tractor traffic in the bulb fields is confined to headlands during the picking period, maintenance of the quality of buds and foliage beforehand demands several tractor passes through the crop for pesticide applications. With each of these high value crops the period of suitability for harvest is very short, and the markets are very volatile, so once harvesting is under way, it proceeds relentlessly, regardless of ground conditions.

Soil damage through surface compaction and rutting of the ground by wheels can be considerable. In daffodil fields there is daily trampling along the furrows as the buds are picked, with repeated movement of tractors removing the crop, rutting the headlands, and in wet weather turning the soil there into a slurry of mud and water. In cauliflower and cabbage fields tractors rut the furrows in the crop as they follow the cutting of the crop heads. Headlands and gateways can become particularly badly rutted and slurried. Cultivation for potatoes is often done under less than ideal conditions during the field capacity season, and results in compaction, degradation of soil structure, and decrease in infiltration capacity, even under good conditions.

A major objective of cultivations for potatoes is to produce an open, loose, fine tilth with minimal consolidation to encourage well formed, clean tubers. De-stoning/de-clodding machines are now used routinely for this and are very effective. However, the fine aggregates and fragments produced are left in a very loose state and are easily moved by rain impact and runoff. The custom of double cropping, common in the favoured areas, also presents an increased risk of soil damage through badly-timed working.

3. RUNOFF AND EROSION

Use of ridges and intensive, often poorly timed land work (from the point of view of soil surface state and infiltration capacity) result in much runoff and soil erosion on what are generally regarded as stable soils. They are rated as having an insignificant erosion risk in the National Soil Map legend (Mackney et al., 1983) and only small risk of accelerated erosion by Evans (1990). However, such soil movement has been active for long periods, as witnessed by the prevalence of 'one-sided' hedge banks in the district. In south west England traditional field boundaries are earth banks 1-2m high, but in these intensively cultivated areas colluvium has often built up to the level of the embankment.

Finely worked soils for seed beds or for ridging are susceptible to erosion before the crop cover has developed and before consolidation of the soil has taken place. Rilling of furrows between daffodil and potato ridges often takes place in the winter months. Where furrows follow the slope, the rills only coalesce on the headland at the lower end of the field. However, irregular slopes may cause ridges locally to follow the contour, and behind them runoff can pond. Eventually this may overtop the ridge and erode breaches in successive ridges downslope. Such breaches are also common where headlands are planted at right-angles to the field crop.

On freshly cultivated ground, and from ridge flanks in particular, movement may be limited to structural aggregates or fragments, which are often redeposited a short way down to form small fans of rolled soil clasts at the end of furrows in potato and daffodil crops. The soil in these deposits suffers little or no modification in particle size, other than a decrease in stone content. Rills do not usually form at this stage.

More vigorous runoff disperses the soil, and may also form rills. Deposition of transported material then begins to reflect soil particle size and composition, with colluvium at the site of deposition being banded or laminated. Sorting occurs into laminae of limited particle size range, often with enrichment or depletion of organic matter. Laminae of rolled aggregates are also present in some colluvium. Separation of stones by size is usual, although on the man-made Ludgvan soil series differential sorting of various anthropogenic materials has been observed. For example, low density predisposes coal cinders to travel relatively long distances, while pottery shards are also readily transported.

Local conditions of ground configuration and runoff velocity and amount regulate the quantity, kind, proportion and location of the colluvium deposited. Even after deposition of aggregates and coarser soil fractions has begun, runoff leaving the site is turbid with suspended soil fines and organic matter. Once concentrated into channels, the runoff is likely to enter watercourses (including ditches and land and road drains), polluting them with soil materials, and any associated nutrients and adsorbed agrochemicals.

Although attention has been concentrated on land in ridges, rilling of more conventionally cultivated ground also occurs. Small rills following tramlines in cereal fields can be observed, at times, coalescing in right-angle turns onto headland tramlines, or on steeper slopes as dendritic systems in the main body of the crop. On newly drilled leys rills can also be present and may start along wheelings or cultivation marks on gentle slopes, then similarly coalesce in dendritic patterns as gradients increase.

On soil surfaces that have suffered rain impact and wetting, surface slaking and capping often seals surface pores and fissures and increases the probability of runoff. Particular soil characteristics and weather conditions may modify such caps. Some crack and fissure, allowing enhanced infiltration, others spall off very small aggregates on drying. There can be preferential movement of fine soil particles and organic matter without obvious rilling of the soil surface, both on the flanks of ridges or on flattened ground.

Runoff, erosion and deposition do not necessarily occur in all fields with the susceptible crops but they are common. Some of their characteristics are summarised in Table 1. The amounts of erosion and deposition are for clearly measurable quantities within or adjacent to the site concerned. Systematic measurement was undertaken of cross-sectional area and length for rills, and of depth and area extent for deposition.

The diversity of circumstances, including slope form, field shape and distance to watercourses, means that measurements of amounts eroded are often different from those deposited. Difficulties can arise in measurement of fresh colluvium disturbed by traffic (as on daffodil headlands), in assessment of contributions by inter-rill or inter-furrow sheet erosion, and in determining amounts of soil material removed in suspension during runoff.

The last two were addressed through measurement of runoff quantity and suspended solid load from land where there was little or no evidence of rilling. This was done once runoff became conveniently channelled, as in cultivation furrows or wheelings on headlands or in gateways. Measurements (Table 2) were made during warm frontal rainfall, which is more predictable in its occurrence and duration and more uniformly distributed.

Intensity of precipitation was modest ($1.4 - 2.5$mm.h^{-1}) on each occasion. Average return periods of such rainfall in west Cornwall are likely to be about 14 days during the field capacity period. In each case slopes were modest and soils intrinsically permeable.

With a mean rainfall intensity slightly over 2mm.h^{-1}, runoff from daffodils, potatoes, asparagus and beans (winter-sown) sometimes exceeded 50% of the rainfall. On already moist soils substantial runoff was observed within 20 minutes of the onset of such rain. Suspended solids loads were varied, as might be expected from the range of cropping and soil surface conditions studied.

These amounts of solids suspended in runoff are modest compared with overall erosional movement of soil at several of the sites. They suggest that rainfall more intense than 2mm.h^{-1} caused the soil erosion and deposition noted in Table 1. However, rainfall of about 2mm.h^{-1}, and therefore runoff of the rates in Tables 1 and 2, are likely to persist for totals of around 70-100 hours in the average field capacity season. Applying this to the hourly suspended solid losses in Table 2 indicates movement from fields as large as 3t.ha^{-1}yr^{-1}(but more typically around 100kg) from unexceptional rainfall rates.

An important aspect of soil erosion is the sorting and winnowing that takes place. These processes were recognised by Massey and Jackson (1952). Table 3 illustrates the modifications to soil redeposited as colluvium within the field or its immediate environs. Silt and organic carbon amounts are substantially increased, but enhancement of clay is negligible. Stones are absent from the deposits.

Table 1. Known soil erosion and deposition in the Mount's Bay catchment and examples of runoff early in 1991.

SITE AND AREA	SOIL SERIES AND SLOPE	EROSION (m^3)	DEPOSITION (m^3)	RUNOFF (mm.h^{-1})	RAINFALL (mm.h^{-1} and date)
DAFFODILS					
Playing Field 2.3ha	Denbigh* and Ivybridge* 2-4°	0.10	25.2	0.22	
Tollhouse 1.8ha	Ludgvan (ungleyed) 3°	1.87	7.3	1.44	2.3 for 5.75 hours (22/2)
Tregilliowe 3.2ha	Denbigh and Trusham 2-4°	None observed	3.0	0.70	
Longlane 2.8ha	Denbigh* 3°	0.47	None observed	0.66	2.5 for 7.25 hours (18/3)
EARLY POTATOES (with polythene)					
Tregadjack 0.8ha	Denbigh 0.6°	4.0	6.4	0.99	1.4 for 5.25 hours (15/3)
Tolver 1.4ha	Ludgvan (ungleyed) 2-11°	88.9	Uncertain, much onto road	--	--
EARLY POTATOES (without polythene)					
Tolver 0.3ha	Ludgvan (ungleyed) 2-11°	13.6	Uncertain, much onto road		
WINTER CEREALS					
Treassowe 0.7ha	Denbigh 0-6°	0.35	1.4	0.48	1.4 for 5.25 hours (15/3)
Tregurtha 0.8ha	Denbigh 1-3°	No erosion or deposition observed		0.08	2.5 for 7.25 hours (18/3)
LEY PASTURE					
Trannack 0.9ha	Denbigh 3°	8.1	Little fresh deposition	0.78	1.4 for 5.25 hours (15/3)

Table 1 continued

SITE AND AREA	SOIL SERIES AND SLOPE	EROSION (m^3)	DEPOSITION (m^3)	RUNOFF (mm.h^{-1})	RAINFALL (mm.h^{-1} and date)
ASPARAGUS					
Townfield 2ha	Denbigh* 3°	Little recognisable	47.1	1.50	2.3 for 5.75 hours (22/2)
BEANS					
Polgoon 0.4ha	Denbigh 0-4°	No erosion or deposition observed		1.40	2.3 for 5.75 hours (22/2)
CAULIFLOWER/CABBAGE					
Varfell 1.0ha	Ludgvan 2°	No erosion or deposition observed		0.30	2.3 for 5.75 hours (22/2)
Ludgvan Leaze 3.9ha	Denbigh 2°	No erosion or deposition observed		0.17	1.4 for 5.25 hours (15/3)
Tregurtha 4.5ha	Denbigh 2°	No erosion or deposition observed		0.26	2.5 for 7.25 hours (18/3)

*Indicates man-made topsoil phases.

NB. With the exception of Playing Field, Tollhouse and Tregilliowe (all daffodils planted since 1989) and Tolver, all these sites have received aldrin in the past.

Erosion and deposition amounts are over-winter accumulations until late February 1991.

Table 2. Rates of runoff and suspended solids losses.

	RUNOFF (mm.h^{-1})	SOIL SERIES AND SLOPE	SUSPENDED SOLIDS LOSS (kg/ha/hour)
22/2/91 (2.3 rain for 5.75 hours)			
Daffodils	0.22	Denbigh, Ivybridge; 2-4°	1.70
	1.18	Ludgvan (ungleyed); 3°	4.39
	0.70	Denbigh; 3°	1.04
Broccoli	0.30	Ludgvan; 2°	--
Potatoes (no polythene)	1.43	Denbigh; 2-5°	--
Beans	1.40	Denbigh; 0-4°	30.96
Asparagus	1.51	Denbigh; 3°	1.77

15/3/91 (1.4mm/hr rain for 5.25 hours)			
Cabbage	0.17	Denbigh, Ivybridge; 2°	0.576
Potatoes (polythene)	0.99	Denbigh; 0-6°	5.64
Potatoes (no polythene)	0.49	Denbigh; 2-5°	8.09
Cereals	0.48	Denbigh; 0-6°	0.555
Ley	0.78	Denbigh; 3°	0.396

18/3/91 (2.5mm/hr for 7.25 hours)			
Daffodils	0.66	Denbigh, Trusham; 2-4°	--
Cabbage	0.26	Denbigh; 1-4°	--
Cereals	0.08	Denbigh; 1-3°	0.068

4. PESTICIDE MOVEMENT TO WATER

The work described above demonstrates how runoff and erosion coupled with damage associated with winter cropping and consequent ill-timed land work can result in movement of soil material from naturally well drained, loamy soils. This is especially relevant because between the early 1960s and 1989 aldrin was applied to several hundred Cornish fields, principally for the control of narcissus bulb fly. In recent years approval was limited to this and the killing of wireworm in pastures broken for potato cropping. Previously its use also included dipping of brassica roots on transplanting, and even addition to general purpose mineral fertilisers.

Discovery of large amounts of dieldrin in eels in the Newlyn Coombe River (Milford, 1989) led to the investigations described in this paper. Although this work resulted in the final withdrawal of the pesticide, its degradation product dieldrin is very persistent. Edwards (1973) gave a half-life for aldrin of 0.3 years as epoxidation progresses into dieldrin, which in turn was given a half life of 2.5 years. Work in the Hayle catchment (Harrod, 1991) suggests that a figure of 4-5 years is more representative of dieldrin degradation under west Cornish conditions.

Both aldrin and dieldrin are effectively insoluble in water. The principal means by which the residues reach the aquatic environment is by transportation on soil particles (organic matter and clay), to which they are strongly bonded. The link between soil erosion and movement of pesticide residues can be demonstrated by comparing the aldrin and dieldrin contents of soil and water samples in runoff, water courses and colluvium with source areas of soil material. Table 4 compares aldrin and dieldrin amounts in soil from the centres of treated fields with those in colluvium deposited downslope. This demonstrates increases of aldrin (mean 14.4 fold) and dieldrin (3.6 fold) in colluvium.

This is often in excess of the enhancement of soil organic matter and silt (Table 3). It suggests that the soil components with which the pesticides are associated are preferentially concentrated by erosion and redeposition. Table 4 also shows that enhancement of aldrin exceeds that of dieldrin, which suggests the soil components with which aldrin and dieldrin are associated have different erosional or depositional properties.

Table 3. Comparison of clay, silt and organic carbon contents of colluvium in seven fields.

	MEAN %	MAXIMUM %	MINIMUM %
CLAY (less than 2µm)			
Field centre	22	29	13
Colluvium	23	31	16
Concentration factor	1.09	1.43	0.75
SILT (2-60µm)			
Field centre	43	56	29
Colluvium	66	76	50
Concentration factor	1.63	1.82	1.28
ORGANIC CARBON			
Field centre	3.4	4.6	2.6
Colluvium	5.0	7.6	3.5
Concentration factor	1.45	1.74	1.27

NB. Concentration factor = colluvium
                            field soil

During moderately steady rainfall (4mm an hour for 5 hours on 14 March 1989) in the Newlyn area (Harrod, 1989), turbid runoff from daffodil fields and water leaving land drains contributed substantial amounts of pesticide residues to watercourses (Table 5). This table also indicates that on 21 April 1989 and 2 May 1989, when flows were small, concentrations were much reduced, though still above detection limits.

Table 4. Aldrin and dieldrin (μg.kg^{-1}) in soil and colluvium in thirteen fields.

	MEAN	MAXIMUM	MINIMUM
ALDRIN			
Field centre	88	300	2
Colluvium	685	2,475	2
Concentration factor	14.4	51.5	0.08
DIELDRIN			
Field centre	264	1,100	10
Colluvium	358	710	20
Concentration factor	3.61	15.8	0.11

NB. Concentration factor = $\dfrac{\text{colluvium}}{\text{field soil}}$

Table 5. Aldrin and dieldrin entering water in the Newlyn catchment (1989).

Site and date of sampling	Aldrin (ng.l^{-1})	Dieldrin (ng.l^{-1})
Titanic Field and adjacent ditch		
Ditch, source end (14/3)	77	70
Ditch, source end (21/4)	3	20
Ditch, source end (2/5)	2	22
Ditch, lower end (14/3)	650	670
Ditch, lower end (2/5)	3	21
Overland flow entering ditch lower end (14/3) (suspended sediment 1,130mg.l^{-1})	960	890
Overland flow (2/5)	4	81
Trereife Stream		
Trereife stream above most bulb fields (14/3)	9	15
Trereife stream runoff from Pathway Field entering stream (14/3)	77	240
Trereife stream drain outfall from Pathway Field (14/3)	60	250
Trereife stream drain outfall from Pathway Field (21/4)	1	8
Trereife stream below bulb fields (14/3)	110	76

NB. Aldrin applied to Titanic Field in 1988 and to Pathway Field in 1987.

Comparison of concentrations in field soil with those in overland flow and suspended solids in the ditch beside Titanic Field indicate concentration of residues in the runoff. In Titanic Field, where aldrin was applied the previous year (1989), soil contents were 300μg.kg^{-1} aldrin and 150μg.kg^{-1} dieldrin, whereas in the suspended solids of water entering the ditch on 14 March 1989 there were 850µg.kg^{-1} aldrin and 788µg.kg^{-1} dieldrin. Aldrin had been concentrated 2.83 times, dieldrin 5.67 times.

Runoff measured in February and March 1991 (Harrod, 1992) had comparable residues (Table 6). This took place under lighter rain, around 2mm.h^{-1}, than the sampling in March 1989 (Table 5), when rainfall was about 4mm.h^{-1}. For the 6 sites represented in Table 6, average date of aldrin application was 5-6 years previously (from a range of 3 to 8 years). These data show that despite sufficient time having elapsed for more than 50% breakdown of dieldrin, there remains a strong potential for concentration under typical weather conditions.

Table 6. Dieldrin entering water in the Mount's Bay catchment, 1991.

	Mean	Minimum	Maximum
Field soil dieldrin µg.kg^{-1}	249	72	932
Dieldrin in runoff entering stream, µg.kg^{-1} of suspended solids	821	300	1,632
Concentration factor	5.79	0.97	22.05
Dieldrin content in receiving streams, µg.kg^{-1} of suspended solids	116	27	272

NB. Data for 22/2, 15/3 and 18/3, 1991

5. CONCLUSIONS

The use of persistent pesticides in agriculture has long been known to present environmental hazards and in the UK has been constrained accordingly since the Cook Report (Cook, 1964), though special cases were exempted, such as the application of aldrin to daffodils. Because of the persistence of the decay product dieldrin, this exception must be regarded as leading to a form of soil degradation likely to last several years.

Data presented here suggest that the intensive winter use of well drained soils on quite gentle slopes in Cornwall also degrades the land by increasing runoff and erosion. This results from ill-timed land work, including cultivations and trafficking which can rut, compact and puddle the soil, the construction of features such as ridges and beds aligned with the gradient and the use of plastic sheeting for forcing.

The intensive winter land use and application of persistent pesticide result in further degradation of the soil by enhancing the pesticide content of colluvium through preferential movement of the soil fractions to which the pesticide residues are bonded. These processes also threaten the aquatic environment, because the residues are held on those soil fractions likely to remain in suspension for at least several hours. This provides a clear example of the close link between management of soils and the wider environment.

Although these conclusions primarily relate to south west England, the potential for similar land degradation exists wherever land work takes place in the field capacity season. In addition to intensively grown horticultural crops and field vegetables, it also has bearing on crops such as maincrop potatoes, maize and sugar beet, which can be harvested in less than ideal conditions in wet autumns.

ACKNOWLEDGEMENTS

This paper is based on work funded by the National Rivers Authority, South West Region. The co-operation, openness and concern over pollution among most farmers and growers in the study area must be recorded. The contribution of the two anonymous referees is also acknowledged.

REFERENCES

COOK, SIR J. 1964. Review of the persistent organochlorine pesticides. HMSO, London.

EDWARDS, C.A. 1973. Persistent pesticides in the environment. CRC Press, Ohio.

EVANS, R. 1990. Soils at risk of accelerated erosion in England and Wales. Soil Use and Management 6:125-31.

FINDLAY, C.D., COLBORNE, G.J.N., COPE, D.W., HARROD, T.R., HOGAN, D.V. and STAINES, S.J. 1984. Soils and their use in South West England. Soil Survey of England and Wales Bulletin 14.

HARROD, T.R. 1989. Pesticide pathways and land use practices in the Newlyn River catchment, Cornwall. Unpublished Report to the National Rivers Authority, South West Region.

HARROD, T.R. 1991. Land use practices and pathways of pesticide residues in the River Hayle basin, Cornwall. Unpublished Report to the National Rivers Authority, South West Region.

HARROD, T.R. 1992. Land use practices and pathways of pesticide residues in the Mount's Bay catchment, Cornwall. Unpublished Report to the National Rivers Authority, South West Region.

MACKNEY, D., HODGSON, J.M., HOLLIS, J.M. and STAINES, S.J. 1983. Legend for the 1:250,000 soil map of England and Wales. Soil Survey of England and Wales.

MASSEY, H.F. and JACKSON, M.L. 1952. Selective erosion of soil fertility constituents. Soil Science Society of America Proceedings 16:353-456.

MILFORD, B. 1989. Organochlorine pesticide residues in freshwater eels. South West Water Environmental Protection Report EP/WZ/89/2.

STAINES, S.J. 1979. Soils in Cornwall: Sheet SW53 (Hayle). Soil Survey Record No. 57.

Chapter 12

NUTRIENT LOSSES IN RELATION TO VEGETATION COVER ON AUTOMATED FIELD PLOTS

V. Andreu, J. Forteza, J.L. Rubio and R. Cerni.
Desertification Research Unit, IATA-CSIC. Jaime Roig 11,
46010 Valencia, Spain

Summary

The sediments produced by several rain events during 1988-1990 have been studied. The four field erosion plots are located on different types of vegetation: natural vegetation (matorral), two shrub species (*Medicago arborea* and *Psoralea bituminosa*), and a bare soil. Sediment and runoff production are related to the quantity and intensity of rain. Some chemical factors that influence loss of fertility are evaluated, relating them to rainfall, vegetation cover, seasonal fluctuations and initial soil status.

For the plots with less vegetation cover, much transport of fine elements can be observed, together with high contents of organic matter, total nitrogen, phosphorus and exchangeable cations (K, Mg, Na and Ca).

1. INTRODUCTION

Erosion-sedimentation processes are basic to the normal development of soil, but these effects can be intensified by anthropogenic activity such as wildfires, inadequate agricultural techniques and inappropriate land use.

Soil erosion can reduce soil fertility by removal of nutrients from the soil surface. This is reflected in a reduction of crop yields and in a decrease in the development of natural vegetation. According to the US Office of Technology Assessment (1982), accelerated erosion reduces soil organic matter, fine clays, water retention capacity of soils and plant rooting depth. Other related factors are the reduction of plant nutrients and degradation of soil structure (USDA, 1981; Frye, 1984; Schertz et al., 1984; Hairston et al., 1988; Francis, 1990).

Soil organic matter is one of the first soil constituents to be removed by erosion since it has relatively low density and is concentrated near the soil surface (Lucas et al., 1977). Organic C loss is likely to have an important effect on the ecosystem because it enhances cation exchange capacity and is a source of mineral N and other plant nutrients (Schreiber and McGregor, 1979; Gilliam et al., 1983; Langdale and Lowrance, 1984; Lowrance and Williams, 1988). Also, the losses of P by soil erosion and surface runoff can be important (Alström and Bergmann, 1988; Kronvang, 1990), leading to the reduction of soil fertility and the eutrophication of lakes and other estuarine systems.

Together with organic matter, C, N and P, other oligoelements and plant nutrients are removed by runoff (Frye, 1984) producing a reduction in plant growth and diversity.

The objective of this paper is to study the effect of erosive rain on loss of soil nutrients and salts, and the physical characteristics of sediments. The study has been made on a set of erosion plots located in a Mediterranean forest area with different vegetation covers.

2. MATERIAL AND METHODS

The experimental plots are situated on land ceded by the Forestry Service of the Valencian Government's Department of Agriculture and Fisheries, near Porta-Coeli (Valencia, Spain). They were sited on a SW-facing forest hillside with scrub cover, developed after a forest fire that occurred in 1984.

The shallow soils belong to the Rendzin-Lithosol association (FAO, 1988), developed on Muschelkalk materials of clayey-sandy marl and conglomerate material. There are frequent outcrops of highly-consolidated conglomerate material. The soils are very stony ($\approx$40%).

Climatically, the area belongs to the dry ombroclimate stage of the thermomediterranean thermoclimate according to Thorntwaite's system. The dominant vegetation type is the *Rosmarinus-Ericion* association. The most abundant species include: *Rosmarinus officinalis, Thymus vulgaris, Stipa tenacissima, Chamaerops humilis.*

The experimental plots measure 40 × 8m and are sited at an altitude of 220m on a slightly concave slope (mean slope is 20%). The four plots are bounded by long bricks (40 × 100cm) following the direction of slope. One plot was left under natural vegetation. Another was kept bare of vegetation by initial manual removal of the vegetal cover and thereafter by herbicide applications. The other plots were planted with shrub species (*Medicago arborea* and *Psoralea bituminosa* respectively). Downslope, the plots narrow from 8 to 2m in order to concentrate runoff. This is collected in 2m-wide collectors which discharge into 2000 litre tanks. These hold all the runoff and sediment, which is then concentrated into a 40 litre tank to ease its collection for analysis (Rubio et al., 1990; Andreu et al., 1991).

Climatic factors were measured by means of a system of sensors (Rubio et al., this volume) which collect data on rainfall intensity, soil and air moisture and temperature, and the evolution of runoff during the storm.

The runoff and sediment collected during each erosive rain event from 1988 to 1990 were analysed to obtain data about soil fertility changes. For this purpose, the most important chemical and physical characteristics (exchange cations, organic matter content, total nitrogen, and mineral nitrogen) were studied.

A multiple correlation matrix, Pearson correlations and stepwise regressions have been applied to study relationships between the different parameters.

3. RESULTS AND DISCUSSION

Soil characteristics for the bare and natural vegetation plots are reported in Table 1. The texture of the soil is silty and its surface status can be seen in Table 2.

Table 1. Physical and chemical characteristics of the soil for the natural vegetation and bare plots.

	PLOT	
	BARE	NATURAL
Humidity (%)	3.52	2.75
Estr. Est. (%)	22.85	34.75
pH	8.00	8.1
E.C. (S/m)	0.81	0.37
CaCO$_3$(%)	45.51	42.30
Org.Mat. (%)	6.26	5.53
Min. N (mol$_c$/kg)	4.77	0.45
Avail. P (mol$_c$/kg)	0.17	0.11
C.E.C. (cmol$_c$/k)	30.90	22.11
Ca (mol$_c$/kg)	28.54	20.18
Mg (mol$_c$/kg)	1.71	1.33
Na (mol$_c$/kg)	0.07	0.07
K (mol$_c$/kg)	0.53	0.58

PARTICLES	PLOT	
Ø(mm)	BARE	NATURAL
0.0-0.002	6.14	7.66
0.002-0.005	6.16	7.52
0.005-0.02	18.55	20.77
0.02-0.05	22.15	22.48
0.05-0.10	13.60	12.60
0.10-0.25	12.09	10.44
0.25-0.50	10.51	9.04
0.50-1.0	5.30	5.46
1.0-2.0	5.50	4.03
GRAVELS (% in weight)		
> 2.0	31.57	42.42

Table 2. Surface status of the soil on the erosion plots.

MEASURES	PLOT					
	BARE			NATURAL		
	1988	1989	1990	1988	1989	1990
Roughness (%)	4.27	4.23	4.01	5.35	5.37	5.20
Shear strength (⋆)	4.14	4.06	3.38	4.43	4.41	3.76
Penetrograph	9.31	9.28	9.23	9.88	9.73	9.77
Stone cover (%)	65	63	65	37	37	35
Bulk density (gcm^{-3})	1.05	0.94	0.98	1.25	1.52	1.11
Crust cover (%)	2.10	1.3	Nd	1.04	0.94	Nd

Nd: Not detectable.   (⋆) : Torvane.

Figure 1 shows the characteristics of the selected rain events. The rain intensity is generally low, except for two episodes in 1989 and 1990. The behaviour of rains seems characteristic of the Mediterranean climate, with episodes of torrential rain in autumn/winter periods. Otherwise rainfall is scarce and of generally low intensity.

The runoff volume and sediment weight are reported in Figure 2. The maximum runoff generated corresponds to the very high rainfall intensity episodes of 16/11/89 (I_{30} = 58) and 09/09/90 (I_{30} = 99), causing overflow of the 2000 litre tanks. The runoff is always higher from the bare plot.

The sediments produced are not excessively high, except in the 09/09/90 event. The plot under natural vegetation always shows less soil loss. The low levels of sediments produced are mainly due to the good soil structure which is due to the high levels of organic matter (around 6%) and structural stability (22-35%). The organic matter content and soil structural stability are closely related, increasing the structural resistance to raindrop impact and splash (De Ploey and Poesen, 1985; Foster et al., 1989).

Figure 1. Characteristics of the erosive rain events during 1988 - 1990.

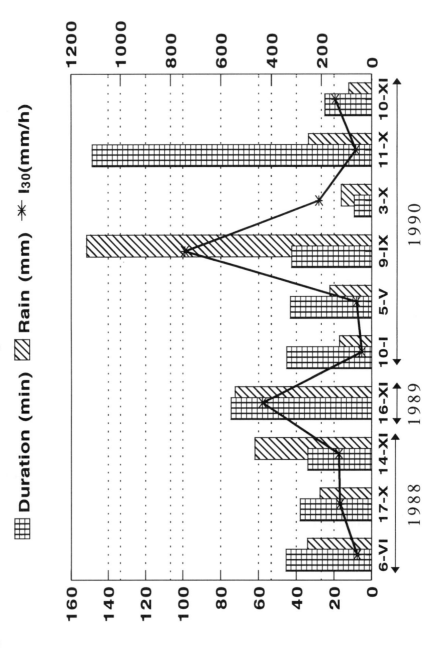

Figure 2. Sediment weight and runoff volume collected from the plots during the erosive rain events (1988 - 1990).

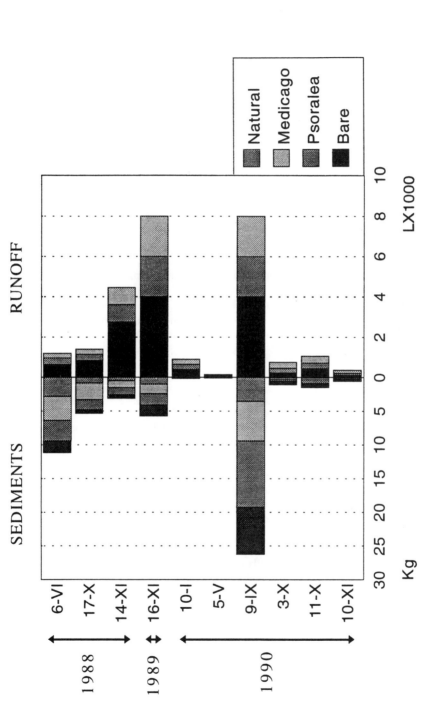

The results of the sediment analysis are presented in Table 3. The calcium carbonate content is high for all samples. The highest content occurs from the bare plot. The event of 17/10/88 (I_{30} = 16.80) shows the highest content of carbonates in sediments for all the plots.

The organic matter content is higher in the sediments from the natural plots, due to the accumulation of plant biomass. The biomass also affects the level of total and mineral nitrogen in the sediments, which increases with vegetation cover. The available phosphorus is highest in the sediments from the bare plot. This is due to P's low mobility and the lack of plant absorption on the bare plot. The values of exchangeable cations in the sediments are similar for all plots. Some differences appear in the case of sodium, with maximum values from the bare plot, due to the more intense effects of evaporation and capillary rise than for the plots with vegetation. High values of available phosphorus and sodium correspond to the lowest intensity event (10/01/90, I_{30} = 5.20). However, magnesium and potassium show highest levels for an event of higher intensity (16/11/89, I_{30} = 58).

With respect to the textural analysis of the sediments, all plots show an enrichment in fine elements, with the proportion of silt and especially clay being higher than in the original soil (Figure 3). The phenomenon is more obvious for the plot without vegetation, with 80% of the sediments being clay and silt particles after the highest intensity rain (09/09/90). This suggests that the erosion processes are very selective in removing the fine fractions of the soil (Lewis, 1981; Morgan, 1986) along with the organic matter associated with them (Fullen, 1991). Indeed, this is reflected by the high organic matter contents in the sediments.

From the statistical analysis of the data, the correlation between runoff and sediments removed is highly significant. No significant correlation with storm duration was found. No significant correlations were observed between the physical and chemical characteristics of the soil and sediments, although organic matter content and the fine fraction of the sediments were related statistically (r = 0.780).

5. CONCLUSIONS

Vegetation has an important influence on soil stabilisation and the avoidance of soil fertility losses due to erosion by water. Losses of nutritional soil components occur in each rainfall event, but these losses increase as the vegetation cover decreases. The Mediterranean regime of rainfall, characterised by low frequency, high intensity rains, favours the removal of salts and organic matter, which are closely associated with other important elements of plant nutrition. The Mediterranean climate favours the capillary rise of salts by influencing evapotranspiration. The precipitated salts are then removed by runoff in the first rains of the humid season.

There are highly significant correlations between rainfall characteristics (volume and intensity) and the quantity of runoff and sediments produced. No other significant relationships have been found, except that between soil organic matter content and the fine fraction of the sediments.

Table 3. Chemical characteristics of the sediments collected.

PLOT		CaCO₃ (%)	O.M. (%)	Nt (%)	Nm (%)	P avail. (mol./Kg)	Mg (mol./Kg)	Na (mol./Kg)	K (mol./Kg)
Natural	Mean	28.91	20.71	692.14	19.94	3.98	1.90	0.34	0.44
	Maximum	35.57	41.56	1062.96	45.82	17.08	4.17	1.02	0.73
	Minimum	25.27	10.66	280.06	7.69	1.00	0.37	0.03	0.12
	Std	2.80	8.92	270.61	11.68	4.59	1.20	0.29	0.17
Medicago	Mean	27.36	17.55	710.04	20.94	4.61	1.70	0.39	0.42
	Maximum	34.02	26.55	1062.18	43.12	18.00	4.21	1.37	0.58
	Minimum	17.09	7.73	372.29	7.45	1.45	0.26	0.06	0.09
	Std	4.64	5.79	232.83	12.23	4.69	1.00	0.38	0.13
Psoralea	Mean	28.09	17.09	692.98	17.77	5.76	1.79	0.29	0.44
	Maximum	33.65	23.45	910.22	36.42	27.33	3.86	1.19	0.64
	Minimum	23.71	10.15	537.21	6.88	1.85	0.42	0.08	0.14
	Std	3.23	4.69	127.25	9.15	7.28	0.90	0.31	0.13
Bare	Mean	33.09	12.07	563.60	14.63	6.58	2.06	0.37	0.53
	Maximum	43.96	22.49	828.19	32.76	34.84	3.12	1.22	0.88
	Minimum	26.13	6.49	298.10	5.96	1.33	0.46	0.07	0.20
	Std	5.31	4.22	150.40	8.73	9.52	0.81	0.31	0.20

O.M.: Organic matter Nt: Total Nitrogen Nm: Mineral Nitrogen P avail.: Available Phosphorus Std: Standard deviation

Figure 3. Textural distribution of clay, silt and sand fractions in the soil and sediments.

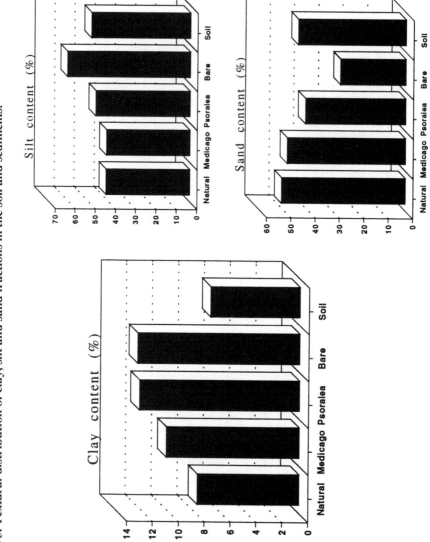

REFERENCES

ALSTRÖM, K. and BERGMANN, A. 1988. Sediment and nutrient losses by water erosion from arable land in South Sweden, a problem with non-point pollution. Vatten 44:193-204.

ANDREU, V., RUBIO, J.L. and CERNI, R. 1991. The Porta-Coeli experimental site. In: Excursion Guide Book. ESSC Workshop 'Soil Erosion and Degradation as a Consequence of Forest Fires'. UIMP, Barcelona-Valencia.

DE PLOEY, J. and POESEN, J. 1985. Aggregate stability, runoff generation and interrill erosion. In: Richards, K.S., Arnett, R.R and Ellis,S. (eds). Geomorphology and soils. George Allen and Unwin, London.

FAO 1988. Soil map of the world. 1:5,000,000 scale. Rome.

FOSTER, I., DEARING, J. and GREW, R. 1989. The North Warwickshire Experimental catchments; studies of historical sediment yield, sediment source and soil erosion. In: Foster, I. (ed.). Field guide to 'Soil erosion and agricultural land'. BGRG Conference, Coventry.

FRANCIS, C. 1990. Soil erosion and organic matter losses on fallow land: a case study from south-east Spain. In: Boardman, J., Foster, I.D. and Dearing, J.A.(eds). Soil erosion on agricultural lands. Wiley, Chichester.

FRYE, W.W. 1984. The effects of soil erosion on soil productivity. In: Harlin, J.M. and Berardi, G.M. (eds). Agricultural soil loss (processes, policies and prospects). Westview Press Inc., Boulder.

FULLEN, M.A. 1991. Soil organic matter and erosion processes on arable loamy sand soils in the West Midlands of England. Soil Technology 4:19-31.

GILLIAM, J.W., LOGAN, T.J. and BROADBENT, F.E. 1983. Fertilizer use in relation to the environment. In: Olsen, R.A. (ed.). Fertilizer technology and use. SSSA, Madison.

HAIRSTON, J.E., SANFORD, J.O., RHOTON, F.E. and MILLER, J.G. 1988. Effect of soil depth and erosion on yield in the Mississippi Blacklands. Soil Sci. Soc. Am. J. 52:1458-1463.

KRONVANG, B. 1990. Sediment-associated phosphorus transport from two intensively farmed catchment areas. In: Boardman, J., Foster, I.D. and Dearing, J.A.(eds). Soil erosion on agricultural lands. Wiley, Chichester.

LANGDALE, G.W. and LOWRANCE, R. 1984. Effects of soil erosion on agroecosystems of the humid United States. In: Lowrance, R. (ed.). Agricultural ecosystems. Wiley, New York.

LEWIS, L.A. 1981. The movement of soil materials during a rainy season in Western Nigeria. Geoderma 25:13-25.

LOWRANCE, R. and WILLIAMS, R.G. 1988. Carbon movement in runoff and erosion under simulated rainfall conditions. Soil Sci. Amer. J. 52:1445-1448.

LUCAS, R.E., HOLTMAN, J.B. and CONNOR, L.J. 1977. Soil carbon dynamics and cropping practices. In: Lockerez, W. (ed.). Agriculture and energy. Academic Press, New York.

MORGAN, R.P.C. 1986. Soil erosion and conservation. Longman, London.

RUBIO, J.L., ANDREU, V. and CERNI, R. 1990. Soil degradation by water erosion: experimental design and preliminary results. In: Albaladejo, J., Stocking, M.A. and Diaz, E (eds). Soil degradation and rehabilitation in Mediterranean environmental conditions. CSIC, Murcia.

SCHERTZ, D.L., MOLDENHAUER, W.C., FRANZMEIER, D.P. and SINCLAIR, H.R. 1984. Field evaluation of the effect of soil erosion on crop productivity. In: Erosion and soil productivity. American Society of Agricultural Engineers 8/85:9-17.

SCHREIBER, J.D. and McGREGOR, K.C. 1979. The transport and oxygen demand of organic carbon released to runoff from crop residues. Prog. Water Techno. 11:253-261.

U.S. DEPARTMENT OF AGRICULTURE 1981. Soil erosion effects on soil productivity: A research perspective. J. Soil Water Conserv. 36:82-90.

U.S. OFFICE OF TECHNOLOGY ASSESSMENT 1982. Impacts of technology on US cropland and rangeland productivity. Congressional Board of the 97th Congress. Library of Congress catalog card 82-600596.

Chapter 13

A MONITORING SYSTEM FOR EXPERIMENTAL SOIL EROSION PLOTS

J.L. Rubio, V. Andreu and R. Cerni
Desertification Research Unit, IATA-CSIC, Jaime Roig 11,
46010 Valencia, Spain

Summary

A field facility for the study of soil erosion processes is described. The experimental set up includes four erosion plots (40 × 8m) with a range of different covers from bare soil to natural matorral. They include two types of selected shrub species of interest for soil conservation in the Mediterranean (*Medicago arborea* and *Psoralea bituminosa*). In order to meet the requirements for the development, evaluation and validation of EUROSEM (European Soil Erosion Model; see Morgan; Quinton; and Rickson, all this volume) a set of sensors has been installed on the plots. The system was prepared to send information in real time, as radio signals via a modem, to a terminal computer allocated in the laboratory. The rain, runoff and sediment characteristics of erosive events are explained.

1. INTRODUCTION

Field studies provide a realistic approach to soil erosion problems. In this sense, field plots are an important tool to acquire valuable data to accomplish many different research objectives (Lal, 1988). Field erosion plots have been used since 1915 (Zachar, 1982) and from this time facilities have been designed, constructed and developed to meet different research objectives. This has resulted in a great variety of equipment depending on the type of study, performance requirements, economics and ecological conditions.

The high cost of setting up and operating erosion plots means that there are only a small number of such sites, distributed widely throuhgout the world (Mutchler, 1963).

The size of plots has varied from the $1m^2$ plots used in 1915 by the US Forest Service (Zachar, 1982) to catchments of many hectares. However, medium size plots as used by the USDA (1979) for the development of the USLE (Wischmeier and Smith, 1965; Wischmeier and Smith, 1978) or the MUSLE (Williams, 1975) are the most widely used and are large enough to represent processes of rill and interrill erosion. Similar plots are used to validate and develop other equations and models.

To validate the EUROSEM (Morgan et al., 1991; Morgan, this volume) it is necessary to obtain real data about the characteristics of the area, rain events, vegetation and soil. To meet this objective, a network of erosion plots has been installed in different regions of Europe, which includes the Porta-Coeli (Valencia, Spain) set of field erosion plots.

The objective of this paper is to describe the automatic system, installed in October 1991, for monitoring rain events, runoff production, soil temperature and soil moisture. These data will be used to acquire a better knowledge of the water erosion processes in a Mediterranean environment and to give data to validate the EUROSEM.

2. SET UP OF THE INSTALLATION

2.1 Area description

The plots are located on a south-west facing forest hillside near Porta-Coeli (Valencia), on terrain assigned by the Forestry Services of the Department of Agriculture and Fisheries of the Valencia Government. All the area is covered by Mediterranean shrub vegetation (*Rosmarino-Ericion* association), with evidence of fire damage in the past.

The dominant lithology of the area is clayey-sand and conglomerates, giving an abundant stony fraction. There are frequent outcrops of highly-consolidated conglomerate material. The shallow soils are Calcic Xerochrept (Andreu et al., 1991).

Climatically, the area has the typical thermomediterranean climate, with a maximum precipitation in autumn (September to November) and a dry period from June to September. The annual precipitation is around 450mm, and an annual mean temperature of 17.2°C. The mean temperature of the dry period is 34°C.

2.2 Field plot characteristics

The four plots measure 40 × 8m, and are located on a 20% slope. They are bounded by hydrophobically treated bricks. The plots are sufficiently wide to minimise edge effects, and yet are large enough for rills to develop. The plots narrow from 8 to 2m width to force the concentration of runoff, which is collected in 2m wide collectors that run into 2000 litre tanks. Inside the tanks there is a smaller tank (30 litres) which retains the sediments. Figure 1 shows the configuration of the plots. Each plot presents a different vegetal cover. One was left with natural shrub vegetation. On two of the plots, two shrub species of interest for soil conservation (*Medicago arborea* and *Psoralea bituminosa* respectively) were planted. On the fourth, the soil was left bare.

A pluviograph and a thermohygrograph were installed in order to obtain daily thermopluviometric data on site, in addition to the automatised equipment measuring these data.

2.3 Automatic monitoring system

The unpredictable occurrence of rain events and the distance (36km) to the experimental site make a direct observation during erosion events in the field difficult. A continuous, fully automated system with radio transmission to send data overcomes these problems. The equipment comprises a set of sensors, a field data collection system, a transmission-acquisition system and equipment to treat and store the data. The set up of the monitoring system appears in Figure 2.

Figure 1. Configuration of the plots and the monitoring system.

1. Pluviograph.	2. Thermohygrograph.	3. Collectors.
4. Sediments.	5. Drainage channel.	6. Cut off channel
7. Network of sensors.	8. Collection-transmission system.	

A: Bareplot
B: Plot with *Psoralea bituminosa*
C: Plot with *Medicago arborea*
D: Plot with natural vegetation

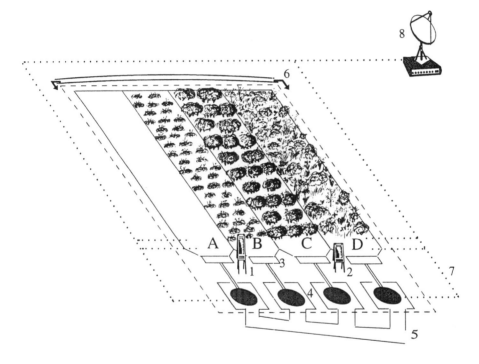

2.4 The sensors

Soil moisture sensors (6513 STARLOG) based in gypsum blocks with a range of soil suction from 0.064-2.24 bars were installed on the bare, naturally vegetated and *Medicago arborea*-planted plots. Soil temperature sensors (ES-060 OMNIDATA) were installed on the bare and the naturally vegetated plots. In the tanks of both plots, level capacitive sensors (LC-20 SYMEN) with a special front sounder were installed to measure runoff discharge during a rain event. A pluviometry sensor (SESYBER PLU-002B) was placed near the plots to record the rain events.

Figure 2. Diagram of the set up of the monitoring system.

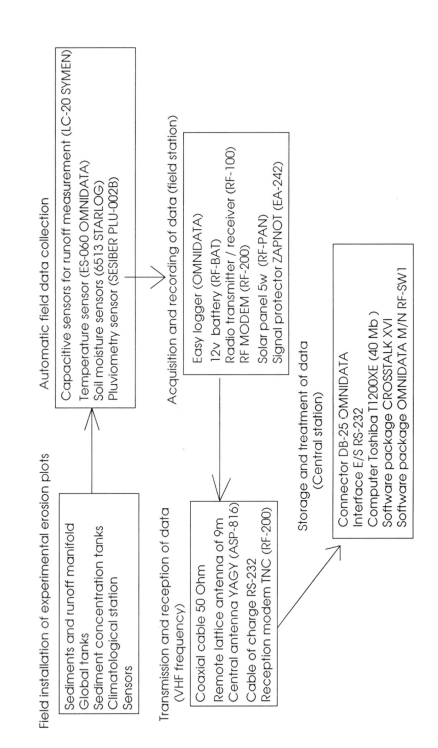

Field installation of experimental erosion plots

Automatic field data collection

Sediments and runoff manifold
Global tanks
Sediment concentration tanks
Climatological station
Sensors

Capacitive sensors for runoff measurement (LC-20 SYMEN)
Temperature sensor (ES-060 OMNIDATA)
Soil moisture sensors (6513 STARLOG)
Pluviometry sensor (SESIBER PLU-002B)

Acquisition and recording of data (field station)

Easy logger (OMNIDATA)
12v battery (RF-BAT)
Radio transmitter / receiver (RF-100)
RF MODEM (RF-200)
Solar panel 5w (RF-PAN)
Signal protector ZAPNOT (EA-242)

Transmission and reception of data
(VHF frequency)

Coaxial cable 50 Ohm
Remote lattice antenna of 9m
Central antenna YAGY (ASP-816)
Cable of charge RS-232
Reception modem TNC (RF-200)

Storage and treatment of data
(Central station)

Connector DB-25 OMNIDATA
Interface E/S RS-232
Computer Toshiba T1200XE (40 Mb)
Software package CROSSTALK XVI
Software package OMNIDATA M/N RF-SW1

2.5 Acquisition and recording of data

All the signals of the sensor are stored in a data logger (Easy logger OMNIDATA) characterised by 12 analogic entry channels of 12 bits of resolution, with four voltage input ranges for two types of signals: 'DC volt' ranges for measuring typical DC voltages, and 'AC resistance' for sensors that requires AC excitation. All the signal lines of the sensors are protected by ZapsNot model FEA-PZ located in a ZapNot Bus Truck. In case of a system fault, the data logger has a RAM of 8K for maintaining the information.

2.6 Transmission/Reception of data

This system consists of an RF MODEM (RF-200) of 1200 baud CMOS TCM 3150 with a processor 84C00-06 8 bit CPU (Z-80), used to change the signal of the sensors from radio frequencies for transmission and vice versa, sustained by a radio transceiver (RF-100). There are two sets - one in the field, and one in the laboratory. Both are connected to an antenna YAGY (ASP-816) attached to a remote lattice antenna of 9m.

2.7 Treatment and data storage

The reception system is connected by a DB-25 OMNIDATA with an interface E/S RS-232 to a portable computer that contains the software communications package CROSSTALK XVI and the software package OMNIDATA M/N RF-SW1, which allows control of the functioning, time of measurement and conversion parameters of the field plot sensors from the laboratory.

The system runs on a 12V (RF-BAT) battery and a solar panel of 5W (RF-PAN) in the field. This battery is charged continuously.

3. FIRST RESULTS OBTAINED

The equipment was installed in the fall of 1990 and began to operate in January 1991. During this year all the parameters covered by the sensors were measured continuously, especially during each rain event. Unfortunately, 1991 was so dry and any rainfall was of low intensity.

The system works well, recording and sending data without any radio interference. Figure 3a reports the monthly distribution and the I_{30} of the rains of 1991, and Figure 3b presents the climatological characteristics of the erosive rain events monitored in that year. In both figures, the low rates of rainfall intensity can be observed and compared with the amount of rainfall and storm duration.

The quantities of runoff generated and sediment transported are shown on Figure 4, illustrating the two really erosive events. These two events correspond to the rainfall of 19th and 20th of February, and even then, only small amounts of soil were removed. In Figure 5, the data correspond to the event on the 20th February, in which the characteristic behaviour of the rain events of that year (long duration and low rainfall intensity) can be observed together with the histogram of runoff production related to the intensity (I_{30}) and quantity of rain over time.

Figure 3. Monthly distribution of rainfall and I_{30} during 1991 (A) and climatological features of the erosive rainfall events of that year (B).

(A)

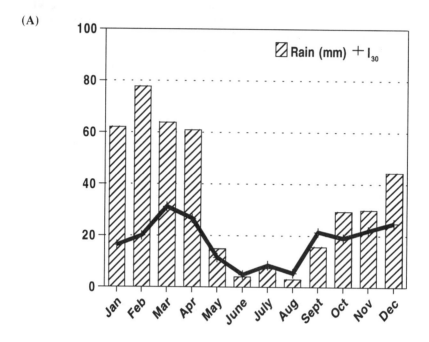

(B)

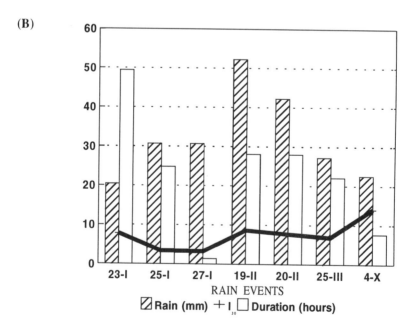

Figure 4. Erosive characteristics of the monitored rainfall events of 1991. A) Runoff generated. B) Sediments.

(A)

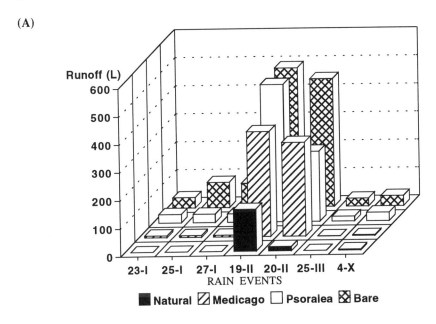

(B)

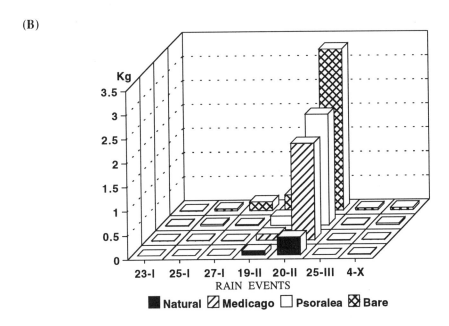

Figure 5. Rainfall and runoff characteristics of the event of 20th February.

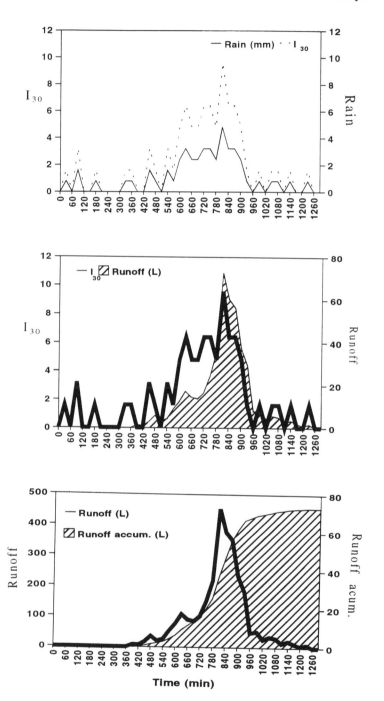

4. DISCUSSION AND CONCLUSIONS

In the development of models, like EUROSEM, and in the research of erosion processes the use of sensors or other automatic monitoring equipment has proved to be an important tool. In this example, the functioning of the system and the transmission/reception has been satisfactory under all conditions.

The data obtained for 1991 reflect the tendency of the rainfall to be of low erosivity, with only two rain events (19th and 20th February) during this period being strong enough to produce measurable runoff. In both cases there was very little sediment produced.

REFERENCES

ANDREU, V., CERNI, R., and RUBIO, J.L. 1991. The Porta-Coeli experimental sites. In: Excursion Guide Book. ESSC Workshop, Valencia, Spain, 3-7 September 1991. UIMP, Valencia.

LAL, R. 1988. Soil erosion research methods. SWCS-ISSS, Ankeny.

MORGAN, R.P.C., QUINTON, J.N. and RICKSON, R.J. 1991. EUROSEM: a user Guide. Silsoe College, Cranfield.

MUTCHLER, C.K. 1963. Runoff plot design and installation for soil erosion studies. ARS-41-79, Agricultural Research Service. US Department of Agriculture, Washington, DC.

US DEPARTMENT OF AGRICULTURE 1979. Field manual for research in agricultural hydrology. Agriculture Handbook No. 224, Washington.

WILLIAMS, J.R. 1975. Sediment yield prediction with the universal soil loss equation, using a runoff energy factor. In: Present and perspective technology for predicting sediment yield and sources. Proceedings, Sediment Yield Workshop. Oxford, Mississippi, 28-30 November 1972. ARS-S-40. US Department of Agriculture, Washington, DC.

WISCHMEIER, W.H. and SMITH, D.D. 1965. Predicting rainfall erosion losses from cropland east of the Rocky Mountains. USDA Agric. Handbook No. 282. US Government Printing Office, Washington, DC.

WISCHMEIER, W.H. and SMITH, D.D. 1978. Predicting rainfall erosion losses. Agriculture Handbook No. 537. US Department of Agriculture, Washington, DC.

ZACHAR, D. 1982. Soil erosion. Elsevier, Amsterdam.

Chapter 14

PRELIMINARY INVESTIGATION OF RUNOFF USING A RAINFALL SIMULATOR

V. Sardo[1], P. Vella[1] and S.M. Zimbone[2]

[1]Istituto di Idraulica Agraria, Via Valdisavoia 5, 95123 Catania, Italy
[2]Istituto di Genio Rurale, Piazza S. Francesco 4, 89061 Reggio, Calabria, Italy

Summary

Field investigations are being carried out in Sicily with a rainfall simulator to develop and validate the European Soil Erosion Model (EUROSEM) (Morgan et al., 1991) for the prediction of soil erosion by water. The simulator consists of a number of self-compensating sprinklers installed at a distance of 2.0 × 2.5m, at a height of 2.0m, over four sandy-loam plots. Each plot measures 22.0 × 2.5m, with a slope of 9%, following the USDA standard (Wischmeier and Smith, 1978). In this paper the results of the preliminary analysis and the steps for the calibration of the rainfall simulator are presented. Simulations show that runoff is influenced significantly by the antecedent conditions (such as moisture content of the soil at depth). There is considerable variability between plots.

1. INTRODUCTION

Soil erosion is influenced by a number of factors interacting in a complex way. The large spatial and temporal variability of such factors makes it very difficult to develop prediction models of universal validity. Some of these factors are difficult to control and field investigations are often undermined by variability which can impair data quality.

Under the EUROSEM project, the Istituto di Idraulica Agraria in Catania has installed a rainfall simulator covering 750m ^{2} in order to study the hydrological response of large-scale plots. This paper describes the experimental equipment, the plots and the results of some field calibration tests for the simulated rainfall.

2. MATERIALS AND METHOD

Four adjacent plots, 22m long and 2.5m wide, on a 9% slope have been installed (Figure 1). Each plot is hydrologically isolated, with steel sheets protruding 40cm above the soil surface. Above the plots, runoff is diverted by a furrow. Runoff from the plots is conveyed downslope into tanks.

The soil is a sandy-loam, derived from a weathered mixed calcareous-volcanic parent material. The main physical characteristics of the soil for each plot are reported in Table 1. The plots had a uniform covering of weeds, mainly of the genera *Oxalis* and *Avena*. Plant cover, evaluated by photographic methods, ranged from 80 to 90%.

Table 1. Main physical characteristics of the soil.

Plot	Particle fractions (%) (*)				Median D50 (mm)	Consistency Limits				Infiltrability (**) (mm.hr^{-1})
	Gravel	Sand	Silt	Clay		Liquid limit (%)	Plastic limit (%)	Plasticity index	Shrinkage limit (%)	
1	7	56	32	5	0.10	34	22	12	19	22.6
2	10	45	40	5	0.15	35	22	13	19	31.5
3	7	58	30	5	0.12	33	22	11	19	24.9
4	20	45	30	5	0.35	33	22	11	19	25.0

(*) ASTM classification   (**) As measured with a double ring infiltrometer

Table 2. Main results of the rainfall/runoff tests.

Plot	ri = 24.6				ri = 33.5				ri = 48.7			
	rad = 30		rad = 60		rad = 30		rad = 60		rad = 30		rad = 60	
	rud	c	rud	c	rud	c	rud	c	rud	c	rud	c
1	40	0.58	64	0.58	41	0.61	67	0.70	38	0.62	70	0.68
2	39	0.22	60	0.32	39	0.33	65	0.46	35	0.38	65	0.49
3	39	0.46	64	0.55	40	0.58	67	0.70	36	0.59	68	0.70
4	40	0.32	63	0.43	40	0.44	67	0.55	37	0.48	66	0.53
Mean values	40.2	0.39	62.7	0.47	40.0	0.49	66.5	0.60	36.5	0.52	67.2	0.60
iv	7.5		13.0		8.5		13.4		11.7		19.5	

where ri = rainfall intensity (mm.h^{-1}); rad = rainfall duration (min); rud = runoff duration (min); c = runoff coefficient; iv = mean infiltrated volume (mm), as a difference between rainfall and runoff volumes.

Figure 1. Erosion plots and rainfall simulator

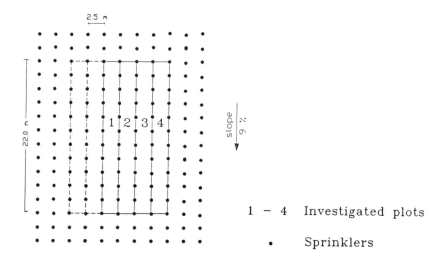

1 – 4   Investigated plots

•   Sprinklers

Rainfall is simulated with a number of sprinklers installed at a distance of 2 × 2.5m, at a height of 2m. They are slow-rotating, self-compensating sprinklers (DAN 2200, Israel), with a nominal flow rate of 240 l.h^{-1}, within operating pressures ranging from 2.5 to 4 × 10^{-2} kPa. The number of simultaneously operating sprinklers can be reduced to simulate different rainfall intensities. The simulator has been designed to give rainfall intensities ranging from about 48mm.h^{-1}, when all the sprinklers operate, to about 12mm.h^{-1} when only 25% operate.

In order to determine the characteristics of the artificial rainfall, some tests have been carried out to analyse the drop size distribution and the spatial distribution of rainfall intensity. The drop size distribution has been determined using the flour pellet method (Hudson, 1964) for a single sprinkler and when 25, 50, 66 and 100% of the sprinklers operate. The spatial distribution of rainfall intensity has been evaluated by collecting rainfall in cans located on the plots. The positions and number of cans varied in relation to the number of operating sprinklers. The uniformity of the distribution has been evaluated through the Christiansen coefficient and the coefficient of uniformity.

In order to ensure constant initial moisture conditions, a preparatory storm of 150mm in 12.5 hours was applied to the soil. During this rainfall event the plots showed different hydrological responses, with runoff coefficients ranging from 0.08 to 0.24.

Twenty minutes after the pre-wetting, tests were carried out with 50, 66 and 100% sprinkler operation. Two sets of tests were run, over 30 and 60 minutes. Between the tests the rainfall simulator was stopped for 20 to 35 minutes. Sediment concentration was determined by sampling water in the pipes just before they discharged into the tanks.

3. RESULTS

The distribution of rainfall intensity and the median volume drop diameter for a single sprinkler are shown in Figure 2. The rainfall intensity varies according to the distance from the sprinkler, with mean values of 1.4 and 1.2 mm.h⁻¹ upslope and downslope, respectively. The median volume drop diameter increases with distance from the sprinkler. The maximum value is about 2.5mm (at a distance of 7m). At 1 and 1.5m from the sprinkler the size of the raindrops is less than 0.71mm.

Figure 2. Rainfall intensities and median volume drop diameter for a single sprinkler.

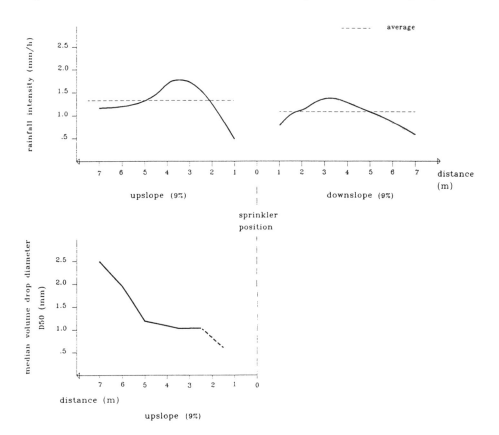

The values of the median volume drop diameter (D_{50}), determined when 25, 50, 66 and 100% of the sprinklers were operating were 1.98, 2.04, 2.17, 2.48mm respectively. There was a high degree of overlap when multiple sprinklers were used. Values of D_{50} do not show a significant variability in drop size with rainfall intensity. This is probably due to the fact that rainfall intensity varies with the number of operating sprinklers, not a change in the water pressure of the sprinklers. The comparison with the results from a single sprinkler suggests that some drops may have coalesced as a consequence of overlapping.

The mean rainfall intensity for 50, 66 and 100% of sprinklers operating were 24.6, 33.5 and 48.7mm.h[-1] respectively, with coefficients of variation up to 3%. Christiansen coefficients and coefficients of uniformity ranged from 97 to 98%.

Figure 3. Hydrographs for different rainfall intensities and durations.

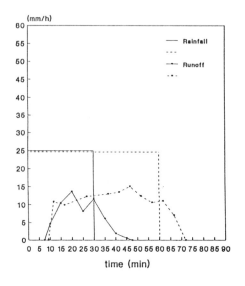

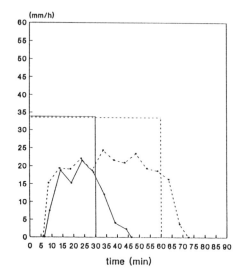

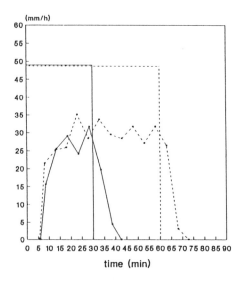

The results of the tests are reported in Table 2. During tests wind velocity was less than 1m/s and no rill formation occurred. There was high variability in the runoff coefficient between the plots. Greater variations were observed for the 30 minute test at 24.6 mm.h^{-1}, in which the runoff coefficient ranged from 0.22 (plot 2) to 0.58 (plot 1). For each plot the runoff coefficient increased with rainfall intensity and duration. Duration of runoff for each single test varied by up to 8% between the plots. Mean runoff duration ranged from 36.5 to 40.2 minutes and from 62.7 to 67.2 minutes, respectively, for the 30 and 60 minute tests.

Infiltration rate increased with rainfall intensity, as observed by Flanagan et al. (1988) (Table 1). As expected (Boers et al., 1992; Flanagan et al., 1988) the mean infiltrated volumes were far lower than those measured with the double ring infiltrometer.

Figure 3 shows the hydrographs, averaged over the four plots for the two rainfall durations and at the three intensities. These hydrographs show that the ponding time (5 to 10 minutes) decreases with increasing rainfall intensity. They also show the absence of significant differences between the times of concentration under the different rainfall conditions, and a slight decrease in the falling limb for the 60 minute tests.

The values of sediment concentration were affected by errors due to the deposition along the collecting pipes of larger soil particles. In order to eliminate this, arrangements to reduce pipe length and increase pipe gradient are being undertaken.

4. CONCLUSIONS

The simulator operates with a high spatial uniformity in terms of rainfall intensity and drop size. A more accurate characterisation of drop size distribution would be advisable. Drop diameters should be controllable and modified to simulate more realistic, natural conditions (Vories and Von Bernuth, 1990).

Preliminary tests showed the need for controlling initial soil moisture conditions. Consequently, before starting the rainfall-runoff tests, 150mm of water was applied over about 12 hours. Tests under such controlled conditions still showed considerable variability in runoff and sediment under different rainfall intensities and durations. This suggests the need for a better understanding of the effects of initial conditions.

REFERENCES

BOERS, T.M., VAN DEURZEN, F.J.M.P., EPPINK, L.A.A.J. and RUYTENBERG, R.E. 1992. Comparison of infiltration rates measured with an infiltrometer, a rainulator and a permeameter for erosion research in S.E. Nigeria. Soil Technology 5:13-26.

FLANAGAN, D.C., FOSTER, G.R. and MOLDENHAUER, W.C. 1988. Soil pattern effect on infiltration, runoff and erosion. Transactions of the ASAE 31:414-420.

HUDSON, N.W. 1964. The flour pellet method for measuring the size of raindrops. Research Bulletin No. 4. Dept. of Conservation and Extension, Rhodesia.

MORGAN, R.P.C., QUINTON, J.N. and RICKSON, R.J. 1991. EUROSEM - A user guide. Silsoe College, UK.

VORIES, E.D. and VON BERNUTH, R.D. 1990. A laboratory study of the erodibility of disturbed soil. Transactions of the ASAE 33:1497-1502.

WISCHMEIER, W.H. and SMITH, D.D. 1978. Predicting rainfall erosion losses - a guide to conservation planning. USDA Agriculture Handbook No. 537.

Chapter 15

THE EFFECTS OF SLOPE LENGTH AND INCLINATION ON THE SEPARATE AND COMBINED ACTIONS OF RAINSPLASH AND RUNOFF

C. Kamalu

Silsoe College, Cranfield University, Silsoe,
Bedfordshire, MK45 4DT, UK

Summary

Under certain conditions of slope length and inclination, interactions between rainfall and runoff are observed to increase erosion rates.

Laboratory work on a standard sand and a sandy loam soil has tested this interaction under varying slope lengths and inclinations. Results showed the rainfall-runoff interaction was most pronounced on short slope lengths (about 1.5m) and intermediate slope inclinations (5%). Tests on 1.5m, 2m and 2.5m slope lengths showed the 1.5m slope was often subjected to the greatest erosion rates. Also, different surface processes (either runoff, rainsplash or their combination) became dominant under different conditions. Generally, runoff showed the greatest sensitivity to slope variation, becoming clearly dominant on steeper slopes. It was the least erosive form on gentle slopes. An interactive combination of rainfall and runoff was marginally dominant over other erosive processes on gentle to intermediate slopes.

1. INTRODUCTION

Under certain conditions of slope length and inclination, interactions between rainsplash and runoff are observed to increase erosion rates. This was a conclusion reached from a series of experiments undertaken at the University of Birmingham on 1.5m length model road shoulders constructed with sand and sand/clay materials, at a fixed slope of 5% (Kamalu, 1991). These tests clearly revealed that the various erosional processes (rain, runoff and their combination) applied as a fixed mass rate of water (90kg mass over 1 hour), yielded varied amounts of soil, reflecting their different levels of efficiency in causing erosion.

Further laboratory work on a standard sand at Silsoe College has also studied both the rainfall-runoff interaction and the comparative erosion of rainfall, runoff and their combination. In these studies varied slope lengths (1.5m, 2m and 2.5m) and inclinations (3.5%, 5%, 7% and 9.4%) were used. Rainfall and runoff were applied as a fixed mass rate of water in the following proportions:

i)   100% rain (110mm.hr^{-1});
ii) 100% discharge (5.5, 4.6, 3.67 l.min^{-1} depending on slope length);
iii) 82% rain (90mm.h^{-1}) plus 18% discharge (1, 0.84, 0.67 l.min^{-1});

iv) rain only (90mm.h^{-1});
v) discharge only (1, 0.84, 0.67 l.min^{-1});

Plainly, (iv) and (v) split the combined run of (iii) into its separate rain and runoff components. The proportion of rain to runoff was chosen such that a measurable amount of erosion occurred for most runoff-only runs under varied slope lengths and inclinations. The chosen rainfall intensities were similar to those for tropical storms.

In initial tests on standard sand, relatively small coefficients of variation of about 7% between replicates were obtained when four and, later on, three replicates were used. The number of replicates was thereafter reduced to 2 per treatment and the data was analysed as a 2-block, 2^4 factorial set of experiments.

2. MATERIALS AND METHODS

Figure 1 is a sketch of the soil bed (1m × 2.5m), rain simulator and runoff simulator used. Experiments were conducted on a standard D14 Leighton Buzzard Lower Greensand and a sandy loam, both of fairly uniform grading with most particles in the 0.2mm - 0.6mm size range. The soil was saturated before each run to eliminate variations of soil moisture between replicates. Soil profiles are given in Table 1.

Table 1. Particle size analysis

Standard Sand		Sandy loam soil	
Mesh Size (microns)	% Retained	Particle Size (microns)	% Weight
1000	0.025	600 - 2000	5.57
710	0.33	212 - 600	76.41
500	15.43	106 - 212	8.72
355	48.70	63 - 106	1.67
250	27.06	2 - 63	1.59
180	6.24	>2	7.61
150	0.75		
125	0.68		
90	0.63		
63	0.13		

The rotating nozzle rain simulator discharged rain drops from a 2/3 HH 40 size nozzle at a constant pressure of 8psi. This produced drops of approximately 2mm median diameter which, under the influence of both pressure and gravity, hit the bed (1.8m below) at about 95% of their terminal velocity. Rainfall intensity was altered by adjusting the angle of the nozzle.

Figure 1. Apparatus used

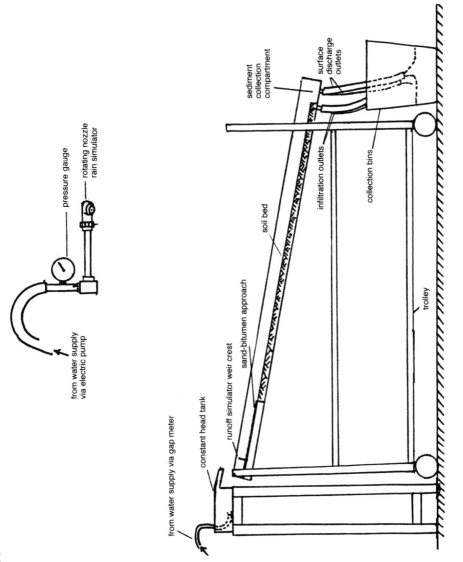

pressure gauge

rotating nozzle
rain simulator

from water supply
via electric pump

from water supply via gap meter

constant head tank

runoff simulator weir crest

sand-bitumen approach

soil bed

sediment
collection
compartment

surface
discharge
outlets

infiltration outlets

collection bins

trolley

The runoff was simulated by flow from a weir built into the top of the soil tray. The rate of flow was controlled by a gap meter. A sand-bitumen surface immediately downstream of the weir introduced runoff onto the soil bed.

The soil bed length was shortened by covering the soil surface at the bottom of the slope with a steel sheet (1m × 0.5m or 1m × 1m), placed flush with the upstream soil surface. The soil bed slope was adjusted by means of a scissor jack supporting the rig at its downstream end. Outlets are provided from the base of the soil tray to allow collection of water infiltrated through the soil bed. A sediment collection compartment exists at the downstream end and outlet pipes from this section discharge water and sediment into collecting bins.

Each experiment ran for 30 minutes. At the end of each run the total volumes of water discharged from the slope and infiltrated water from the base of the soil tray were collected. The collected sediment was then placed in an oven to dry for 24 hours at 105°C. The weight of dry soil was determined per unit area of soil bed to account for adjustments of slope length and per litre of water to eliminate the effects of fluctuations in the amount of water supplied for each replicate.

3. RESULTS AND DISCUSSION

3.1 Effect of slope inclination

Runoff appears to be the process most sensitive to a change of slope inclination. Beyond some threshold inclination, runoff becomes the dominant erosive process. Test results show that threshold slope inclinations at which runoff becomes the dominant erosive process on standard sand, lie in the region of 7-8%. These compare with 10% (Quansah, 1985; Bubenzer et al., 1966) and 9% (Moss et al., 1979).

For the sandy loam these thresholds appear to grow steeper for the longer slope lengths. On the 2.5m shoulder runoff is not dominant even at the 9.4% slope. However, given that all other results are consistent with this trend, it is reasonable to assume that the runoff will eventually become dominant, but at a much steeper slope.

One also observes a generally sharper rise in the erosion rates beyond a certain threshold. For both of the soils this threshold lies in the region of 5-7%.

The sharp increase in erosion rates generally was accompanied by an increase in the intensity of rilling. For instance, on the 1.5m standard sand shoulder where the greatest erosion rates occurred, it was observed that at the 3.5% slope inclination only 10 rills were formed at the tail of the bed under 100% applied runoff. At 7% this multiplied to 50 rills. Beyond 7% the further intensification of the rilling no longer involved the multiplication of the number of rills, but the concentration of the flow into fewer rills which then widened and deepened. At the 9.4% slope inclination, for instance, the flow concentrated into two wide channels of about 8cm width and up to 3cm depth.

For the case of 100% rain on a 2m standard sand slope no rilling occurred until the slope reached 7%, at which point just two rills were discernible. At 9.4%, 4 rills were formed.

Certain authors note (for instance, Moss et al., 1979; Meyer and Monke, 1965) that at gentle slopes, rainfall tends to suppress rill formation by wearing down the rill side, making them wider and shallower. This has the effect of decreasing depths of runoff flow, decreasing flow velocity and hence detachment.

3.2 Effects of slope length

Generally, there was a trend whereby erosion rates were often greatest on the shortest shoulder, but rarely so on the longest. There is often a significant difference between the erosion rates on a 1.5m length compared with 2m and 2.5m slopes. At 7%, erosion rates in general are significantly greater on the 2.5m compared with the 2m length slope (Table 2). This pattern occurs again on the standard sand under the action of various erosional processes. However, except for the case of rainfall/runoff combined at 7%, the difference is never statistically significant.

Table 2. Table of means - slope × length

Slope length	Slope steepness			
	3.5%	5%	7%	9.4%
1.5m	2.33	3.55	6.1	16.79
2m	1.62	2.85	6.29	15.86
2.5m	1.5	2.73	6.93	16.9

For the 100% discharge curves for tests on standard sand erosion was not significantly to be significantly greater on the 1.5m shoulder compared with the 2m and 2.5m shoulders until an inclination of 9.4%. In the case of rainfall/runoff combined, which is the most effective erosional process on gentle and intermediate slope gradients, the trend occurs on the 3.5-7% slopes and the differences between the erosion rates on the 1.5m and 2m length slopes are always significant. They are also significant between these lengths for rainfall at the 5% slope. To explain these observations, we note that at gentle and intermediate slope inclinations, rainfall and rainfall/runoff combined detachment are dominant over runoff detachment, due to the runoff not having sufficient velocity to detach soil particles. Thus, under the action of rainfall or rainfall/runoff combined at gentler slopes, transport capacity of the flow may have been reached within a few decimetres of travel distance; especially on a sandy soil lacking in cohesion and which is easily dispersed and detached (Rauws and Govers, 1988; Smith et al., 1954). It is debatable whether the erosion rate levels off after reaching transport capacity, or falls as the test results indicate. A fall implies deposition, which would explain the greater erosion rates on the 1.5m compared with the 2m and 2.5m shoulders. Furthermore, there will be some increase in flow depth with distance downstream in the case of rainfall induced flow. This may have the effect of cushioning raindrop impact and limiting the amount of detachment downstream on longer slopes.

For the tests on sandy loam, at gentle slopes the differences between the erosion rates due to the different processes were not significant. The overall pattern was less consistent than for the standard sand. Nevertheless, the most consistent trend observed was that the greatest soil losses under the 100% discharge treatment occurred on the shortest shoulder.

3.3 Effects of rainfall-runoff interaction

Examining the table of means (Table 3), the interaction effect seems most pronounced at a slope inclination of 5%, and it is clearly significant. The clear trend is that rainfall/runoff combined is dominant at gentle and intermediate slopes. This is because (i) rainfall on gentle slopes has the capacity to detach plenty of material, but not to transport it, (ii) runoff on gentle slopes acts as sheetflow with the capacity to transport a lot of material, but not to detach it; whilst (iii) rainfall/runoff combined has both the capacity to detach and transport a lot of material. Quansah (1985) notes that at certain slope inclinations rainsplash and runoff contribute equally to detachment. To explain the pronounced effect of rainfall/runoff combined at this slope, a certain combination of rainfall and runoff in the region of 5% slope steepness may be optimum for effectiveness in eroding soil.

Table 3. Table of means - slope × rainfall - discharge treatment (see page 143/44)

Slope steepness	Rainfall = $110mm.hr^{-1}$	Discharge = $3.67 / 4.6 / 5.5 \ l.min^{-1}$	Rainfall = $90mm.hr^{-1}$ + Discharge = $0.67 / 0.84 / 1.0 \ l.min^{-1}$	Discharge = $0.67 / 0.84 / 1.0 \ l.min^{-1}$	Rainfall = $90mm.hr^{-1}$
3.5%	2.465	1.595	2.555	0.0333	2.435
5%	4.12	2.084	4.54	0.892	2.897
7%	8.32	6.895	8.75	1.3	6.84
9.4%	19.29	22.94	19.16	5.51	15.7

In the particular case of the 1.5m standard sand shoulder inclined at 5%, a 100% increase in erosion due to the $90mm.h^{-1}$ of rainfall was obtained by adding just $0.67 \ l.min^{-1}$ of runoff. This is a similar result to that obtained under similar conditions at Birmingham University (Kamalu, 1991).

In contrast, no significant interaction is observed to occur on the 2m and 2.5m standard sand shoulders. This masking of the interaction effect by increased slope length can be observed in the table of means (Table 4). We find that at 1.5m the interaction effect is positive and significant (if we disregard both slope and soil type). The possibility that the rainfall/runoff combined is able to reach transport capacity within a short distance may also explain the masking of the interaction effect observed on the shortest slope. The curtailed ability of the flow to detach once it has travelled far enough to reach transport capacity, and the increase of flow depth downstream, serving to cushion the impact of raindrops, may both explain the absence of the rain-runoff interaction effect at longer lengths.

Table 4. Table of means - rainfall - discharge treatment × length

Slope length	Rainfall = 110mm.hr^{-1}	Discharge = 3.67 / 4.6 / 5.5 l.min^{-1}	Rainfall = 90mm.hr^{-1} + Discharge = 0.67 / 0.84 / 1.0 l.min^{-1}	Discharge = 0.67 / 0.84 / 1.0 l.min^{-1}	Rainfall = 90mm.hr^{-1}
1.5m	8.8	9.53	9.72	1.69	6.21
2m	8.74	7.93	7.95	1.53	7.11
2.5m	8.11	8.16	8.55	2.59	7.58

4. CONCLUSIONS AND RECOMMENDATIONS

Runoff became the dominant erosive process beyond some slope inclination threshold which appeared to differ with soil type. This threshold was found to be more sensitive to length on the sandy loam compared with the standard sand. Rainfall/runoff combined was the dominant erosive process on a narrow shoulder (1.5m), of gentle to intermediate slope inclination (3.5%-7%). Rainfall-runoff interaction was observed to occur under these conditions, but this appeared to be masked at longer lengths. The shortest slope length (1.5m) had the greatest erosion rates, whilst the difference between the erosion rates of the 2m and 2.5m shoulders was comparatively small. Both the standard sand and the sandy loam have a threshold inclination (between 5 and 7%) beyond which erosion rates increase more rapidly with slope steepness.

REFERENCES

BUBENZER, G.D., MEYER, L.D. and MONKE, E.J. 1966. Effect of particle roughness on soil erosion by surface runoff. Transactions American Society of Agricultural Engineers. pp.562-564.

KAMALU, C. 1991. Soil erosion on road shoulders: The effects of rainfall-runoff interaction. IECA Proceedings. pp.249-263.

MEYER, L.D. and MONKE, E.J. 1965. Mechanics of soil erosion by overland flow. Transactions American Society of Agricultural Engineers 8:572-577,580.

MOSS, A.J., WALKER, P.H. and HUTKA 1979. Raindrop simulated transport in shallow flows. Sedimentary Geology 22:165-184.

QUANSAH, C. 1985. Rate of soil detachment by overland flow with and without rain, and its relationship with discharge, slope steepness and soil type. In El-Swaify, S.A., Moldenhauer, W.C. and Lo, A. (eds). Soil erosion and conservation. Soil Conservation Society of America.

RAUWS, G. and GOVERS, G. 1988. Hydraulic and soil mechanical aspects of rill generation on agricultural soils. J. Soil Science 39:111-124.

SMITH, R.M., HENDERSON, R.C. and TIPPIT, O.J. 1954. Summary of soil and water conservation research from the Blackland Experiment Station, Temple, Texas. Bulletin No. 781. Texas Agricultural Experimental Station.

Chapter 16

DATA HANDLING FOR SOIL MANAGEMENT AND CONSERVATION: THE SOTER APPROACH

E.M. Bridges, N.H. Batjes and L.R. Oldeman
International Soil Reference and Information Centre,
PO Box 353, 6700 AJ Wageningen, the Netherlands

Summary

A soils data base which is capable of storing and handling information at different scales of operation is presented. The digital World Soils and Terrain Database (SOTER) has been developed at the International Soil Reference and Information Centre (ISRIC) for use at a global scale and for individual continents or countries. Trials of this system have been carried out in several parts of the world and it is now ready for a wider range of applications including studies of soil degradation. At ISRIC a computer program for water erosion prediction hazard has been developed, which draws upon the information stored in the SOTER database. Compilation of databases and the development of database handling systems for soil information are currently an important area of development in soil science.

1. INTRODUCTION

There is an intense and increasing pressure on national land and water resources, leading to soil erosion and degradation, pollution of soils, water and the environment, and a permanent loss of productive capacity. In all countries there is a pressing need, acknowledged by most governments, for a system which can store detailed information on natural resources in such a way that these data can be accessed and combined immediately and easily. In this way, each combination of land, soil, water and population within a country can be rapidly analysed and classified in terms of potential use, in relation to food requirements, socio-economic factors and environmental impact or conservation. Such a system is a prerequisite for policy formulation, development planning, efficient use of resources and for implementation of development programmes.

The lack of such a system at international and national levels has, until now, been one of the most important constraints to the solution of fundamental problems and to the efficient use of natural resources. This has been felt both by the individual countries, and by aid donors frustrated at the meagre results of their contributions. Now, however, due to the rapidly falling real costs of computer hardware and software, and the equally rapid development over the past two to five years of easy-to-use computer programmes, these systems can be provided in a relatively short time at what is, comparatively speaking, an insignificant cost. In a typical case such a system would consist of the following:

1. A computerised database containing available and in most cases very detailed information on topography, soils, climate, vegetation, land use, population (human, domestic

animals and wildlife) and infrastructure, and the whole range of socio-economic factors such as food requirement attitudes, skills, costs of input and availability of markets.

2. A geographical information system or GIS, which ties each item of information to its precise geographical location, but which is able to display each separate type of information as a separate layer or overlay. This makes it possible to display or print maps of any combination of information required virtually instantaneously.

3. A set of crop yield models which can calculate the level of production which could be obtained from each and any combination of soil and climate in the region or country, at a number of different input levels, management systems or rates of erosion.

4. Various environmental impact models, which, for example, allow the calculation of erosion rates for a given land unit, use and production system.

A similar system has been proposed by an ad hoc meeting of soil scientists from the European Community and developing countries (Stoops & Cheverry, 1992). Systems with the above capabilities can now operate on common desk-top, office type computers with optional peripheral equipment such as digitising tables and colour printers. Hardware and software for the most basic system would cost in the order of US$50,000, and would be suitable for district planning at the local level. Such a system would require only a few weeks of training to operate. At the national level, where greater capacity and higher quality of product is required, total equipment costs today would be of the order of US$150,000. In this case training of local staff would last a matter of months.

2. SOTER

Development of computerised natural resource inventories and a system for land evaluation, especially for use in developing countries, has been carried out by the Food and Agriculture Organisation (FAO) since the early 1970s, under the Agro-Ecological Zones (AEZ) programme. Considerable experience has been acquired (FAO, 1978-1981). However, there was a conspicuous absence of any uniform digital database concerning the soils of the world. Consequently, in 1986 ISRIC was asked by the International Soil Science Society to develop a methodology for a World Soils and Terrain Digital Database - SOTER (ISSS 1986a;b). The SOTER procedures would allow knowledge of the spatial distribution and attributes of world soils to be updated and refined using a uniform methodology. This is necessary because much of the information presented on the Soil Map of the World (FAO/Unesco 1971-1981) was collected prior to the 1970s. Now the map is partly out of date (Sombroek 1990). The following paragraphs briefly indicate the nature of the methodology.

In SOTER, geo-referenced map units are linked to a digital database specifying the key attributes of the components of these mapping units. A brief summary of the features of the SOTER system includes:
- a resolution equivalent to an average map scale of 1:1M;
- compatibility with global databases of other environmental resources;
- capable of updating and purging obsolete and/or irrelevant data;
- accessible to international, regional and national decision and policy makers;

- transferable and applicable to developing countries, for national database development at larger scales (up to 1:100,000).

2.1 Content and Structure

The general approach in SOTER is to review all existing soil and terrain data in a geo-referenced area - whether or not registered on official soil maps - and to supplement this information with remote sensing data where appropriate. The collected data are then rearranged according to the procedures of the SOTER manual (van Engelen and Pulles, 1991). Underlying this approach is the grouping and mapping (polygon based) of land areas showing a distinctive, and often repetitive pattern of land form, slope, parent material and soils. Areas so obtained are delineated on the base map as SOTER units. Whereas individual SOTER units are characterised in the database in terms of their terrain and soil components, these cannot be displayed on a 1:1M scale map (Fig. 1). Each SOTER unit is geo-referenced and considered unique with respect to its constituent soil and terrain characteristics.

Figure 1. Schematic representation of a terrain unit and its terrain and soil components.

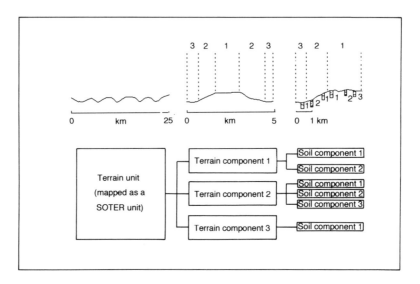

Areas with different lithologies within similar landforms are separated as different terrain units which display similar patterns of surface form, slope, micro-relief and parent materials. The next step includes the delineation of terrain components, each of which is divided into a number of soil components (Fig. 2). These soil components are represented by actual soil profiles described from the area and linked to the SOTER system as shown schematically in Figure 3. The respective attribute data are specified using either descriptive terms or numerical values, as appropriate (van Engelen & Pulles, 1991).

Figure 2. Example of a SOTER unit, as represented on a map and its legend, and its constituent terrain units, terrain components (TC) and soil components (SC) as specified in the database.

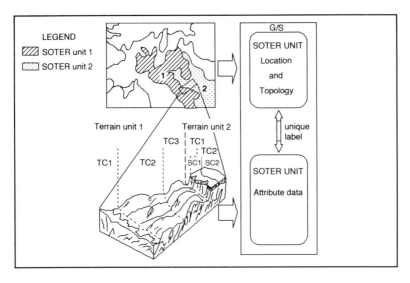

Figure 3. Structure of the SOTER database (1:M stands for one to many relations; M:1 for many to one relation).

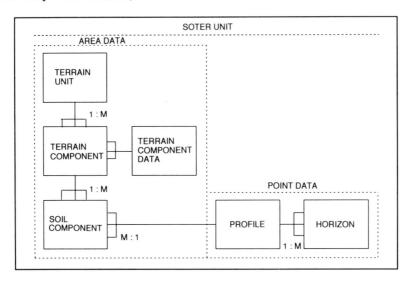

The chemical, physical and morphological attributes of the soils, for instance horizonation, organic carbon, cation exchange capacity, texture, structure, stoniness and moisture holding capacity, are considered to be representative for the whole area typified by the major soil under consideration. Initially, the soils of each soil component will be represented by a small number of profiles. In the long term, the intention is to increase the number of profiles as rapidly as possible. This will enable identification of soil attributes and subsequently calculation of the range of properties within each soil component.

The terrain characteristics and key attributes of profiles can readily be incorporated in the relational database management system of SOTER, which is an improvement on the traditional methods of small scale soil cartography (e.g. FAO/Unesco 1971-1981). The SOTER attribute files are defined into 'structures', compatible with widely used commercial software database management systems such as dBASE, INFO and ORACLE.

Extensive trials of the SOTER system have been carried out in Latin America (LASOTER) and in North America (NASOTER). Interest in soil data base handling systems and requests for technical assistance to develop them were received by ISRIC from six neighbouring countries in West Africa, from Austria, Hungary, Czechoslovakia, and from several Middle Eastern and Central American countries. The number of requests to date is indicative of the demand for, and importance attached to the land resources database, land evaluation, and land use planning system which SOTER can provide (Oldeman and van Engelen, 1992). A workshop on the use of SOTER software for participants from Argentina, Brazil and Uruguay took place in March-April, 1992 (van Engelen & Peters, 1992).

Worldwide coverage by SOTER at 1:1M scale will require from 10 to 20 years to complete. However, at the global level, there remains an urgent need to generate rapid, accurate and reliable information to help solve a multitude of soil-related environmental problems. The response to this situation is that work has begun upon a framework of data which will enable representative soil information to be made available in a digital format at least for broad-based studies at the global scale.

3. RECENT DEVELOPMENTS

Arising from the work in the LASOTER area, experiments have been carried out using the database to map information on features of soil erosion by water. A number of single feature maps have been produced using GIS in association with SOTER. These examples include maps of those areas where recent overland flow and sediment deposition have been observed, together with maps of topsoil loss, severity of terrain deformation and erosion hazard. These maps record the visual interpretation of the surveys involved in the LASOTER study.

Following this work, a computer program for water erosion hazard assessment linked to SOTER has been developed by ISRIC staff. This program, entitled SOTER Water Erosion Assessment Program (SWEAP) has the possibility of using either the Universal Soil Loss Equation (USLE) or the Soil Loss Estimation Model of South Africa (SLEMSA). The system works on time steps of one month in which seasonal dynamics of crop cover and rainfall erosivity are accommodated. The results are interpreted in terms of erosion hazard units rather than quantified estimates of potential soil loss (van den Berg, 1992).

4. THE NEED FOR GLOBAL LAND RESOURCE DATABASES

International agencies such as the World Resources Institute, the World Bank, the UNEP, FAO, CGIAR and IGBP have expressed the need for quantified basic information on land resources. This is required to assess the potential productivity of the land, to monitor the status, risk and rate of soil degradation, to develop action plans for conservation or rehabilitation of the land and to improve our understanding of global change. Two programmes are discussed in this context that are of a global nature and that could have benefited tremendously if a World Soils and Terrain Database had been available.

4.1 Global Assessment of Soil Degradation (GLASOD)

Although soil degradation is recognised as a serious and widespread problem, its geographical distribution and the total areas affected are only roughly known. Sweeping statements about soil erosion undermining the future prosperity of mankind do not help planners who need to know where the problem is serious and where it is not. This feeling was also expressed by UNEP which indicated the need to produce a scientifically credible global assessment of the status of human-induced soil degradation in the shortest possible time (ISSS, 1987). In 1987 ISRIC undertook to carry out a qualitative assessment of the state of human induced soil degradation (GLASOD) for the United Nations Environment Programme (UNEP). The objective of GLASOD was defined as 'strengthening the awareness of policy-makers and decision-makers of the dangers resulting from inappropriate land and soil management' so as to create 'a basis for the establishment of priorities for action programmes' (Oldeman et al., 1991).

The GLASOD map shows four categories of human induced soil degradation processes: water erosion (two types of degradation: Wt, loss of topsoil; Wd, terrain deformation/mass movement), wind erosion (Et, loss of topsoil; Ed, terrain deformation; Eo, overblowing), chemical deterioration (Cn, loss of nutrients and/or organic matter; Cs, salinization; Ca, acidification; Cp, pollution), and physical deterioration (Pc, compaction; Pw, waterlogging; Ps, subsidence of organic soils). Additionally, three types of 'stable terrain' and six types of 'wasteland' have been depicted on the map. The severity of the soil degradation process is characterised by the degree to which the soil is degraded and by the relative extent of the degraded area within a given physiographic unit. The human-derived causative factors of soil degradation are depicted on the map using letter codes. They include deforestation, overgrazing, over-exploitation of the vegetative cover for domestic use, agricultural activities and (bio)industrial wastes.

The GLASOD map has been digitised, and the descriptive legend incorporated in a 'soil degradation data base'. This linkage permits the generation of a range of area statistics (Oldeman et al., 1991; Table 1), some of which have been published in the 1992-1993 World Resources Report (World Resources Institute, 1992). Additionally, a range of derived thematic maps have been prepared for publication in UNEP's new Atlas of Desertification (UNEP, 1992). However, it is vitally important that the information contained in the GLASOD map is viewed in its correct context, and at the scale of the original survey. Taking the maps and the information held in the database out of context, and reproduction at a more detailed scale than the original map cannot be justified, on the basis of the information which was used to compile the map. Equally, the scale used in cartographic representation of the extent of degradation must be interpreted carefully, as a superficial impression may be gained that any one aspect of degradation is more extensive than it really is.

Table 1. Extent of human-induced soil degradation for the world expressed in million hectares (Source: Oldeman et al., 1991).

Type	Light	Moderate	Strong	Extreme	Total
Wt Loss of Topsoil	301.2	454.5	161.2	3.8	920.3
Wd Terrain Deformation	42.0	72.2	56.0	2.8	173.3
W WATER	343.2	526.7	217.2	6.6	1093.7 (55.6%)
Et Loss of Topsoil	230.5	213.5	9.4	0.9	454.2
Ed Terrain Deformation	38.1	30.0	14.4	-	82.5
Eo Overblowing		10.1	0.5	1.0	11.6
E WIND	268.6	253.6	24.3	1.9	548.3(27.9%)
Cn Loss of nutrients	52.4	63.1	19.8	-	135.3
Cs Salinization	34.8	20.4	20.3	0.8	76.3
Cp Pollution	4.1	17.1	0.5	-	21.8
Ca Acidification	1.7	2.7	1.3	-	5.7
C CHEMICAL	93.0	103.3	41.9	0.8	239.1(12.2%)
Pc Compaction	34.8	22.1	11.3	-	68.2
Pw Waterlogging	6.0	3.7	0.8	-	10.5
Ps Subsidence organic soils	3.4	1.0	0.2	-	4.6
P PHYSICAL	44.2	26.8	12.3	-	83.3(4.2%)
TOTAL	749.0(38.1%)	910.5(46.4%)	295.7(15.1%)	9.3(0.5%)	1964.4(100%)

4.2 World Inventory of Soil Emission Potentials (WISE)

This project is being carried out by ISRIC within the framework of the Netherlands National Research Programme on Global Air Pollution and Climate Change (NOP MLK). Terrestrial ecosystems, and soils in particular, are important sources of a number of 'greenhouse gases' such as water vapour, carbon dioxide, methane, nitrous oxide and nitric oxide. The soil conditions and processes which regulate production and emissions of these radiatively active trace gases from soils are incompletely understood (e.g. Bolin et al., 1986; Bouwman, 1990).

A soil database developed from the 1992 version of the Soil Map of the World is being compiled at ISRIC, in close consultation with FAO and national soil survey organisations where necessary. This will enable the quantification of the soil conditions and processes which regulate gaseous production and emissions from soils. It is then proposed to make estimates of potential methane production for broadly defined natural and man-influenced ecosystems (Batjes and Bridges, 1992a).

It is anticipated that the WISE database could also be used to model N_2O emissions from soils and to refine estimates of the soil carbon pool (Batjes and Bridges, 1992b). Additionally, it could form the basis of an assessment of the susceptibility of soils to specified types of environmental degradation at the global and regional levels, including the vulnerability of soils to accumulation and sudden release of heavy metals, pesticides and other toxic substances following climatic changes (Batjes and Bridges, 1991).

5. CONCLUSIONS

Internationally, policy and decision makers increasingly recognise that land resources must be preserved for future generations. By implication this means that attention and support should be given to the development of improved environmental information systems to provide fast and accurate information upon which sound decisions can be based. Implementation of worldwide programmes, such as the International Geosphere and Biosphere Programme (IGBP), require soil scientists, biologists, climatologists, land use specialists and information technologists to work in close collaboration to develop and implement effective soil-geographical information systems for use at different scales (Zobler, 1989; Bliss, 1990). As far as soil resources are concerned, the usefulness of SOTER has recently been confirmed by field trials and its methodology officially endorsed by FAO, UNEP and the International Society of Soil Science.

Data handling systems such as SOTER in association with geographical information systems can provide a framework for handling soil data for quantifying the varied problems of soil conservation. Although the pressing need at the moment is seen to be the identification of susceptible areas and assessment of the amount of soil erosion actually taking place, an increasing pressure exists for the provision of soil information for land evaluation, crop production potentials and the assessment and solution of soil chemical pollution problems. This work is envisaged at the macro-scale. More detailed follow-up studies are needed.

REFERENCES

BATJES, N.H. and BRIDGES, E.M. 1991. Mapping of soil and terrain vulnerability to specified chemical compounds in Europe at a scale of 1:5M. Proceedings of an International Workshop organised in the framework of the Chemical Time Bomb Projects of the Netherlands Ministry of Housing, Physical Planning and Environment and the International Institute for Applied Systems Analysis (Wageningen, 20-23 March 1991). International Soil Reference and Information Centre, Wageningen.

BATJES, N.H. and BRIDGES, E.M. (eds). 1992a. World inventory of soil emission potentials. Proceedings of an international workshop organised in the framework of the Netherlands National Research Programme on Global Air Pollution and Climatic Change (NOP). WISE Report No. 2. International Soil Reference and Information Centre, Wageningen.

BATJES, N.H. and BRIDGES, E.M. 1992b. A review of soil factors and processes that control fluxes of heat, moisture and greenhouse gases. Technical Paper 23. International Soil Reference and Information Centre, Wageningen.

BLISS, N.B. 1990. A hierarchy of soil databases for calibrating models of global climate change. In: Bouwman, A.F. (ed.). Soils and the greenhouse effect. John Wiley, Chichester.

BOLIN, B., BÖÖS, B.R., JÄGER, J. and WARRICK, R.A. (eds). 1986. The greenhouse effect, climate change, and ecosystems. SCOPE 29. John Wiley, Chichester. pp.311-325.

BOUWMAN, A.F. (ed.) 1990. Soils and the greenhouse effect. John Wiley, Chichester.

FAO 1978-1981. Reports of the agro-ecological zones project. World Soil Resources Reports No. 48; Vol.1.: Africa; Vol.2: Southwest Asia; Vol.3: South and Central America; Vol.4: Southeast Asia. Food and Agriculture Organization, Rome.

FAO/UNESCO 1971-1981. Soil Map of the World. 1:5,000,000. Volumes 1 to 10, Unesco, Paris.

ISSS 1986a. Proceedings of an international workshop on the structure of a digital Soil Resource Map annex Data Base (20-24 January 1986, ISRIC, Wageningen).

Baumgardner, M.F. and Oldeman, L.R. (eds). SOTER Report No. 1, International Society of Soil Science, Wageningen.

ISSS 1986b. Project proposal 'World Soils and Terrain Digital Database at a scale of 1:1M (SOTER)'. Baumgardner, M.F. (ed.). International Society of Soil Science, Wageningen.

ISSS 1987. Proceedings of the Second International Workshop on a global soils and terrain digital database. (van de Weg, R.F. (ed.)). SOTER Report No. 2. International Soil Reference and Information Centre, Wageningen.

OLDEMAN, L.R. and VAN ENGELEN, V.W.P. 1992. A world soils and terrain digital database (SOTER): An improved assessment of land resources for sustained utilization of the land. Paper prepared for the ISSS/SSSA Conference on Operational Methods to Characterise Soil Behaviour in Space and Time, Ithaca, New York. Working Paper and Preprint 92/06, International Soil Reference and Information Centre, Wageningen.

OLDEMAN, L.R., HAKKELING, R.T.A. and SOMBROEK, W.G. 1991. World map of the status of human induced soil degradation: An explanatory note (Second revised edition). Published in co-operation with Winand Staring Centre, International Society of Soil Science, Food and Agriculture Organization of the United Nations, International Institute for Aerospace Survey and Earth Sciences. International Soil Reference and Information Centre, Wageningen/ United Nations Environment Programme, Nairobi. (Note: The explanatory note accompanies three maps sheets at an average equatorial scale of 1:15M).

SOMBROEK, W.G. 1990. Geographic quantification of soils and changes in their properties. In. Bouwman, A.F. (ed.). Soils and the greenhouse effect. John Wiley, Chichester pp.225-244.

STOOPS, G. and CHEVERRY, C. (eds). 1992. New challenges for soil research in developing countries: a holistic approach. Proceedings of a workshop funded by the European Community life sciences and technologies for developing countries (STD 3 Programme). ENSAR, Rennes.

UNEP 1992. World Atlas of Desertification (Middleton, N.J. and Thomas, D.S.G. (eds.)). United Nations Environment Programme. Edward Arnold, London.

VAN DEN BERG, M. 1992. SWEAP. A computer program for water erosion assessment applied to SOTER. SOTER Report No. 7. International Soil Reference and Information Centre, Wageningen.

VAN ENGELEN, V.W.P. and PETERS, W.L. 1992. Manual del curso de Entrenamiento Lasoter SIT. Discussion Paper and Preprint 92/3. International Soil Reference and Information Centre, Wageningen.

VAN ENGELEN, V.W.P. and PULLES (eds). 1991. The SOTER Manual: Procedures for small scale digital map and database compilation of soil and terrain conditions. Working Paper and Preprint 91/3. International Soil Reference and Information Centre, Wageningen.

WORLD RESOURCES INSTITUTE. 1992. World Resources 1992-1993: A guide to the global environment. World Resources Institute in collaboration with the United Nations Environment Programme and the United Nations Development Programme. Oxford University Press, New York.

ZOBLER, L. 1989. A world soil hydrology file for global climate modelling. In. Proceedings of the International Geographic Information Systems (IGIS) Symposium. Association of American Geographers, Arlington.

Part II:

PREDICTION AND MODELLING

Chapter 17

MODELLING REGIONAL SOIL EROSION SUSCEPTIBILITY USING THE UNIVERSAL SOIL LOSS EQUATION AND GIS

S. Jäger

Geographisches Institut, Universität Heidelberg, Im Neuenheimer Feld 348,
W-6900 Heidelberg, Germany

Summary

An investigation of the regional soil erosion susceptibility in Baden-Württemberg, Germany has been carried out using readily available data. A raster based Geographic Information System was used to process and analyse the data. The Universal Soil Loss Equation (USLE) is used. Its adaptation to the macro-scale is demonstrated for each factor. The results are presented as generalised maps of soil erosion susceptibility.

The evaluation model is based on a quantitative classification directed toward conservation policy. Four classes of soil erosion susceptibility are derived by comparing tolerated and computed erosion values. Approximately 3% of the state's arable land shows a very high susceptibility, 63% is highly susceptible, 13% moderately susceptible, and 21% is classified as slightly susceptible. The input data contained some errors, especially in the topographic variables of slope length and slope gradient, as measured on topographic maps by the State Agronomy Agency. A comparison of these manually derived LS-factors with factors derived from a Digital Elevation Model (DEM) shows weak correlation. In this study, this comparison was carried out using only three available DEMs. As a result of this investigation a DEM for the entire state has been made available. The LS values of the USLE will be derived from the DEM for the entire state in a forthcoming project.

1. INTRODUCTION

Soil erosion has been a well-known problem for centuries in Baden-Württemberg (approximately 36,000 km^2, Figure 1), especially in the loess covered areas of the state. The main geomorphic units are the Rhine Graben in the west, flanked by the mainly forested uplands of the Black Forest and Odenwald. In the middle part, the Mesozoic strata form dominant cuestas. South of the Danube, the landscape is dominated by hilly ground and end moraines. Loess is a common sediment in the vicinity of the Rhine Graben and near the Alps.

Several qualitative and quantitative investigations of soil erosion have been carried out in the area (Dikau, 1986; Murschel-Raasch, 1991, Quist, 1987). The recently enacted State Soil Conservation Law (Staatsministerium, 1991) is the first of its kind in Germany. The law now permits the delineation of problem areas with different hazards, such as soil contamination from heavy metals, acidification and soil erosion. Also, restrictions on special crop rotations or tillage methods can be applied. Therefore, to satisfy legal constraints, a

161

quantitative knowledge of soil erosion rates is needed. Thus, several projects aim to develop appropriate models for the assessment of soil erosion at different scales. They are financed by the Ministry of Agriculture and the Ministry of the Environment. The work described in this paper was conducted in 1990 and 1991. It is a contribution to the macro scale assessment of soil erosion susceptibility of the state's arable land.

Figure 1. Overview map of the area of investigation. Major cities are Stuttgart (S), Mannheim (M), Heilbronn (H), Ulm (U) and Basel (B).

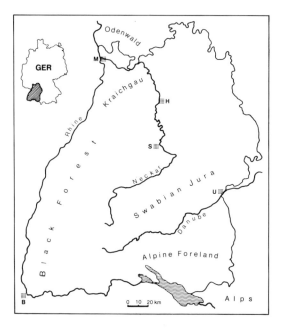

The term 'susceptibility' is used here in order to distinguish this investigation from 'hazard' and 'risk' mapping, terms which are often used interchangeably. According to the United Nations Disaster Relief Co-ordinator, hazard is defined as the probability of recurrence of a specific damaging phenomenon. The term risk takes into account the extent of socio-economic damage of that specific phenomenon (UNDRO, 1991). In contrast, susceptibility maps delineate areas with different potential for soil erosion.

2. THE MODEL

The model used for the present study is the Universal Soil Loss Equation (Wischmeier and Smith, 1978). Limitations of the model are well known and described in several publications (e.g. Bork, 1988; Schwertmann et al., 1987; Morgan, 1986). Its empirical nature means that the USLE has to be adapted to each region as boundary conditions vary. For southern Germany this work has been done by Schwertmann et al. (1987). Another difficulty with the

USLE is a scale problem. The equation has been developed to estimate average annual erosion rates at the field scale. It is therefore not usable for regional evaluation without modification. Several deterministic models have been developed over the last few years. Although they have become more efficient due to advances in computer technology, these models do not provide the broader view required by governmental agencies. On the other hand, there have been several attempts to evaluate regional soil erosion susceptibility with the USLE. The most sophisticated work has been presented by Auerswald and Schmidt (1986) with a collection of maps showing the distribution of the USLE factors in the neighbouring state of Bavaria. The investigation presented here is intended to complete the overview of soil erosion distribution and susceptibility in south west Germany.

3. METHOD

To overcome the scale problem mentioned above, several restrictions have to be made to remain within the assumptions of the model (Wischmeier, 1976). The first step was to divide the area into grid cells of 2 × 2km. This grid size was chosen as slope angle data was available at that scale. One major condition was to use data already available from several state agencies. The approach was to define the dominant value for each USLE factor in each grid cell and calculate the soil erosion rate and susceptibility based on those values. Given the grid size, it is not possible to derive site specific erosion rates. Raster overlay procedures in the low cost Geographic Information System (GIS) IDRISI (Eastman, 1989) were used to perform the calculations.

4. RAINFALL AND RUNOFF FACTOR, R

To avoid the very time-consuming procedure of digitising series of rainfall records, R and other more easily determined and observed rainfall parameters have been correlated. This is a common practice for a regional assessment of R, for example in West Africa, in Peninsular Malaysia, in Southern Africa (summarised by Morgan, 1986) and in Bavaria (by Rogler and Schwertmann, 1981). Data of the Environmental Protection Agency of Baden-Württemberg were used, which is based on the regression equation of Rogler and Schwertmann (1981):

$$R = -1.48 + 0.141*Ns$$

where:

R   = mean annual erosivity in $kJ.m^{-2}.mm.h^{-1}$
Ns = mean rainfall amount (mm) in summer (May - October).

It is assumed that the transfer of this regression equation is possible because rainfall patterns of Bavaria and Baden-Württemberg are very similar and the sensitivity of erosion to R is low compared to C and LS (Auerswald, 1987). The result of applying the above equation is documented in the report of the Environmental Protection Agency of 1987 (Landesanstalt für Umweltschutn Baden-Württemberg, 1987) as an isoerosivity map. To obtain the required raster data the isolines were digitised and the grid values were subsequently calculated by interpolating between the isolines. The classified result is shown in Figure 2.

Figure 2. Rasterised erosivity map (R factor) of Baden-Württemburg.

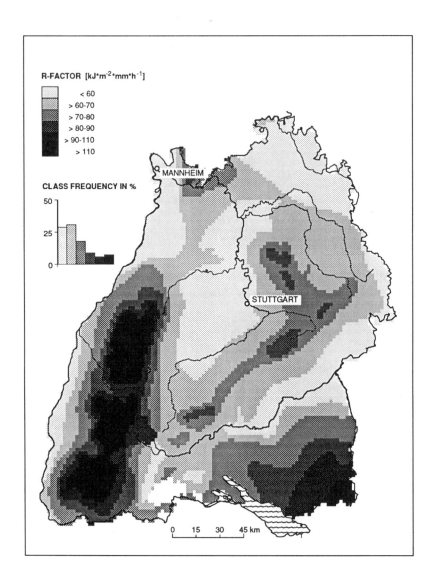

5. SOIL ERODIBILITY FACTOR, K

Baden-Württemberg's loess areas are among the most erosion-susceptible regions in Europe. The adaptation of the K-factor to German conditions has been carried out by Schwertmann et al. (1987) and the following equation is used in Germany (AG. Bodenkunde, 1982):

$$K = 2.17 * 10^{-6} M^{1.14} (12\text{-}OS) + 0.043 (A\text{-}2) + 0.033 (4\text{-}D)$$

where:
M : silt + very fine sand content (in %)
OS : percentage of organic matter
A : the class of aggregate size
D : the permeability class.

Usually, all these parameters have to be measured in the laboratory or field. However, a method developed by Auerswald (1986) was applied (Table 1), in which regression analysis was used. The method is based on soil taxonomy data that evaluates soils according to productivity.

Table 1. Evaluation scheme to derive K-factors from the soil taxonomy

Soil texture class	Origin	Soil condition	
		≤3	≥4
S	D	0.12	0.07
Sl	D	0.14	0.13
lS	D	0.23	0.20
SL	D	--	0.27
sL	D	0.44	0.38
	Loe	0.52	--
L	D	0.48	0.48
	Loe	0.53	0.55
LT	D	00	0.37

S	Sand	Sl	Sand, slightly loamy	lS	Loamy sand
SL	Sand, very loamy	sL	Sandy loam	L	Loam
LT	Heavy loam	D	Diluvial	Loe	Loess

N.B. Soil condition reflects the influence of climatic conditions, water supply, topographic conditions and land use. It can vary from 1 (very good soil condition) to 7 (very poor soil condition) (after Auerswald, 1986).

K can be estimated with sufficient accuracy from the soil's origin, texture and condition which are available for all of Germany (Auerswald, 1986). Soil taxonomy data on a municipal level is available in digitised form from the Ordnance Survey of Baden-Württemberg. The taxonomy only evaluates arable soils, but this is considered to be an advantage as only a small number of large-scale soil maps are available. Since the boundary of each municipality is also available in digital form, K values can be assigned to these as a mean K-factor. Subsequently the grid format of the data is established by vector to raster conversion in the GIS. The resulting K-factor map is shown in Figure 3.

Figure 3. Rasterised erodibility map (K factor); arable land only.

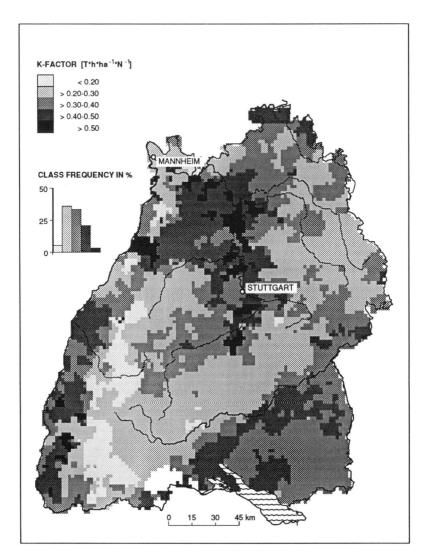

As expected, highest values are found in the loess regions of the northern part of the state and near the foothills of the Rhine Graben area. The foreland of the Alps, which is mainly covered by ground and end moraines, also shows high K values. The map appears to overestimate the erodibility of the sand and gravel soils that dominate this region, but in fact this is correct because agriculture is dominated by dairy-farming and only the very productive soils are tilled. Thus, the total percentage of arable land is only slightly above 10%.

6. TOPOGRAPHIC FACTOR, LS

The combined factor LS reflects the influence of topography on soil erosion. Compared with the other factors it is the most sensitive, because in Baden-Württemberg its possible variation is greater than that of any other factor, as shown in Table 2 (see Auerswald, 1987 also).

Hence, accuracy in the determination of this factor is very important. The equations to calculate LS are derived empirically, but were recently confirmed using deterministic assumptions (Moore and Burch, 1986). Data for the LS factor were gathered by the State Agronomy Agency by manual measurements from topographic maps at the 1:25,000 scale.

Table 2. Descriptive statistics of the USLE factors in Baden-Württemberg.

USLE factor	Mean	Standard deviation	Minimum	Maximum	Range in %
R	66.45	13.60	45.00	185.00	411
K	0.36	0.08	0.10	0.55	550
L	2.59	0.60	1.04	3.29	316
S	0.90	0.94	0.03	9.28	30933
C	0.14	0.04	0.04	0.32	800
A	8.00	9.27	0.00	117.00	

Only typical values of slope length and gradient on arable land were measured using a grid overlay. In order to meet the requirements of the scale of this investigation, a grid resolution of 2km was chosen.

Values of slope length ≥600m appeared unreasonable (Jäger, 1990). Thus data for a 100km^2 area, corresponding to 25 2 × 2km grid cells, were compared with LS-factors derived from a 50m Digital Elevation Model.

A generalisation algorithm, using a data window of 2 × 2km size placed sequentially over the 50m, was then applied to produce a grid size of 2km. In each 2 × 2km window the modal value of slope length and gradient were calculated and assigned. The results were then compared with the manually measured values (Figure 4).

Correlations between the automatically determined values and the manually measured values are very low. The manually LS data are generally higher, mostly due to slope length values. These showed no correlation to the DEM data at all (Jäger, 1990). It cannot be proved without doubt that the DEM-derived values are more correct. However, because of the much higher number of data points on which the calculations were based, errors in the DEM cannot be responsible for the poor correlation, and although only 100 km^2 were compared, significant errors in the manually-derived data seem probable.

Slope length data have therefore been statistically corrected. The most frequent (modal value) slope length in each 1° slope class of the manual slope data has been calculated. The resulting values have been interpreted as the typical slope length of that class of slope gradient. Thus, extreme values and the use of a mean slope length value for the entire state

167

could be avoided. A mean value would not reflect the real situation, because slope length is assumed to be different due to historical reasons and due to the fact that slope length decreases with increasing slope gradient (Auerswald and Schmidt, 1986). Table 3 shows the mean slope length in Baden-Württemberg for each slope gradient class and the derived L and LS values. The LS map is shown in Figure 5.

It can be seen from Table 3 that slope length values decrease with increasing slope angles, whereas LS values, and therefore erosion, increase.

Figure 4. Comparison of topographic factors in 2 × 2 km grid cells, measured manually on maps and calculated with a DEM, using a generalised algorithm.

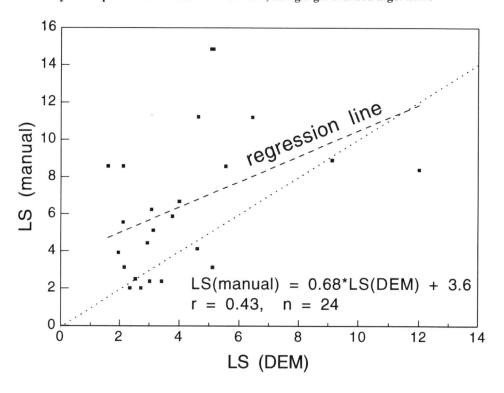

Table 3. Distribution of the most frequent slope length in each slope gradient classes and corresponding L and LS values in Baden-Württemberg.

Slope gradient in degrees	Mode of slope length in m	Number of measurements	L	LS
1	240	183	2.05	0.34
2	240	411	2.05	0.62
3	180	382	2.85	1.38
4	240	445	3.30	2.31
5	180	356	2.85	2.74
6	180	454	2.85	3.59
7	130	116	2.43	3.86
8	180	187	2.85	5.61
9	100	40	2.13	5.06
10	110	12	2.23	6.31
11	65	64	1.79	5.69
12	38	28	1.31	5.04

7. COVER AND MANAGEMENT FACTOR, C

This is the most complex factor in the USLE because it must take into account a variety of crop rotations and management practices. C also depends on the distribution of erosive rains during the year. Thus, relative soil erosion rates for each crop and the rainfall distribution must be known if the method is to be applied (Schwertmann et al., 1987). A detailed investigation of crop rotations including corn has been carried out by Auerswald et al. (1986). For a state-wide determination of the C-factor detailed data about the crop rotations are not available. C can be calculated from the spatial distribution of the crops (Auerswald and Schmidt, 1986), assuming that:
1. the crop rotation through time can be characterised by the distribution of crops in space at a specific date for a specific area, and
2. a linear relationship between the portion of a single crop in a crop rotation and the C-factor for that crop rotation exists.

Crop distribution data are available from the State Statistics Agency for each municipality. The weighted C-factor for each crop is determined, partly according to Bavarian experience as follows:

Corn	0.40	Barley	0.10
Potato	0.30	Vines	0.10
Sugar beet	0.30	Vegetables	0.10
Rape	0.10	Legumes	0.40
Others	0.06		

169

Figure 5. Map of the topographic factor, LS, based on slope length and gradient; arable land only.

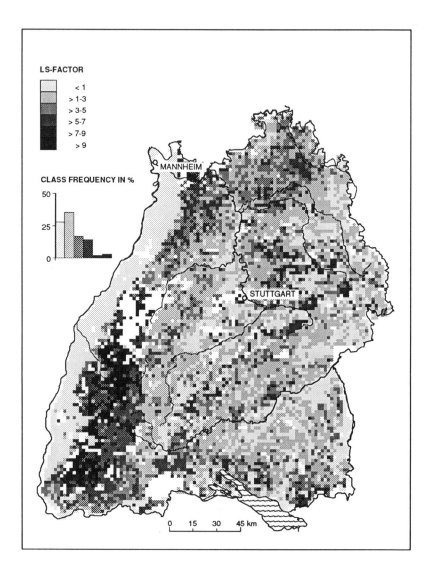

Although vineyards do not form part of normal crop rotations, they are included here because they dominate on the foot slopes of the Black Forest and the Odenwald, and soil erosion rates are high in those areas. The value of 0.1 is chosen arbitrarily. Its low value was selected to account for the high percentage of grass covered vineyards. To obtain the required raster data the same procedure as for the K values was used. The C-factor map is shown in Figure 6.

Figure 6. Rasterised map of cover and management factor, C; arable land only. Areas with less than 10% tilled land are not evaluated and are left white on the nap.

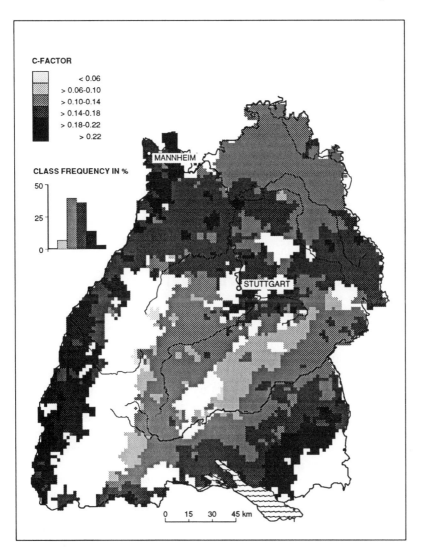

8. PROTECTION FACTOR, P

The erosion prevention factor, P was set to unity because generally, no protection measures are used in the state.

9. THE SUSCEPTIBILITY MODEL

Several approaches exist for evaluating susceptibility to soil erosion. Often the calculated erosion rates are subdivided into arbitrary classes marked as high, moderate and low. For

171

the proposed assessment, a different method was applied following Heimlich and Bills (1984). Their classification method is especially useful for conservation policy. The procedure is to subdivide the USLE factors into their natural, unchangeable components (R, K, S) and the anthropogenic, changeable (C, P, L) parts. In contrast to Heimlich and Bills (1984), L is put into the second group because it is considered to be changeable by means of drainage or paths. The classification determines whether it is possible to reduce soil erosion to tolerable values with common management practices, and whether high CxPxL values inevitably lead to high erosion rates. Therefore the tolerable soil loss (T) has to be defined first.

10. TOLERABLE SOIL LOSS

Different values for tolerable soil loss, T, can be defined, according to whether an ecologic or economic point of view is taken. Furthermore, T can be related to off-site or on-site damages (Schertz, 1983). For the proposed assessment, an evaluation model developed by Auerswald and Schmidt (1986) has been applied. It defines T as being linearly related to the depth of the soil profile, as estimated from soil taxonomy data. This approach allows soil erosion rates of up to $10t.ha^{-1}yr^{-1}$ on deep loess soils. The data were converted to raster format in the same manner as the K and C factors.

11. THE CLASSIFICATION PROCEDURE

To classify the susceptibility it is first necessary to determine the lowest and highest values of CPL in the state. The minimum and maximum CPL for common tillage methods in Baden-Württemberg are 0.04 and 0.91 respectively. The next step is to calculate the value of T/0.91 (defined as RKS_{min}) and T/0.04 (defined as RKS_{max}) for each raster element. RKS_{min} gives the upper limit for the RKS value of a raster element which is not to be exceeded if common tillage methods are to be applied. Any raster element with a RKS value below RKS_{min} cannot have erosion rates above T and is classified as low susceptibility. RKS_{max} gives the upper limit of the natural factors that should not be exceeded by a raster element. Raster elements with a RKS value above RKS_{max}, given any common tillage method, have erosion rates above T and therefore are classified as very highly susceptible. The remaining area is then divided into two further classes. If the calculated erosion rate is above T the raster element is classified as highly susceptible. This class implies that erosion rates can be reduced below tolerable values using common tillage methods. Moderately susceptible areas have calculated erosion rates below T. For this class, where unfavourable tillage or crop rotations are used, there is the risk of erosion rates rising above T. This classification method meets the requirements of policy-makers because effects of conservation laws can be modelled. It is a numerical approach to a regional assessment of susceptibility and therefore it can be checked at any time.

12. RESULTS

Two maps are presented as the results of the investigation. They are taken from a series of evaluation maps (Jäger, in prep.). The soil erosion map is shown in Figure 7.

Figure 7. Modelled soil erosion rates; arable land only. Soil erosion rates are calculated by multiplying the unclassified data of the factors, R, K, L, S and C for each raster element.

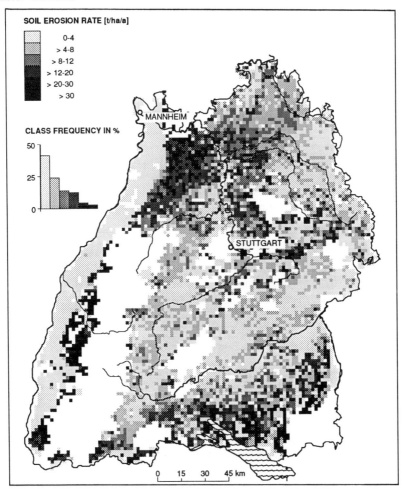

Raster elements with less than 10% arable land are not classified so as not to give too much weight to high erosion rates in those areas. The soil loss ranges from 0 to $117t.ha^{-1}yr^{-1}$. In the Rhine Graben and other areas with gentle slopes there is minor erosion. For the well-known problem areas in the loess regions the model produced good results. The high values of more than $30t.ha^{-1}yr^{-1}$ obtained in the forelands of the Alps are the result of a combination of high R and C values. However, the percentage of arable land is low in this area, with values between 10 and 20%. In agreement with the situation in Bavaria (Auerswald and Schmidt, 1986) erosion rates are high in areas with a lot of corn-fed cattle; corn being the crop with the highest C-factor.

One major difficulty in regional assessments is the validation of model results. Only few data on soil erosion rates have been published so far. These values, based on different methods, are compared with the calculated values presented in this study (Table 4). Although similarities can be observed, there is insufficient data to confirm the calculated results. Test plot measurements vary in duration and do not cover a long enough period to be compared with the modelled long-term average soil losses.

The soil erosion susceptibility map (Figure 8) clearly shows the most sensitive areas. It is obvious that these areas do not necessarily correspond to areas with high erosion rates because soil loss is considered relative to T. Approximately 63% of the evaluated raster elements are classified as highly susceptible. This is a high percentage but high susceptibility shown by the model means that soil erosion rates below T can be achieved by applying suitable tillage methods or crop rotations. Only 3% is classified as very highly susceptible. Most of these elements are located at the transition from the Rhine Graben to the uplands, and exhibit steep slopes under vineyards. Errors here are assumed to be high because of the poor data for the dominating LS factor and because of the unknown C factor for vineyards. Twenty-one percent is classified as low susceptibility and approximately 12% is moderately susceptible, implying that poor management will make erosion rates rise above T.

Table 4. Calculated soil erosion rates compared with data already published.

Author	Duration	Region	Method	Soil erosion rates	
				Measured $t.ha^{-1}.yr^{-1}$	Calculated $t.ha^{-1}.yr^{-1}$
Bleich 1978	1974-67	Gäue	Test plots	12	0-8
Dikau 1986	1979-81	Kraichgau	Test plots and mapping	25-165	12-30
Eichler 1974	--	Kraichgau	One event	Up to 400	8-30
Murschel-Raasch 1991	1978-88	Kraichgau	CREAMS Model	9-14	8-30
Ostendorf & Zürl 1964	--	Filder	*	15	0-8
Quist 1987	1978-86	Kraichgau	Test plots	30-120	8-30

* Estimation of long-term soil erosion rates by measuring the height of steps at the transition from forest to arable land. A bulk density of 1.5 $g.cm^{-3}$ was used to calculate erosion rates.

13. CONCLUSIONS

At field scale empirical models such as the USLE are being replaced more and more by physically based process models. Nevertheless the USLE remains useful for regional evaluation. This is the case in developed countries where basic data are already available in digital form and can be easily processed in a GIS. The results can serve as an important basis for the determination of areas where soil conservation should be emphasised. The GIS approach is useful for evaluation because different scenarios can be developed in a very short time. It would be no problem, for example, to calculate new susceptibility classes with a new T model. However, reliability of the results is highly dependent on the quality of the basic data. The highest accuracy should be observed for determining the topographic factor LS, as it has the highest influence on the results. Manual measurements on topographic maps can be replaced by automated procedures based on Digital Elevation Models.

Figure 8. Soil erosion susceptibility map, according to the classification of Heimlich and Bills (1984); arable land only.

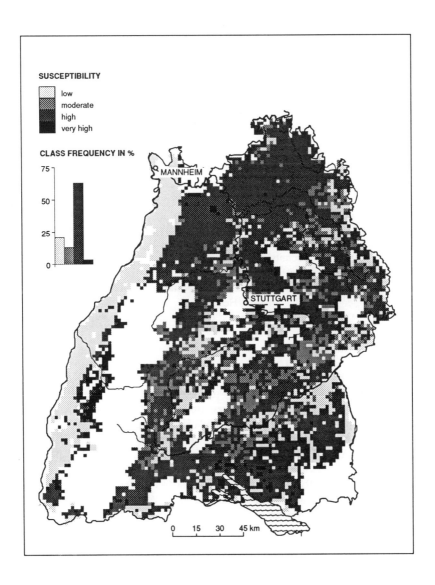

ACKNOWLEDGEMENTS

The financial support by the Ministry of Agriculture for the work presented is gratefully acknowledged. Many thanks to Richard Dikau for helpful comments, to Sabine Klein for the drawing, Clemens Ritter for the assistance in processing the computer maps and to Stephen Smyth for helping me with the translation.

REFERENCES

AG BODENKUNDE 1982 Bodenkundliche Kartieranleitung; E. Schweizerbart'sche Verlagsbuchhandlung, Hannover.

AUERSWALD, K. 1986. Einstufung der Bodenerodibilität (K-Faktor) nach dem Klassenbeschrieb der Reichsbodenschätzung für Südbayern, Zeitscheift für Kulturtechnik und Flurbereinigung 27;344-351.

AUERSWALD, K. 1987. Sensitivität erosionsbestimmender Faktoren, Wasser und Boden 1;34-38.

AUERSWALD, K. and SCHMIDT, F. 1986. Atlas der Erosionsgefährdung in Bayern. Karten zum flächenhaften Bodenabtrag durch Regen. GLA Fachberichte 1; Geologisches Landesamt, München.

AUERSWALD, K., KAINZ, M. and VOGL, W. 1986. Vergleich der Erosionsgefährdung durch Maisfruchtfolgen (C-Faktor), Bayerisches Landwirtschaftliches Jahrbuch 63/1:3-8.

BLEICH, K.E. 1978. Erosion von Böden infolge Bodennutzung, Daten und Dokumente zum Umweltschutz, Sonderreihe Umwelttagung 22; Stuttgart.

BORK, H.-R. 1988. Bodenerosion und Umwelt, Verlauf, Ursachen und Folgen der mittelalterlichen und neuzeitlichen Bodenerosion, Bodenerosionsprozesse, Modelle und Simulation. Landschaftsgenese und Landschaftsökologie, Heft 13; Abteilung für Physische Geographie, Technische Universität, Braunschweig.

DIKAU, R. 1986. Experimentelle Untersuchungen zu Oberflächenabfluss und Bodenabtrag von Messparzellen und Landwirtschaftlichen Nutzflächen, Heidelberger Geographische Arbeiten 81; Geographisches Institut, Heidelberg.

EASTMAN, R. 1989. IDRISI, a grid based Geographic Analysis System; Clark University, Graduate School of Geography, Worcester, Massachusetts.

EICHLER, H. 1974. Bodenerosion im Kraichgauer Löss, Kraichgau 4.

HEIMLICH, R.E. and BILLS, N.L. 1984. An improved soil erosion classification for conservation policy. Journal of Soil and Water Conservation 39:261-266.

JÄGER, S. 1990. Modellierung der regionalen Bodenerosionsgefährdung Baden-Württembergs auf Basis der Allgemeinen Bodenabtragsgleichung. Unpublished M.Sc. Thesis, Geographisches Institut der Universität Heidelberg.

JÄGER, S. In prep. Bodenerosionsatlas Baden-Württemberg.

LANDESANTSTALT FÜR UMWELTSCHUTZ BADEN WÜRTTEMBERG 1987. Umweltbericht; Landesanstalt für Umweltschutz, Karlsruhe.

MOORE, I.D. and BURCH, J. 1986. Physical basis of the length-slope factor in the Universal Soil Loss Equation. Soil Science Society of America Journal 50:1294-1298.

MORGAN, R.P.C. 1986. Soil erosion and conservation. Longman.

MURSCHEL-RAASCH, B. 1991. Die Entwicklung eines Informationssystems zur Reduzierung der Erosion und des Stoffaustrages am Beispiel von Ackerböden in Kraichgau. PhD Dissertation, Stuttgart-Hohenheim.

OSTENDORFF, E. and ZÜRL, K. 1964. Auffällige Erosionsschäden in Württemberg, ihre Ursachen, Bekämpfung und Sanierung, Jahresheft des Vereins für vaterländische Naturkunde in Württemberg 118/119:87-146.

QUIST, D. 1987. Bodenerosion - Gefahr für die Landwirtschaft im Kraichgau. Karichgau 10:42-62.

ROGLER, H. and SCHWERTMANN, U. 1981. Erosivität der Niederschläge und Isoerodenkarte Bayerns. Zeitschrift für Kulturtechnik und Flurbereinigung 22:99-112.

SCHERTZ, D.L. 1983. The basis for soil loss tolerance. Journal of Soil and Water Conservation 38:10-14.

SCHWERTMANN, U., VOGL, W. and KAINZ, M. 1987. Bodenerosion durch Wasser: Vorhersage des Abtrags und Bewertung von Gegenmassnahmen. Ulmer, Stuttgart.

STAATSMINISTERIUM BADEN-WÜRTTEMBERG 1991. Gesetz zum Schutz des Bodens, Gesetzblatt für Baden-Württemberg 16:434-440.

UNDRO 1991. Mitigating natural disasters. Phenomena, effects and options. United Nations Disaster Relief Co-ordinator. United Nations, New York.

WISCHMEIER, W.H. 1976. Use and misuse of the Universal Soil Loss Equation. Journal of Soil and Water Conservation 31:5-9.

WISCHMEIER, W.H. and SMITH, D.D. 1978. Predicting rainfall erosion losses, A guide to conservation planning. USDA Agriculture Handbook 537, Washington, D.C.

Chapter 18

COMPARISON OF OBSERVED AND COMPUTED SOIL LOSS, USING THE USLE

P.P. Tomás and M.A. Coutinho

Instituto Superior Técnico, Av. Rovisco Pais, 1096 Lisboa Codex, Portugal

Summary

The main objectives of the present study are (i) to analyse the data from the Vale Formoso Experimental Erosion Centre (VFEEC); (ii) to characterise rainfall regimes, especially erosivity in South Portugal; (iii) to apply the Universal Soil Loss Equation (USLE) to the VFEEC plots and compare observed and computed soil loss; and (iv) to calibrate the USLE if necessary and possible.

Sixteen erosion plots (22 metres long and 8 metres wide) were set up in 1962, and have undergone several crop rotations since then. Water and sediment from the plots are collected about 20 times a year. Two meteorological stations collect precipitation and wind data. Data from the plots (runoff, soil loss, vegetation cover and management practices) and from the meteorological stations were processed into different data bases. Records from recording rain gauges were digitised. Several methodologies for determining rainfall characteristics were used and the USLE factors for each plot and year were also computed. The soil loss for each plot and year was calculated with the USLE and results were compared with observed soil loss.

The most important conclusions of this study were (i) there is a significant difference between the estimates of rainfall erosivity for southern Portugal using different methodologies; (ii) the USLE over-predicted soil loss from plots - the ratio between computed and observed soil loss is about 10; (iii) the USLE adjustment to the data, with linear and non-linear models, was not fully satisfactory; and (iv) there is a sequence of high and low soil loss values from the plots from year to year. The soil loss from odd hydrologic years (e.g. 1963/64) is greater than the soil loss from even hydrologic years (e.g. 1964/65), for plots under identical conditions. This hypothesis was verified for plots cultivated with wheat, using the χ^2 test with $\alpha = 5\%$.

1. INTRODUCTION

In 1960 a programme was established by the Direcçao-Geral de Hidráulica e Engenharia Agrícola (General-Directorate of Hydraulics and Agricultural Engineering), DGHEA, to collect erosion data in the cultivated areas of southern Portugal. In autumn 1961 an experimental erosion centre was established east of Beja, on the Vale Formoso farm, which is on red schist soils, prone to erosion by water.

The programme for the Centre was designed to fulfil the following main objectives:

- adaptation and validation of the Universal Soil Loss Equation (USLE) to Portuguese conditions;
- establishment of methodologies to obtain adequate values for the inputs to the USLE;
- evaluation of soil conservation measures.

Extensive and valuable data have been gathered (Araújo, 1974; Ferreira et al., 1982), but these main objectives have only been partially achieved. Some discrepancies still persist when comparing plot values with USLE estimates, which limits the acceptance of this methodology.

The regime of precipitation in Vale Formoso varies greatly, both annually and seasonally (and even within the seasons). This regime is different from that used to derive the USLE. Rainfall erosivity in Portugal is relatively higher compared with an equivalent amount of precipitation in the US. We do not know if this higher erosivity corresponds with a higher rate of erosion.

The limitations of applying the USLE to Mediterranean conditions has led to a modified methodology being developed for use of the equation in California (SCS, 1977).

This paper shows an important improvement in the comprehension and analysis of the erosion regime in southern Portugal. A description of available data and previous studies is made in Coutinho and Tomás (1990).

2. DATA AVAILABLE

2.1 Vale Formoso Experimental Erosion Centre

Vale Formoso Centre is located 11km east of the Guadiana river, at 37°45'30"N and 7°37'W (Figure 1). The brush and green oak natural vegetation of the area was cleared for small grain production in the 1930s. This resulted in accelerated soil erosion, which has led to the shallow red schist soils observed today (Ferreira et al., 1982).

The mean annual rainfall is about 550mm   (480mm   in the last 30 years), with almost 75% of the annual rainfall falling between October and March. During the operation of the Erosion Centre the minimum and maximum recorded precipitation is 220 and 980mm , in the years of 1982/83 and 1989/90 respectively.

The centre has 16 Wischmeier erosion plots and two meteorological stations. Plots are arranged side by side on slopes between 10% and 20%, facing between east and south. Each plot is 8.33m wide and 20.0m long (1/60ha).

Runoff and associated sediment are collected in three calibrated tanks (400 litres each), arranged in series, at the base of each plot. The first tank collects the total water and sediment from the plot. The other two tanks collect 1/11 of the flow from the upper tank. The first tank collects almost all the sediment. Corrections are made to account for evaporation and rainfall in the tanks. On average, 20 observations are made each year.

Figure 1. Location of the Vale Formosa Experimental Erosion Centre.

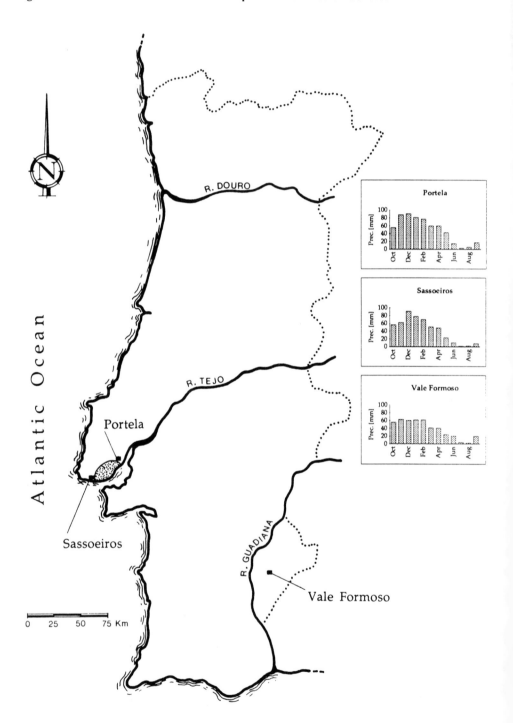

The plots are subject to several representative crop sequences. For this study only the three longest crop rotations were analysed, which correspond to 12 different plots (Table 1).

Table 1. Crop sequences, plot identification and length of time series.

Crop sequence	Duration of a full crop sequence	Size of time series	Identification of plots	Plot years of data	Number of full crop sequences
Wheat-Summer Fallow (W.SF)	2 years	22 years	1, 2 , 10 and 11	88	11
Wheat-Green Manure (W.GM)	2 years	20 years	3, 4, 5 and 6	80	10
Wheat-Green Manure-Wheat-Dry Beans $(W_1.GM.W_2.DB)$	4 years	16 years	8, 9, 12 and 13	64	4

The plots are divided into two groups, each one having a meteorological station equipped with rain gauges and recording rain gauges, an anemometer, evaporimetric pans and thermometers.

Annual values for this study are reported for the hydrological year, from 1st October to 30th September.

3. RAINFALL REGIMES

3.1 Erosivity Indices

Coutinho and Tomás (1988) outline a detailed analysis of three different methodologies for calculating erosivity indices. Two of these are modified versions of the EI_{30} index (Wischmeier and Smith, 1978). The other was based on the Hudson index (Hudson, 1981).

The EI_{30} type indices were calculated following two different approaches: (i) only considering storms with precipitation greater than 0, 5, 10, 12.5 and 15mm; and (ii) calculating the kinetic energy of rainfall with intensities greater than 0, 10 and 25mm.h^{-1} and considering the maximum storm amount over 15, 30 and 45 minutes. The Hudson type index was obtained using only the sum of kinetic energy of precipitation with intensities greater than 0, 5, 10, 15, 20 and 25mm.h^{-1}.

The kinetic energy of precipitation, for all indices, was calculated using the equation proposed by Foster et al. (1981).

$$e_k = 0.119 + 0.0873 \log10(i_m) ; \quad i_m \leq 76mm.h^{-1}$$
$$e_k = 0.283 ; \quad i_m > 76mm.h^{-1} \tag{1}$$

where e_k is the kinetic energy in MJ.ha^{-1}.mm^{-1} and i_m the precipitation intensity in mm.h^{-1}.

All indices were calculated on an annual basis for three meteorological stations (Figure 1); one at Vale Formoso and two in the Lisbon area (Portela and Sassoeiros). The indices calculated for any given station were strongly correlated. The Wischmeier index does not perform very well in climates that have very intense precipitation events and I_{30} may be a calibration parameter.

To calculate the cover and management factor of the USLE, the monthly variation of the EI_{30} index was calculated (Figure 2). From this figure we can see that Portela and Sassoeiros have a rather similar pattern, unlike that for Vale Formoso, due to differences in climate.

Figure 2. Monthly variation of $(EI_{30})_{12}$ index.

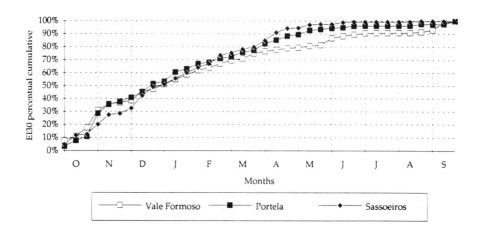

3.2 Intensity-Duration-Frequency Curves

Intensity-Duration-Frequency (IDF) curves are used in the West Coast of the US as the basis of an alternative methodology to determine the rainfall factor R of the USLE (Foster et al., 1981).

These curves were calculated for the three locations, assuming the following equation:

$$i = at^b \tag{2}$$

where i is the maximum intensity of precipitation in $mm.h^{-1}$; t is the duration in minutes; a and b are parameters. The computed parameters for the curves are presented in Table 2 and the IDF curves for a two year period for the three stations are presented in Figure 3.

Table 2. Parameters of IDF curves.

Return period (years)	Portela		Sassoeiros		Vale Formoso	
	a	b	a	b	a	b
2	290.3	-0.673	356.9	-0.755	484.7	-0.802
5	325.1	-0.630	468.2	-0.739	611.5	-0.774
10	343.9	-0.610	531.4	-0.730	718.3	-0.768
20	360.8	-0.595	588.9	-0.723	828.8	-0.766
50	383.4	-0.581	653.6	-0.715	961.7	-0.762
100	399.0	-0.572	698.3	-0.710	1076	-0.764
Series length (years)	14		14		23	
Correlation coefficient	From 0.94 to 0.98					

Figure 3. IDF curve for the 2 year return period.

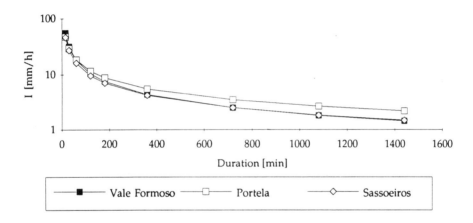

For the western States of the USA, an equation was proposed to estimate R, based on the two year return period, six hour duration precipitation (Foster et al., 1981):

$$R = 0.417 \, P^{2.17}_{[2\,year,6\,hours]} \tag{3}$$

where R is the USLE rainfall factor in MJ.mm.ha^{-1}.hr^{-1}.yr^{-1}. and P is the volume of precipitation in mm. This equation was applied to the three stations and compared with the R factor computed by the Wischmeier methodology. The ratio of R calculated from the Wischmeier methodology and R estimated from equation (3) varies from 1.5 to 2.0.

4. OBSERVED DATA

4.1 Data analysis

Table 3 shows the annual average values of observed soil losses and Figures 4, 5 and 6 show the time series of soil losses for each crop sequence.

Table 3. Annual average observed soil loss.

	Crop sequence W.SF				Crop sequence W.GM				Crop sequence $W_1.GM.W_2.DB$			
Plot	1	2	10	11	3	4	5	6	8	9	12	13
N	22	22	22	22	20	20	20	20	16	16	16	16
$\bar{x}$	0.540	1.330	0.378	0.907	0.534	0.606	0.464	0.324	0.203	0.218	0.851	0.636
s_{N-1}	0.763	2.060	0.648	1.352	0.769	0.815	0.618	0.400	0.324	0.348	1.303	1.093
cv	1.4	1.5	1.7	1.5	1.4	1.3	1.3	1.2	1.6	1.6	1.5	1.7
$\bar{x}W$	0.771	2.375	0.255	0.072	0.869	0.928	0.224	0.227	-	-	-	-
$\bar{x}W_1$	-	-	-	-	-	-	-	-	0.174	0.023	1.234	1.322
$\bar{x}W_2$	-	-	-	-	-	-	-	-	0.051	0.150	1.433	0.701
$\bar{x}SF$	0.309	0.285	0.501	1.742	-	-	-	-	-	-	-	-
$\bar{x}GM$	-	-	-	-	0.200	0.284	0.704	0.227	0.220	0.432	0.037	0.490
$\bar{x}DB$	-	-	-	-	-	-	-	-	0.366	0.267	0.699	0.029

UNITS: $t.ha^{-1}yr^{-1}$

N - series length; $\bar{x}$ - series average; s_{N-1} - series standard deviation; cv - series variation coefficient; $\bar{x}W$ - annual wheat average (crop sequences W.SF and W.GM); $\bar{x}W_1$ and $\bar{x}W_2$ - annual wheat average (crop sequence $W_1.GM.W_2.DB$); $\bar{x}SF$ - annual summer fallow average; $\bar{x}GM$ - annual green manure average; $\bar{x}DB$ - annual dry beans average.

I. Wheat-Summer Fallow

This crop sequence is installed on plots 1 and 2 (10% slope; oriented south) and on plots 10 and 11 (16% slope; oriented east). In any year, the same crop is grown on Plots 1 and 10, and a different crop is grown on both Plots 2 and 11. The analysis of data justifies the following comments.

- Plots with identical slope, orientation and crop have significant differences in soil loss (for example, soil loss for wheat on plots 1 and 2 is $0.77t.ha^{-1}$ and $2.38t.ha^{-1}$ respectively, a ratio of approximately 1:3).

- Figure 4 shows a sequence of high and low soil loss values over time. A statistical analysis was made, using a two-way classification and the χ^2 test (Dixon and Massey, 1983). The series of soil loss for each crop sequence (average of all crops on different plots for the same year) was used. The χ^2 test was significant, for all plots, with $\alpha = 5\%$. The odd years (e.g. 1963/64) show a higher soil loss than that for the even years (e.g. 1964/65).

- Plots 1 and 2 have a greater soil loss than the steeper plots of 10 and 11. This can be related to plot orientation and its effects on the angle of incidence of raindrops during heavy storms, and to different amounts of solar radiation on the plots. This result calls for more analysis.

Figure 4. Annual observed soil loss for Wheat-Summer Fallow crop sequence.

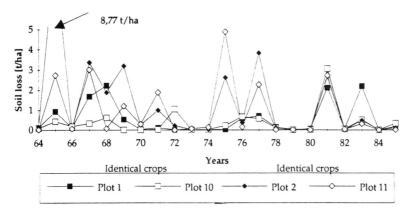

Figure 5. Annual observed soil loss for Wheat-Green Manure crop sequence.

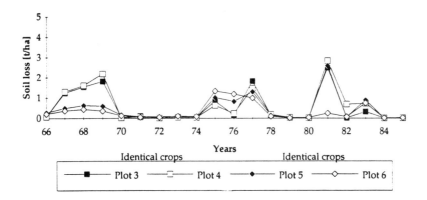

Figure 6. Annual observed soil loss for Wheat-Green Manure-Wheat-Dry Beans crop sequence.

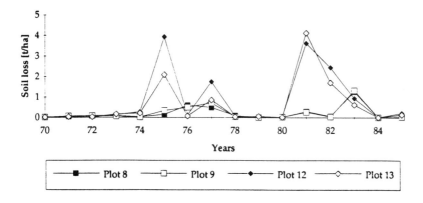

185

- The soil loss series has a rather high standard deviation. For plots 1, 2, 10 and 11, there are 15 extreme years, from the total of 80 plot-years of data. These extreme years produced almost 65% of the total sediment load.

II. Wheat-Green Manure

This crop sequence is installed on plots 3 and 4 (same crop) and on plots 5 and 6 (same crop). All plots have similar slopes and orientations. The main comments on the data are presented below.

- There was a sequence of high and low soil loss values over the years, with the χ^2 test significant for $\alpha = 5\%$, describing the differences of soil loss.

- The standard deviation of the series is also high.

III. Wheat-Green Manure-Wheat-Dry Beans

This crop sequence is installed on plots 8, 9, 12 and 13, with no spatial replicates. The differences between soil losses for different crops still remain, but the χ^2 test was not significant for $\alpha = 5\%$. The standard deviation of the series is high.

5. ESTIMATED EROSION RATES

The USLE is an erosion model designed to compute long-term average soil loss from sheet and rill erosion under specified conditions (Wischmeier and Smith, 1978). The equation groups the variables affecting sheet and rill erosion in 6 generic erosion factors, presented in SI units by Foster et al. (1981).

Rainfall factor (R)

The average R factor for the period from 1963/64 to 1985/86 was calculated:

$$R = 954 \; \frac{MJ.mm}{ha.hr.yr}$$

with a standard deviation of 552, a minimum of 250 and a maximum of 1950 $MJ.mm.ha^{-1}.hr^{-1}.yr^{-1}$.

Soil erodibility factor (K)

From the nomograph presented by Foster et al. (1981), the K factor was based on soil characteristics (Table 4):

$$K_{nomograph} = 0.045 \frac{t.ha^{-1}.hr^{-1}}{ha^{-1}.MJ.mm^{-1}}$$

Table 4. Characteristics of Vale Formoso red schist soil

Texture (%)				Structure	Permeability
Clay	Silt	Sand	O.M.		
14	56	30	1.65	Fine granular	Moderate

Because of the high content of rock fragments in the soil (30% in weight) and because no measured K factor was available, the study of Lima (1990) was used. This author studied the VFEEC soil with a rainfall simulator and corrected the soil loss due to the rock fragments. From the results, a new K factor was calculated:

$$K_{adopted} = 0.016 \frac{t.ha.hr}{ha.MJ.mm}$$

Topographic factor (LS)

The two factors L and S are usually considered as a single topographic factor, LS (Wischmeier and Smith, 1978). Considering S improvement by McCool et al. (1987):

$$LS = \left(\frac{\lambda}{22.13}\right)^m (10.8 \sin(\Theta) + 0.03) \; ; \; s < 9\% \qquad (4)$$

$$LS = \left(\frac{\lambda}{22.13}\right)^{0.5} (16.8 \sin(\Theta) - 0.50) \; ; \; s \geq 9\%$$

where λ is the slope length in metres, m is an exponent related to slope steepness and Θ is the angle of slope.

Cover and management factor (C)

The C factor was established using the methodology and tables presented by Wischmeier and Smith (1978). Each crop is divided into six crop stage periods and the percent erosivity for each period is used as a weighting factor for calculating the annual value of C. The C factor was calculated for each plot and year, based on the agricultural calendar, observed crop development and temporal variation of the EI_{30} index.

Support practice factor (P)

The practice of tillage and planting on the contour was applied to all plots during the study period. The P values for each plot are presented in Table 5.

Table 5. LS and P factors for VFEEC plots

Plot	s [%]	LS [-]	P [-]
1	11.00	1.27	0.6
2	10.25	1.16	0.6
3	9.80	1.09	0.6
4	9.60	1.05	0.6
5	10.45	1.19	0.6
6	11.95	1.42	0.6
7	12.90	1.58	0.7
7x	13.35	1.65	0.7
8	14.90	1.89	0.7
9	15.55	1.99	0.7
10	15.95	2.05	0.7
11	16.05	2.07	0.7
12	16.55	2.15	0.8
13	17.60	2.31	0.8
14	20.15	2.71	0.8
15	20.35	2.74	0.8

s - plot slope
LS - plot topographic factor
P - plot support practice factor

Application of the USLE and discussion of results

The USLE was applied annually (Table 6). The model overpredicts soil loss. The ratio between observed and calculated soil loss varies between 5 and 20, with an average of 10. The following comments are made on the comparison between observed and calculated soil loss values:

- Differences in calculated soil loss for the same crop on identical plots still persists, but they are more attenuated than for the observed soil loss values. The χ^2 test for the sequence of high and low values of calculated losses over the years was not significant.

- The time series of calculated soil losses is smoother than for the observed ones. The coefficient of variation, that was always greater than 1.0 for the observed series, is always less than 1.0 for the calculated series.

- Plots 10 and 11 have a steeper slope than plots 1 and 2 and a greater value of calculated soil loss, as would be expected. However, plots 1 and 2 have a greater value of observed soil loss than plots 10 and 11.

Table 6. Annual average calculated soil loss by USLE

	Crop sequence W.SF				Crop sequence W.GM				Crop sequence W_1.GM.W_2.DB			
Plot	1	2	10	11	3	4	5	6	8	9	12	13
N	22	22	22	22	20	20	20	20	16	16	16	16
$\overline{x}$	4.320	4.926	8.108	10.33	4.113	3.937	4.249	5.071	7.129	6.586	9.038	9.717
s_{N-1}	3.081	4.181	5.814	8.645	2.965	2.856	2.617	3.123	4.697	3.594	5.878	6.309
cv	0.7	0.8	0.7	0.8	0.7	0.7	0.6	0.6	0.7	0.5	0.7	0.6
$\overline{x}$W	5.323	7.803	9.946	16.25	5.507	5.305	4.211	5.025	-	-	-	-
$\overline{x}$W1	-	-	-	-	-	-	-	-	6.756	7.332	11.63	12.95
$\overline{x}$W2	-	-	-	-	-	-	-	-	7.336	5.588	12.05	12.49
$\overline{x}$SF	3.316	2.050	6.270	4.417	-	-	-	-	-	-	-	-
$\overline{x}$GM	-	-	-	-	2.719	2.570	4.288	5.116	10.13	6.094	4.853	7.692
$\overline{x}$DB	-	-	-	-	-	-	-	-	4.297	7.330	7.623	5.740

UNITS: $t.ha^{-1}yr^{-1}$

N - series length; $\overline{x}$ - series average; s_{N-1} - series standard deviation; cv - series variation coefficient; $\overline{x}$W - annual wheat average (crop sequences W.SF and W.GM); $\overline{x}$W$_1$ and $\overline{x}$W$_2$ - annual wheat average (crop sequence W_1.GM.W_2.DB); $\overline{x}$SF - annual summer fallow average; $\overline{x}$GM - annual green manure average; $\overline{x}$DB - annual dry beans average.

6. USLE ADJUSTMENT

A simplified analysis was done to adjust the Wischmeier methodology for estimating erosion losses in southern Portugal. Assuming all the factors in the USLE are applicable, except the erosivity index, a local index was calculated correlating the Wischmeier index with the corresponding erosivity value for the observed soil loss. The approach was:

1. EI_{30} was computed on an annual basis;

2. An equivalent erosivity index (R_{eq}) was calculated from the equation

$$R_{eq} = \frac{A_{obs}}{KLSCP} \qquad (5)$$

where A_{obs} is the observed annual soil loss;

3. A regression model between values of R_{eq} and EI_{30} was established to obtain the local erosivity index (R_{local}), in the form of:

$$R_{local} = \alpha(EI_{30})^\beta \qquad (6)$$

where α and β are the regression parameters.

189

Table 7. Comparison of soil loss values: observed, calculated by USLE and adjusted by regression, for plots and crop sequences.

Crop sequence	Plot	SLobs	SLcal	SLadj	SLcal / SLobs	SLadj / SLobs	α	β	R^2
W.SF	1	0.540	4.320	0.536	0.8	1.0	6.243E-06	2.322	50%
	2	1.330	4.926	0.780	3.7	0.6	2.092E-04	1.652	59%
	10	0.378	8.108	0.536	21.4	1.4	3.766E-05	2.156	29%
	11	0.907	10.331	0.782	11.4	0.9	1.917E-05	2.033	39%
	1,2	0.935	4.623	0.658	4.9	0.7	1.870E-05	2.228	59%
	10,11	0.643	9.220	0.659	14.3	1.0	1.112E-04	1.814	38%
W.GM	3	0.534	4.113	0.411	7.7	0.8	6.136E-07	2.632	66%
	4	0.606	3.937	0.424	6.5	0.7	6.023E-06	2.331	49%
	5	0.464	4.249	0.221	9.2	0.5	2.074E-04	1.812	39%
	6	0.324	5.071	0.166	15.7	0.5	1.224E-03	1.490	34%
	3,5	0.499	4.181	0.299	8.4	0.6	1.988E-05	2.150	53%
	4,6	0.465	4.504	0.285	9.7	0.6	1.029E-04	1.904	44%
W_1.GM	8	0.203	7.129	0.076	35.2	0.4	3.548E-04	1.523	25%
W_2.DB	9	0.218	6.586	0.069	30.2	0.3	3.361E-04	1.528	23%
	12	0.851	9.038	0.297	10.6	0.3	2.550E-05	2.004	28%
	13	0.636	9.717	0.198	15.3	0.3	6.996E-04	1.474	19%
	8,9,12,13	0.477	8.117	0.204	17.0	0.4	2.733E-04	1.642	27%

Units: $t.ha^{-1}yr^{-1}$. SLobs - observed soil loss; SLcal - calculated soil loss by USLE; SLadj - adjusted soil loss; α, β - regression parameters; R^2 - determination coefficient.

Table 7 shows the soil loss values observed, calculated by USLE and adjusted by the local erosivity index and the regression determination coefficient. The regression model is satisfactory only for plots with the highest erosion rates (plots 1, 2, 3 and 4) and also for the average of plots 1 and 2, and of plots 3 and 5. For these selected plots, the annual average adjusted soil loss is about 70% lower than the observed. Nevertheless, the authors suggest the following values for the regression parameters, based on an average from plots 1, 2, 3 and 5:

$$\alpha = 1.93 \text{ E-5}$$
$$\beta = 2.2$$

Figure 7 presents the time series of soil loss values observed, calculated by USLE and adjusted by regression for plot 2. A satisfactory agreement of observed and adjusted data can be confirmed visually.

ACKNOWLEDGEMENT

The authors wish to acknowledge the assistance and support given by several institutions since the beginning of these erosion studies, namely: INMG, DGHEA, University of Évora and JNICT.

Figure 7. Annual soil loss: observed, calculated with the USLE, and adjusted by regression for plot no. 2.

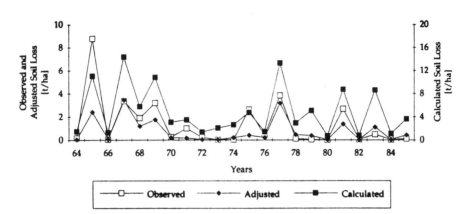

REFERENCES

ARAUJO, E.B. 1974. A sedimentaçao na albufeira de Vale Formoso. Separata da revista de Ciências Agrárias I(II):157-170.

COUTINHO, M.A. and TOMAS, P.P. 1988. Estudo do parâmetro de erosividade da Equaçao Universal de Degradaçao de Solo. 4. SISLUB, Lisboa.

COUTINHO, M.A. and TOMAS, P.P. 1990. Applying the Universal Soil Loss Equation to the southern part of Portugal. Seminar on Interaction between Agricultural Systems and Soil Conservation in the Mediterranean Belt. Oeiras, 4-8 September, Portugal.

DIXON, W.J. and MASSEY, F.J. 1983. Introduction to statistical analysis. International Student Edition. McGraw-Hill Book Company.

FERREIRA, I.M., FERREIRA, A.R. and SIMS, D.A. 1982. Análise preliminar dos dados dos talhoes do Centro Experimental de Erosao de Vale Formoso, Alentejo (Portugal) em termos da Equaçao Universal de Perda de Solo, para os anos de 1962/63 a 1979/80. Dir. Geral Hidráulica e Engenharia Agrícola, Évora.

FOSTER, G.R., McCOOL, D.K., RENARD, K.G. and MOLDENHAUER, W.C. 1981. Conversion of the USLE to SI metric units. Journal of Soil and Water Conservation Nov-Dec:355-359.

HUDSON, N.W. 1981. Soil conservation. Batsford , London.

LIMA, M.P. 1990. Laboratory experiments on water erosion of stony soils. Seminar on Interaction between Agricultural Systems and Soil Conservation in the Mediterranean Belt. Oeiras, 4-8 September, Portugal.

McCOOL, D.K., BROWN, L.C., FOSTER, G.R., MUTCHLER, C.K., and MEYER, L.D. 1987. Revised slope steepness factor for the Universal Soil Loss Equation. Transactions of the American Society of Agricultural Engineers 30(5).

SOIL CONSERVATION SERVICE 1977. Guides for erosion and sediment control. USDA, SCS, Davis, California.

WISCHMEIER, W.H. and SMITH, D.D. 1978. Predicting rainfall erosion losses. USDA, Agricultural Research Service, Handbook 537.

Chapter 19

COMPARISON OF FOURNIER WITH WISCHMEIER RAINFALL EROSIVITY INDICES

M.A. Coutinho and P.P. Tomás
Instituto Superior Técnico, Av. Rovisco Pais, 1096 Lisboa Codex, Portugal

Summary

The Fournier index has been used in Portugal as a rainfall erosivity index for estimating soil loss and to produce maps of erosion risk. Our objectives are to compare the Fournier and Wischmeier rainfall erosivity indices in order to establish the standard use of the Wischmeier index instead of the Fournier index.

This paper is part of a major study of soil erosion phenomena, in particular the understanding and modelling of the fundamental processes. The Universal Soil Loss Equation (USLE) has already been applied to the southern part of Portugal and compared with data from field erosion plots. To apply the USLE, the rainfall erosivity index EI_{30} was calculated for 3 places, for a time series from 14 to 23 years. The Fournier indices were calculated for each place and year, using the classic and modified approach. The erosivity indices (EI_{30} and Fournier) and the annual precipitation have been correlated.

The most important conclusions of this study are:

i.   There is poor correlation between the EI_{30} and the Fournier indices. The best relationship was found between the EI_{30} and the modified Fournier (F_m);

ii.   There is a good correlation between the EI_{30} and the annual precipitation;

iii. To estimate a rainfall erosivity index for areas in Portugal where the EI_{30} is not available, the authors suggest the use of the regression equation EI_{30} - annual precipitation, because they consider this a better estimate and is more meaningful than the determination of the modified Fournier index.

1. INTRODUCTION

The Fournier rainfall erosivity index has been used in Portugal to evaluate soil losses and to produce maps of erosion risk. The aim of this study is to compare the Fournier and Wischmeier indices, in order to use the latter as a benchmark, instead of the former. The work was part of a study of soil erosion phenomena in southern Portugal. Its main objective was the establishment of a soil erosion prediction model.

The Universal Soil Loss Equation (USLE) was the model chosen at the beginning of the study, because of its worldwide application. It has already been applied to the southern part of Portugal (at the Vale Formoso Experimental Erosion Centre (VFEEC)) and compared with data from field erosion plots. A detailed analysis considering the rainfall factor of the USLE has already been conducted (Coutinho and Tomás, 1988).

Three locations were considered for the rainfall index analysis: Portela, Sassoeiros and Vale Formoso (Figure 1). The Fournier indices were calculated for each place and year, using the classic and modified approach. To compute the Wischmeier index EI_{30} (USLE rainfall factor), for the three locations, the standard methodology was used. For both rainfall erosivity index, time series of 14 and 23 years were used.

The Fournier and EI_{30} rainfall erosivity indices and the annual precipitation have been correlated to find the degree of agreement.

2. FOURNIER EROSIVITY

Fournier (1960) sought to explain the distribution of suspended sediment yields at a regional or worldwide scale. Based on extensive data from almost a hundred river basins, an index was selected:

$$F_0 = p^2/P$$

where p = mean rainfall for the wettest month and P = mean annual precipitation, both in mm (Gregory and Walling, 1973).

Several attempts have been made to correlate erosion risks, at the basin or regional scale, by use of this rainfall aggressiveness index. Also, relationships between this index, drainage properties and the Wischmeier erosivity index were analysed by several authors, but agreements were not close (Morgan, 1986).

Several modified Fournier indices have been developed and used to improve the accuracy. Such an index was applied in a simplified methodology used to map erosion risk in Europe under the CORINE programme (Sequeira, 1989).

Other modified Fournier indices have been developed. For instance, a precipitation Concentration Index PCI has been analysed for several meteorological stations in the Mediterranean region (Gabriels, 1990), which provides a better description of erosivity. Although this PCI is a valuable index for regional characterisation it cannot be used as an absolute erosion predictor without a model to support it. In this study, besides the basic Fournier index, F_0, the following two modified indices were considered.

i.   The 1st modified Fournier index is given by:

$$F_{m1j} = p_{Mij}^2/P_j,$$

where p_{Mij} = monthly rainfall value for the wettest month i of the year j, and P_j = annual precipitation. The mean annual value for the index F_{m1} is the average of the annual values F_{m1j};

193

Figure 1. Location of the three meteorological stations.

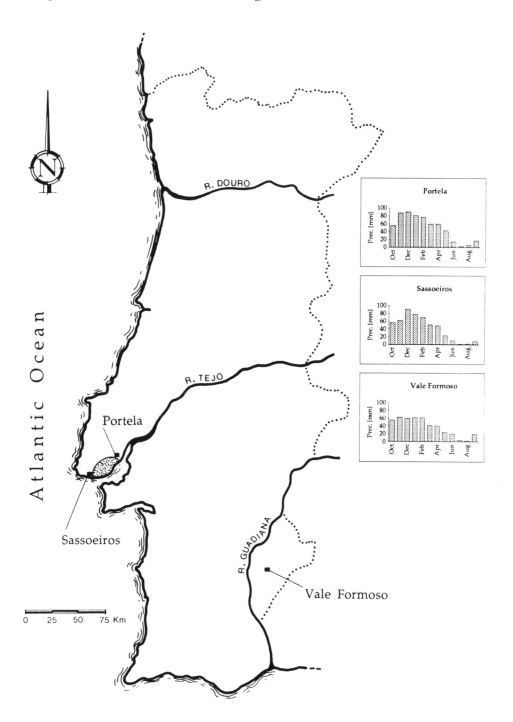

ii.   The 2nd modified Fournier index is given by:

$$F_{m2j} = \sum_{i=1}^{12} p_{ij}^2 / P_j,$$

where p_{ij} = rainfall for the ith month, and P_j = annual precipitation. The mean annual value for the index F_{m2} is the average of the annual values F_{m2j} (Arnoldus, 1980).

Although the 1st modified Fournier index F_{m1} does not differ from the basic index F_0 too much, the annual values and the mean annual values show important differences, because in the considered locations, the wettest month in a given year is not always the same.

Table 1 shows the monthly recorded precipitation, total annual rainfall and values of the Fournier indices.

3.   WISCHMEIER EROSIVITY INDEX

The rainfall factor R of the USLE is obtained by adding the storm erosivity index EI_{30} for all storm events in a year, according to Foster et al. (1981). Rain showers of less than 12.5mm and separated from other rain periods by more than six hours were omitted from the erosivity index computations, unless as much as 6.35mm of rain fell in 15 minutes. The maximum amount in 30 minutes was limited to 63.5mm.h^{-1} (Wischmeier and Smith, 1978).

The kinetic energy of precipitation was calculated using the equation proposed by Foster et al. (1981):

a) for $i_m \leq 76$mm.h^{-1}

$$e_k = 0.119 + 0.0873 \, log10(i_m); \qquad (1)$$

b) for $i_m \geq 76$mm.h^{-1}

$$e_k = 0.283;$$

where e_k is the kinetic energy in MJ.(ha.mm)$^{-1}$ and i_m the precipitation intensity in mm.h^{-1}.

The annual values of EI_{30} for the three locations considered in the study are presented in Table 1.

Table 1. Precipitation and erosivity indexes for all locations. a) VALE FORMOSO.

Year	Oct mm	Nov mm	Dec mm	Jan mm	Feb mm	Mar mm	Apr mm	May mm	Jun mm	Jul mm	Aug mm	Sep mm	Total mm	F_O mm	F_{m1} mm	F_{m2} mm	EI_{30} Mj.mm^{-1} ha^{-1}.h^{-1}
63	13	75	209	55	55	104	13	32	8	0	0	21	584	9.7	74.6	116.3	1055
64	3	35	36	60	47	21	2	3	23	0	0	54	284	4.4	12.8	43.3	282
65	169	145	22	71	106	0	55	11	21	0	0	44	643	32.5	44.2	111.3	1789
66	49	48	2	30	68	19	20	46	44	0	0	2	327	6.9	13.9	45.7	546
67	166	118	8	0	155	39	47	15	16	0	8	1	572	24.2	47.9	121.7	1958
68	49	125	89	114	147	93	65	36	50	22	0	31	820	19.1	26.2	95.7	1700
69	89	103	35	229	9	31	26	25	54	0	0	0	601	17.5	87.1	128.7	1378
70	5	32	47	74	4	28	123	75	26	11	8	0	433	2.4	35.0	71.8	624
71	28	13	32	98	80	65	24	12	0	10	0	28	390	0.4	24.8	61.3	478
72	82	17	58	53	20	19	2	40	35	1	5	0	331	0.8	20.5	50.8	379
73	12	29	42	28	36	73	68	8	21	1	0	0	317	2.6	16.9	48.2	251
74	4	31	17	33	92	105	30	53	28	0	0	0	394	2.5	27.9	67.0	761
75	11	11	68	11	47	49	80	24	34	0	11	49	393	0.3	16.1	51.2	795
76	95	68	162	114	83	8	4	5	35	2	1	9	586	7.9	44.6	104.2	1191
77	122	48	182	38	56	49	58	42	4	0	0	31	630	3.6	52.5	100.6	1946
78	61	60	91	170	132	48	52	4	5	28	0	9	659	5.5	43.9	102.8	13.75
79	157	14	8	24	85	35	41	47	7	5	6	6	434	0.4	56.5	87.2	702
80	45	89	0	2	6	15	41	8	5	0	6	27	237	33.3	33.3	53.4	416
81	12	1	151	59	26	66	26	0	0	16	0	71	444	0.0	51.1	84.6	1349
82	10	83	21	0	20	8	52	11	2	0	17	0	223	30.6	30.6	49.1	384
83	66	198	52	16	7	57	20	38	3	0	16	4	461	84.8	84.8	111.9	1139
84	37	71	15	99	62	0	47	19	23	0	0	2	375	13.6	26.0	62.2	386
85	0	41	55	40	72	25	38	12	29	0	0	72	383	4.4	13.5	51.2	1065
Avg.	56	63	61	62	61	42	41	25	21	4	3	20	457	13.4	38.5	79.1	954

b) PORTELA.

Year	Oct mm	Nov mm	Dec mm	Jan mm	Feb mm	Mar mm	Apr mm	May mm	Jun mm	Jul mm	Aug mm	Sep mm	Total mm	F_0 mm	F_{m1} mm	F_{m2} mm	EI_{30} Mj.mm^{-1} ha^{-1}.h^{-1}
																Erosivity Indices	
			Monthly and annual precipitation														
70	4	138	36	145	14	52	124	128	47	3	10	1	702	1.8	30.0	111.5	1294
71	1	11	52	246	152	120	17	20	0	11	0	45	675	4.0	89.2	153.1	1232
72	147	19	176	161	30	14	65	110	1	0	0	0	722	42.8	42.8	133.2	1167
73	21	109	28	72	70	36	43	27	45	0	0	2	454	1.7	26.3	64.2	739
74	5	46	17	59	115	171	15	32	14	0	0	14	487	0.6	59.7	102.1	711
75	5	45	59	62	61	22	95	1	0	3	43	68	463	7.6	19.4	62.6	1172
76	62	87	24	0	39	32	19	22	24	7	0	9	325	1.7	23.4	49.5	313
77	179	100	174	55	187	49	66	55	23	0	1	5	894	33.9	39.2	135.2	2070
78	30	93	257	149	207	108	77	14	18	3	0	4	959	69.1	69.1	165.7	1867
79	192	16	22	47	55	67	22	54	4	3	22	17	518	0.9	70.8	98.8	415
80	47	70	8	12	34	69	92	35	3	7	0	33	411	0.1	20.6	58.7	583
81	49	1	275	106	70	27	83	7	6	11	4	51	690	109.4	109.4	151.5	2608
82	6	97	29	5	45	3	93	45	8	0	1	0	333	2.6	28.3	69.3	406
83	27	398	116	30	0	78	32	53	16	0	0	0	751	18.0	211.2	245.0	2373
Avg.	55	88	91	82	77	60	60	43	15	3	6	18	599	21.0	60.0	114.3	1211

c) **SASSOEIROS.**

Year	|	Oct mm	Nov mm	Dec mm	Jan mm	Feb mm	Mar mm	Apr mm	May mm	Jun mm	Jul mm	Aug mm	Sep mm	Total mm	|	F_0 mm	F_{m1} mm	F_{m2} mm	EI_{30} MJ.mm^{-1} ha^{-1}.h^{-1}
						Monthly and annual precipitation												Erosivity Indices	
70		0	56	24	122	8	51	101	78	0	0	0	0	439		1.3	33.9	85.3	383
71		2	13	61	168	111	92	18	22	3	1	0	25	515		7.2	54.5	105.3	519
72		138	33	156	73	23	14	0	0	0	7	0	16	460		52.8	52.8	110.4	631
73		11	80	78	72	65	29	60	19	45	0	0	0	460		13.3	13.9	62.9	564
74		3	31	11	34	125	132	10	18	0	0	0	0	364		0.3	48.0	98.4	428
76		0	0	183	93	68	32	12	1	30	0	0	7	426		78.5	78.5	114.3	548
77		209	50	135	72	141	36	63	34	8	0	0	9	759		24.2	57.8	127.2	1264
78		61	55	233	180	154	91	65	3	9	1	0	1	852		63.9	63.9	152.2	1653
79		193	14	5	37	67	58	46	15	0	1	25	9	469		0.1	79.1	105.7	704
80		52	77	11	12	35	64	109	32	7	0	0	0	398		0.3	29.6	67.9	687
81		21	0	165	112	72	16	46	4	6	10	8	37	496		54.6	54.6	99.1	797
82		18	109	33	7	34	0	94	39	11	1	3	0	350		3.0	34.1	71.7	365
83		30	292	89	38	13	52	12	35	13	0	1	3	579		13.7	14.0	172.4	2298
Avg.		57	62	91	78	71	51	49	23	10	2	3	8	505		24.1	57.5	105.6	834

4. COMPARISON OF EROSIVITY INDICES

The following comparisons were made:

i. A comparison of the computed Fournier indices at the three locations shows good agreements only between F_{m1} and F_{m2} (Table 2);

Table 2. Correlation coefficients between erosivity indices and annual precipitation.

	VF	PO	SS
F_0 vs F_{m1}	0.48	0.31	0.20
F_0 vs F_{m2}	0.35	0.47	0.40
F_{m1} vs F_{m2}	0.85	0.88	0.86
F_0 vs EI_{30}	0.20	0.75	0.21
F_{m1} vs EI_{30}	0.50	0.64	0.76
F_{m2} vs EI_{30}	0.81	0.80	0.86
P vs F_{m2}	0.82	0.78	0.70
P vs EI_{30}	0.86	0.80	0.71

Legend: VF - Vale Formoso; PO - Portela; SS - Sassoeiros

ii. The Fournier indices have been correlated with EI_{30}. A good correlation only exists between F_{m2} and EI_{30} (Table 2);
iii. The correlation between F_{m2} and EI_{30} and the annual precipitation P are presented in Table 2. Good agreement was obtained for both indices.

Table 3 shows the best regression equations found for EI_{30} vs. F_{m2}, EI_{30} vs. P, and F_{m2} vs. P at the three studied locations.

Table 3. Regression equations for the relations between indices and annual precipitation.

	VF	R^2	PO	R^2	SS	R^2
EI_{30} vs P	$0.04105.P^{1.626}$	74	$0.01710.P^{1.729}$	75	$0.01567.P^{1.731}$	65
EI_{30} vs F_{m2}	$15.718.F-289.8$	66	$11,196.F-69.3$	65	$15.457.F-798.3$	74
F_{m2} vs P	$0.26374.P^{0.929}$	74	$0.05364.P^{1.193}$	76	$0.73061.P^{0.797}$	49

Legend: R^2 - determination coefficient

5. CONCLUSIONS

The most important conclusions of this study can be summarised as follows:

i.   There is a poor correlation between the EI_{30} and the Fournier indices. The best relationship was found between the EI_{30} and F_{m2};

ii. There is a good correlation between the EI_{30} and annual precipitation P;

iii. To estimate a rainfall erosivity index, in southern areas of Portugal, where EI_{30} is not available, it is suggested to use the regression equation EI_{30} vs. P because this index represents a rainfall erosivity factor more closely than the modified Fournier index;

iv. The Fournier type indices are not strong erosivity predictors due to the time and spatial framework used in their estimation. At the regional scale they can provide useful information. However, extrapolation to different climatic regions, namely the Mediterranean, must be done carefully and should be supported by field data. In cases where field data is available, this should be correlated with a more meaningful index.

REFERENCES

ARNOLDUS, H.M.J. 1980. An approximation of the rainfall factor in the Universal Soil Loss Equation. In: Assessment of erosion. De Boodt, M.M. and Gabriels, D. (eds). John Wiley, Chichester.

COUTINHO, M.A. and TOMAS, P.P. 1988. Estudo do parâmetro de erosividade da Equaçao Universal de Degradaçao de Solo. 4° SISLUB, Lisboa, Portugal.

FOSTER, G.R., McCOOL, D.K., RENARD, K.G. and MOLDENHAUER, W.C. 1981. Conversion of the USLE to SI metric units. J. of Soil and Water Conservation. p.355-359.

FOURNIER, F. 1960. Climat et erosion: la relation entre l'erosion du sol par l'eau et les precipitations atmospheriques. Presses Universitaires de France.

GABRIELS, D. 1990. A provisional method for calculating the rain erosivity in EEC countries with reference to the Mediterranean region. In Rubio, J.L. and Rickson, R.J. (eds). Strategies to combat desertification in Mediterranean Europe. Commission of the European Communities, Report EUR 11175EN/ES.

GREGORY, K.J. and WALLING, D.E. 1973. Drainage basin form and process. A geomorphological approach. Edward Arnold Ltd.

MORGAN, R.P.C. 1986. Soil erosion and conservation. Longman.

SEQUEIRA, E.M. 1989. Programa CORINE. Projecto transfronteiriço Algarve/Andaluzia, E. Agricola Nacional, Oeiras, Portugal.

WISCHMEIER, W.H. and SMITH, D.D. 1978. Predicting rainfall erosion losses. USDA, Agr. Res. Serv. Handbook 537, USA.

Chapter 20

RAINFALL EROSIVITY FOR SOUTH-EAST POLAND

K. Banasik and D. Górski
Department of Hydraulic Structures,
Warsaw Agricultural University - SGGW,
ul. Nowoursynowska 166, PL-02-766 Warsaw, Poland

Summary

Rainfall erosivity for four stations in south-east Poland has been calculated from 29 years of
rain gauge data. The annual erosivity index varies at each station significantly. The ratio of
annual maximum and minimum erosivity index at the stations is 6.7 at Pulawy, 10.2 at
Sandomierz, 24.6 at Limanowa, 9.0 at Lesko, and the average annual erosivity at the
stations is 64.1, 66.4, 96.8 and 84.3 $(MJ.ha^{-1}) \cdot (cm.h^{-1})$ respectively. About two thirds to
three quarters of the erosion hazard corresponds to the summer months of June-August.

1. INTRODUCTION

The Universal Soil Loss Equation (USLE), developed by Wischmeier and Smith (1978),
and various erosion sedimentation models are being used to develop soil conservation
programmes and identify best management practices. Rainfall erosivity is a key input
parameter to the USLE, in which estimated soil loss is computed as a product of six factors:

$$A = R \cdot K \cdot L \cdot S \cdot C \cdot P \tag{1}$$

where A is soil loss per unit area, R is the rainfall erosivity factor, K is the soil erodibility
factor, L is the slope-length factor, S is the slope-steepness factor, C is the cover and
management factor, and P is the supporting conservation practice factor. South-east Poland
is a region of relatively high soil erosion intensity, with the Carpathian Mountains in the
south, and loess soils in the south-east. A few attempts of applying the USLE to this region
have been made (Banasik, 1985; Madeyski and Banasik, 1989; Piest and Ziemnicki, 1979),
but only with indirect estimation of the R-value. These indirect methods are based on the
relationship between rainfall erosivity and some rainfall characteristics such as the 2-year
rainfall depth of given duration (Madramootoo, 1988; Foster et al., 1982; Ateshian, 1974),
as estimated for other regions. These may, however, give inaccurate results.

2. METHODOLOGY AND DATA USED

The rainfall erosivity factor, R, is defined as the average annual value of the rainfall erosion
index, EI. This is obtained by summing the EI values for individual storms for a number of
years. The minimum length of record recommended by Wischmeier and Smith (1978) is 22
years. The sum of the EI values is then divided by the number of years of record. The
value of EI for a given storm is the product of total storm kinetic energy, E, times the

maximum 30-minute rainfall intensity during the storm (I_{30}). Rain gauge data for the four locations in south-east Poland (Pulawy, Sandomierz, Limanowa and Lesko) over 29 years (1960-88) have been used to estimate rainfall erosivity. The estimation has been carried out in two phases. First, all the continuous records have been used for the computation of the EI values and identification of the relationship P vs. EI (rainfall depth versus erosion index of the rainfall). Second, for periods without such records (because rainfall recorders had not been installed or were faulty) EI values were computed from the measured rainfall depths, taking account of the P vs. EI relationship established in the first phase.

Rainfall of less than 12.7mm which was separated from other rainfall events by more than 6 hours has not been included in the computation of EI, unless as much as 6.3mm of rain fell in 15 minutes (Wischmeier and Smith, 1978).

In the first step of the estimation, after having selected the erosive rainfall events, the following values were computed:

a) the kinetic energy of rainfall in the time step over which intensity is considered to be constant from the energy equation (Wischmeier and Smith, 1978) converted to the metric units (Pasák et al., 1983):

$$E_{ki} = (206 + 87 \cdot \log I_i) \cdot P_i \qquad (2)$$

where E_{ki} is in $J.m^{-2}$, I_i is the rainfall intensity in the ith increment in $cm.h^{-1}$, P_i is the depth of rainfall for the ith increment of storm hyetograph in cm;
b) the kinetic energy of the whole rainfall event (E_k) by summing the E_{ki} values for all the time increments of the rainfall event;
c) erosivity of the rainfall

$$EI = 1/100 \cdot E_k \cdot I_{30} \qquad (3)$$

where EI has units of $MJ.ha^{-1} \cdot cm.h^{-1}$ (here the EI-value is equal to that which is expressed in $kJ.m^{-2}.m.h^{-1}$, as in Schwertmann et al. (1987)), I_{30} is the maximum 30-minute rainfall intensity during the storm in $cm.h^{-1}$, and 1/100 is the conversion factor for the units;
d) annual and monthly erosivity, by summing EI values for individual storms;
e) mean monthly and annual values of rainfall erosivity, and
f) parameters of the relationship

$$EI = a \cdot P^b \qquad (4)$$

Values of the parameters for the four locations are shown in Table 1.

Where rain recorders did not work, the erosivity index EI for individual storms has been estimated according to regression equation 4. The values were summed for each month, year, and then with values estimated in the first step.

Table 1. Parameters of the regression equation for rainfall erosivity $(MJ.ha^{-1})\cdot(cm. h^{-1})$ as a function of rainfall depth (mm) $[EI=a\cdot P^b]$.

Station	Parameters of regression equation		Number of rainfall events	Correlation coefficient
	a	b		
Pulawy	0.153	1.205	159	0.54
Sandomierz	0.175	1.076	221	0.44
Limanowa	0.291	0.896	323	0.47
Lesko	0.143	1.128	306	0.55

3. RESULTS AND CONCLUSIONS

Analysis of the rainfall records from the period 1960-88 and other rain data have shown that the average number of erosive rainfall events in a year were 7.5 at Pulawy, 9.9 at Sandomierz, 13.0 at Limanowa and 11.7 at Lesko. The percentage of rainfall with continuous records in the total estimated erosivity was 79% at Pulawy, 89% at Sandomierz, 91% at Limanowa, and 94% at Lesko. Mean annual erosivity indices at the stations are shown in Figure 1, with the number of erosive rainfall events in a year in Table 2. The values of EI increase (from 64 at Pulawy to 97 at Limanowa) with the increase in elevation (Table 2). In the case of Pulawy, the values of average annual EI, previously obtained with the use of indirect estimates, ranged from approximately 30 to 100 (Banasik, 1985; Piest and Ziemnicki, 1979).

Table 2. Mean annual erosivity of rainfall in southeast Poland.

Erosivity estimated with	Stations			
	Pulawy(142)[1]	Sandomierz(217)	Limanowa(414)	Lesko(386)
	$(MJ.ha^{-1})\cdot(cm.h^{-1})$			
Rainfall with continuous records	50.6	58.9	87.9	79.2
Use of regression equation	13.6	7.4	9.0	5.1
Total - EI	64.1	66.4	96.8	84.3
Mean number of erosive rainfalls per year	7.5	9.9	13.0	11.9

[1] Numbers in parentheses are elevation of the stations in metres above sea level.

There are significant variations in annual erosivity from year to year for each of the stations, as shown in Fig. 2. The following ratios of maximum and minimum annual erosivity were 6.8 for Pulawy, 10.2 for Sandomierz, 24.6 for Limanowa and 9.0 for Lesko. The seasonal distribution of EI data had a peak during the summer months: in July at Pulawy, in August at Sandomierz and in June at Limanowa and Lesko (Fig. 3).

Figure 1. Location of the stations with rainfall erosivity factor, R.

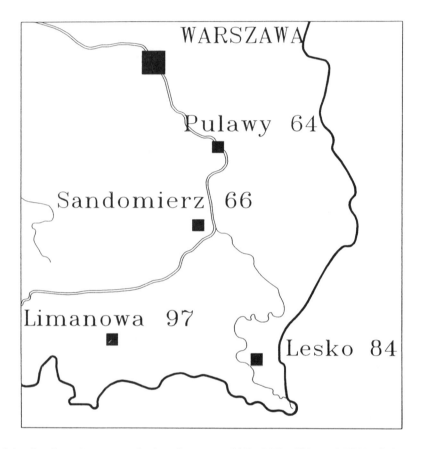

Erosivity for June-August at the locations was 64%, 74%, 69% and 73% of the annual values, respectively.

Rainfall at Limanowa and Lesko could cause more soil loss not only because of the higher annual EI value, but also because of the seasonal peak of EI in June, when some crops (e.g. potatoes) are in the early stage of growth and provide only limited soil protection against erosion.

Figure 2. Annual variation of rainfall erosivity at the four sites in Poland.

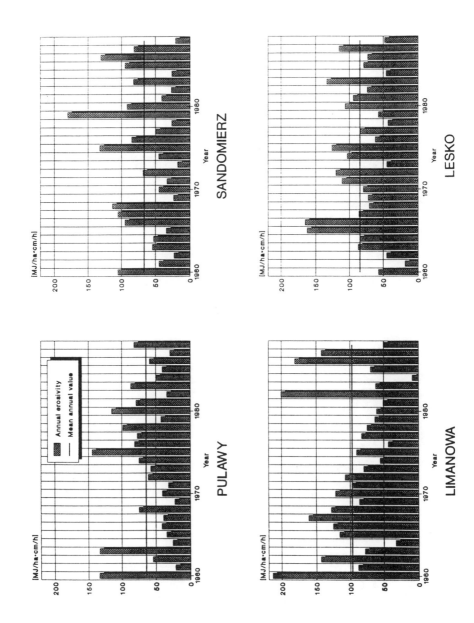

Figure 3. Mean seasonal distribution of EI values.

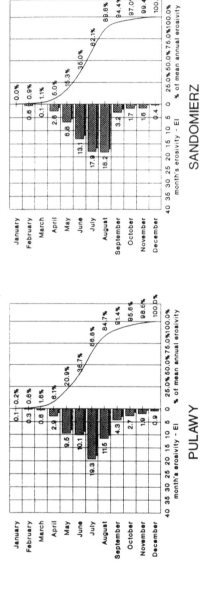
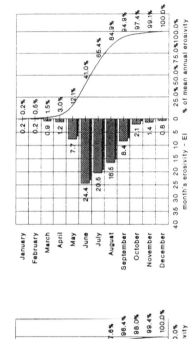

ACKNOWLEDGEMENT

The investigation has been supported by the project No. KBN-SGGW-10 and the joint research project No. PL-ARS-195.

REFERENCES

ATESHIAN, J.K.H. 1974. Estimation of rainfall erosion index. Journal of the Irrigation and Drainage Division. Vol.100, IR3:293-307.

BANASIK, K. 1985. Applicability of the Universal Soil Loss Equation for predicting sediment yield from small watersheds in Poland. Proceedings of the International Symposium on Erosion, Debris Flow and Disaster Prevention. Tsukuba, Japan. p.85-88.

FOSTER, G.R., LOMBARDI, F. and MOLDENHAUER, W.C. 1982. Evaluation of rainfall-runoff erosivity factors for individual storms. Transactions of ASAE 25:124-129.

MADEYSKI, M. and BANASIK, K. 1989. Applicability of the modified Universal Soil Loss Equation in small Carpathian watersheds. Catena Supplement 14:75-80.

MADRAMOOTOO, C.A. 1988. Rainfall and runoff erosion indices for eastern Canada. Transactions of the ASAE 31:107-110.

PASAK, V., JANECEK, M. and SABATA, M. 1983. The protection of agricultural land against erosion. Ministerstvo Zemëdëlství a Vyzivy CSR, Metodiky 11 (in Czech).

PIEST, R.F. and ZIEMNICKI, S. 1979. Comparative erosion rates of loess soils in Poland and Iowa. Transactions of the ASAE 22:822-827,833.

SCHWERTMANN, U., VOGL, W. and KAINZ, M. 1987. Bodenerosion durch Wasser; Vorhersage des Abtrages und Bewertung von Gegenmassnahmen. Verlag Eugen Ulmer, Stuttgart.

WISCHMEIER, W.H. and SMITH, D.D. 1978. Predicting rainfall erosion losses - a guide to conservation planning. Agriculture Handbook 537, USDA-ARS.

Chapter 21

RELATIVE ENERGIES OF SIMULATED RAINSTORMS

H.H. Becher
Lehrstuhl für Bodenkunde, Technical Universitat München,
DW 8050 Freising-Weihenstephan, Germany

Summary

The effects of simulated rainstorms are only comparable if the energies and temporal and spatial distribution of the simulated rainfall produced by different simulators are similar. Using a drop impact transducer connected to a high resolution X-t-recorder, the relative energies and distributions of rainfall simulated by five different field simulators at almost the same rainfall intensity were monitored on runoff plots during an extensive field study on soil erosion. Evaluation of the impact pulses registered indicated that the rainfall energies of all simulators varied over time and position as a function of simulator characteristics, the general setting and the weather conditions. Inspite of this variability, differences between the different rainfall simulators were significant in many cases. Furthermore, the temporal distributions in energies of the simulator using Perrot-nozzles and of that using plastic tubes and a fine wire gauze as drop formers had continuous drop impacts of changing energy. The latter was also the most sensitive to wind. The rainfall simulators with moving nozzles (Veejet 80100) as dropformers, in contrast, showed characteristic periodicities in their energy distribution. It is hypothesised that the different periodicities of jets of water drops of these simulators affect (i) the redistribution of water in aggregates on the soil surface and thus (ii) the breakdown of soil aggregates, which in turn influences the susceptibility of the soil to erosion.

1. INTRODUCTION

In areas with irregular distribution of rainstorms, it is common practice to use rainfall simulators in soil erosion studies. These simulators, however, differ considerably in design and may therefore give different soil losses for similar soils and rainfall amounts (Neff, 1979). It is therefore necessary to compare the effect of different simulators on soil loss and infiltration.

In the present study six simulators were compared in 1988 on the same field. The simulators, site and experimental conditions have been described in detail by Auerswald et al. (1992) and Kainz et al. (1992). Briefly, drops were formed by different methods: three simulators used Veejet 80100 nozzles, one used Perrot-nozzles, and two used an arrangement of capillary tubes of soft plastic or bore holes in hard plastic pipes, together with a fine wire gauze placed at 4m and 7m, respectively, above the soil surface.

As the soils and their treatments on the different plots were virtually the same (Auerswald et al., 1992), differences observed in surface conditions after rainfall application were

208

essentially due to differences in rainfall energies generated by the different simulators. To obtain information on the energy distribution over time on the plots, drop impacts were continuously monitored at several positions during the runs. These runs were carried out on plots of equal dimensions (1.0 × 4.5m) and at almost the same rainfall intensity (Table 1) for at least one hour.

Table 1. Rainfall intensities (mm.h[-1]) during the monitored runs and kinetic energy relative to natural rains of the same intensity (K.E. = 28J.(m[2].mm[-1])).

Rainfall simulators using	Intensity	Relative kinetic energy
Veejet-80100 nozzles		
Weih. rotating	62.3	0.8[1]
Weih. rectangular	66.3	0.8[1]
Bonn	68.7	0.8[1]
Perrot-nozzles		
Giessen	57.8 (left)	0.8[2]
	72.5 (right)	
Plastic tubes with gauze		
Trier	72.8 (left)	1.0[3]
	68.5 (right)	

[1] Meyer (1958)
[2] Bunza et al. (1985)
[3] Hassel and Richter (1988)

2. METHODS

The monitoring and evaluation of the relative energies of five of the six simulators (Table 1) are reported in more detail elsewhere (Becher, 1990). The transducer was placed at different positions, about 50cm apart along the slope, near the inside edge of each plot in most cases, starting at the top of the plot, for a duration of 4 minutes. Drop impacts on the active part (area 44cm[2]) of a modified loudspeaker induced changes of voltage in the millivolt range, lasting for a few milliseconds. These changes were registered by an X-t-chart recorder (150mm.min[-1]) as straight lines (= pulses) perpendicular to the time axis. The very quick changes of voltage together with the high recording speed enabled identification of each impact. Due to the small area of target, the probability of two impacts at the same time was zero (see also Kowal et al., 1973). The length of the registered lines ranged from 2mm to almost 100mm, and were subdivided into 5mm intervals giving 20 classes of relative kinetic energy. The total time of registration per position was subdivided into 20 second time intervals (Fig. 1), which enabled a sufficient number of replicates for counting the pulses within each class. This counting was done for each position, thus also permitting variations in pulse distribution with time to be identified.

Figure 1. Frequency distributions of mean pulses over time and all positions for all rainfall simulators.

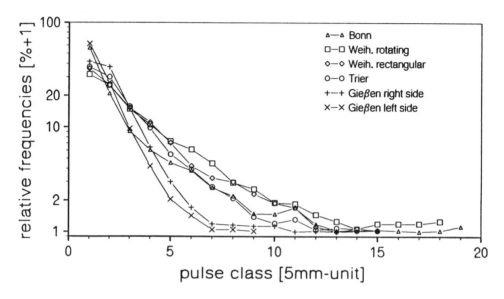

3. RESULTS AND DISCUSSION

At the same rainfall intensity, the simulators producing small drops generate many more impacts per unit area and time than those producing larger drops. Distributions of absolute frequencies of pulse are therefore less informative than those of relative distributions. Counts per pulse class related to total number of counts (i) per time interval was used to evaluate variation over time, or, in most cases (ii) per position, to evaluate variation over the plot, to assess total energy and that of each drop. For this reason, only their relative distributions are discussed. In Fig. 2 these are presented in a logarithmic scale after adding 1 to the percentages of frequency. This is necessary because the non-transformed relative frequencies are dominated by high percentages of low pulses that masked the small percentages of high pulses. The transformation thus makes differences in the range of high pulses clearer.

The evaluation of the peak class distributions indicates that none of the simulators has the same frequency distribution of pulses at different positions along the slope or over time (Becher, 1990), although such consistency would be desirable. The simulator with Perrot-nozzles, although raining continuously, differs from all others by having many low pulses, with only a few medium and no high pulses. It thus produces only low rainfall energy at the same rainfall intensity as the other simulators (Fig. 2). Fig. 2 also demonstrates the observed differences in energy distribution between the left and right sides under this simulator. The largest numbers of high pulses were observed under the Trier simulator. Although raining continuously (Fig. 1), this simulator showed a very large variation over time, which is due to its sensitivity to wind. This sensitivity results in a loss of pulses for classes >6. This loss is due to the fact that under windy conditions, the primary drops leaving the plastic tubes drift leewards. Thus only relatively small drops hit the transducer, whereas larger drops falling vertically under calm conditions drift outside the transducer position or even outside the plot in strong wind gusts. The Bonn simulator used the Veejet 80100 nozzles at almost the same pressure of 42.2kPa (Meyer, 1958) as the Weihenstephan simulators. Nevertheless, it gave an overall pulse frequency distribution different from those of the two Weihenstephan simulators, both of which had very similar overall pulse frequency distributions (Fig. 2). The reason for this deviation is not clear.

Figure 2. X-t-recorder output of the Trier simulator demonstrating the aperiodicity of impacts.

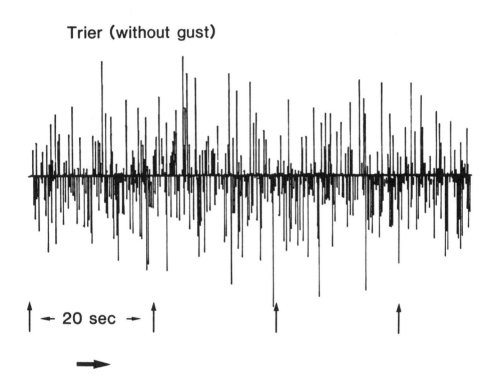

In contrast to the Giessen and Trier simulators, the simulators with moving nozzles show a distinct variation over a very short time period, giving more or less pronounced periodicities of high pulses within the 20 second time interval used for counting (Fig. 1). These high mean pulses may have been produced when the centre of the jet passed over the transducer. These periodicities therefore depend on the transducer position relative to the nozzles (and thus the jet of water drops), and on the type of simulator, namely ones with nozzles on booms rotating or going back and forth and horizontally oscillating nozzles (Kainz et al., 1992). The smallest and largest time intervals observed between impacts are given in the second and third columns of Table 2, and the effects of two (Bonn) or three or four (Weihenstephan, rotating) adjacent moving jets on the pulse distributions are given under the

Table 2. Time intervals (sec) for moving-nozzle simulators at different positions downslope (cm).

Position	Interval s*	l*	Periodicity	Impacts at	Total interval
Weihenstephan, rotating boom simulator					
200	1.2	8.8		2.0; 3.6; 8.4	14.0 (?)
250			8.0 + 8.0 + 1.6		17.6
300	3.2	9.2		ca. 8.8	17.6
350			1.6 + 1.6 + 3.2 + 6.8		13.2
Weihenstephan, rectangular simulator					
0	Decreasing		2.4→2.0→1.6→1.2		
50			1.0	1.0(-1.2)	1.0
100			1.2	1.2	1.2
150			1.6	1.6	1.6
175	2.0	2.2	2.0	2.0	2.0 (ca.)
200	3.6	4.0	Change between 3.6 and 4.0		3.8 (ca.)
Bonn					
-50	7.6	9.6		8.0	8.0
0	0.8	6.8		1.2; 6.0	8.0 (ca.)
50	1.2	8.0		8.0	8.0
100			2.0 + 6.0		8.0
150			1.6 + 6.4		8.0
200	1.2	6.0		1.6; 6.0	8.0 (ca.)
250			8.0	8.0	8.0
300			1.6 + 6.4		8.0
330			1.6 + 6.4		8.0

*s = smallest, l = largest
for further explanation, see text

heading 'periodicity'. Where no clear periodicity was observable the mean interval is given under 'impacts at'. The total time interval for recurrence of impacts is listed in the last column. This total time interval should be 16 seconds for the rotating boom simulator, but at the 200cm position this interval was not exactly identifiable. At the other positions, it clearly differed from the expected time. Each change in speed of nozzle oscillation on the rectangular simulator from Weihenstephan is identifiable as the change of the impact interval from 2.4 seconds to 1.2 seconds, and also of the total time interval. Table 2 also points out that the boom of the Bonn simulator moves most constantly with a total recurrence interval of exactly 8 seconds.

The periodicities also indicate that, in contrast to the Giessen and Trier simulators, the soil surface is hit by a powerful jet of water drops for a short duration followed by a longer spell of repose. This behaviour in performance of these rainfall simulators may have implications for soil erosion monitoring under rainfall simulation. It is hypothesised that after each jet of drops, there is a different relaxation time at each spot underneath the rainfall simulator. During this period, water is redistributed within the surface soil aggregates and within the uppermost soil surface layer itself, at least at the beginning of the rainfall simulation. During the relaxation time, matric potential decreases, so that water menisci form again in the pores at the surface of aggregates and of the soil mass. These menisci draw together soil particles in a manner that the arrangement of the soil particles can partially withstand the impact forces (Al-Durrah and Bradford, 1981) of the next jet of drops. The shorter the relaxation time, the less water can redistribute, and the more soil aggregates break down because they are almost water saturated (Farrell and Larson, 1972). This increases the susceptibility of the soil surface to erosion below the rainfall simulator. This effect should be most pronounced for the rectangular simulator from Weihenstephan with its short total interval of only 1-2 seconds in contrast to the Bonn and rotating simulators (Table 2). Nevertheless, the partial aggregate breakdown occurring during rainfall simulation seals the surface soil layer; the degree of sealing depending on soil initial conditions and properties (e.g. Collins et al., 1986). Thus water cannot redistribute between jets of water drops, and the effect of different relaxation times of the simulators on soil loss will diminish with time of simulation. The results of Farres (1980), although no relaxation times are given, seem to confirm this hypothesis.

ACKNOWLEDGEMENT

This report is based on a project sponsored by the Bundesministerium für Forschung und Technologie under grant No. 0339274A. The author is responsible for the content of this publication. Dr. E. Murad is gratefully acknowledged for revising the English style. The referees are also acknowledged for their comments.

REFERENCES

AL-DURRAH, M. and BRADFORD, J.M. 1981. New methods of studying soil detachment due to water drop impact. Journal of the Soil Science Society of America 45:949-953.

AUERSWALD, K., KAINZ, M., SCHRÖDER, D. and MARTIN, M. 1992. Comparison of German and Swiss rainfall simulators - Experimental set up. Zeitschrift für Pflanzenernährung und Bodenkunde 155:1-5.

BECHER, H.H. 1990. Comparison of German and Swiss rainfall simulators - Relative drop energies and their distribution in time and space for simulated rains. Zeitschrift für Pflanzenernährung und Bodenkunde 153:409-414.

BUNZA, G., DEISENHOFER, H.-E., KARL, J., PORZELT, M. and RIEDL, J. 1985. Beiträge zu Oberflächenabfluss und Stoffabtrag bei künstlichen Starkniederschlägen. I. Der künstliche Starkniederschlag der transportablen Beregnungsanlage nach Karl und Toldrian. DVWK-Schriften 71:1-35. Verlag P. Parey, Hamburg.

COLLINS, J.F., SMILLIE, G.W. and HUSSAIN, S.M. 1986. Laboratory studies of crust development in Irish and Iraqi soils. III. Micro-morphological observations of artificial-formed crusts. Soil Till. Res. 6:337-350.

FARRELL, D.A. and LARSON, W.E. 1972. Dynamics of the soil-water system during a rainstorm. Soil Science 113:88-95.

FARRES, P.J. 1980. Some observations on the stability of soil aggregates to raindrop impact. Catena 7:223-231.

HASSEL, J. and RICHTER, G. 1988. Die Niederschlagsstruktur des Trierer Regensimulators. Mitteilungen der Deutschen Bodenkundlichen Gesellschaft 56:93-96.

KAINZ, M., AUERSWALD, K. and VÖHRINGER, R. 1992. Comparison of German and Swiss rainfall simulators - Utility, labour demands and costs. Zeitschrift für Pflanzenernährung und Bodenkunde 155:7-11.

KOWEL, J.M., KIJEWSKI, W. and KASSAM, A.H. 1973. A simple device for analysing the energy load and intensity of rainstorms. Agric. Meteor. 12:271-280.

MEYER, L.D. 1958. An investigation of methods for simulating rainfall on standard runoff plots and a study of the drop size, velocity and kinetic energy of selected spray nozzles. USDA-ARS, Purdue University, Special Report No. 81.

NEFF, E.L. (ed.). 1979. Rainfall simulator workshop. Proceedings of the Rainfall Simulator Workshop, March 7-9, 1979. Tucson. ARM-W-10.

Chapter 22

RELATIONSHIPS BETWEEN SOIL AND AGRONOMIC PARAMETERS AND SEDIMENT YIELDS

D.J. Mitchell

School of Applied Sciences, University of Wolverhampton,
Wolverhampton, West Midlands, WV1 1SB, UK

Summary

The 1:250,000 Soil Association Maps of England and Wales provide a wide range of soil parameters which can be used in large basin erosion research. Using 56 soil associations in the Wye Catchment, information on available water capacity, soil depth, soil particle size and stoniness have been examined spatially and volumetrically for the River Wye and nine sub-catchments. The relative percentages of sand, silt and clay in the A/O horizon have been also used to construct a clay ratio map (Bouyoucos, 1935), indicating soil erodibility. Based on the soil associations, the average catchment available water capacity (AWC) and the ratio of AWC to total volume of soil have been calculated for each sub-catchment. In the Wye Basin stocking ratios, derived from the Ministry of Agriculture, Fisheries and Food's agricultural parish returns, provide a more significant relationship with sediment yield than arable/pasture ratios. The coincidence of high poaching risk and dense stocking ratios results in high sediment yields. These soil and agronomic parameters could form the basis of a map of national soil erosion potential.

1. INTRODUCTION

It is widely accepted that soil type and farm practice have significant influence on erosion rates and sediment yield. However, to date, numerical parameters have been poorly developed, especially for larger catchments. Detailed soil and agronomic variables are used in statistical models for predicting sediment yield from runoff plots and in small catchment studies, but the application of these factors in the analysis of larger catchments becomes exceedingly vague and generalised.

In order to assess the use of soil and agronomic parameters, suspended sediment yield data for a seven year period (1974-80) from the Wye Catchment and nine sub-catchments are analysed (Figure 1). The River Wye rises on Plynlimon at 677m and joins the Severn Estuary near Chepstow, draining an area of 4,180km^2. Geologically, the catchment can be divided into two contrasting areas, namely a relatively impermeable upper area, above Erwood, of Lower Palaeozoic mudstones and shales, and a lower, more permeable, area of Upper Palaeozoic marls and sandstones.

Figure 1. Location of the Wye and Farlow catchments.

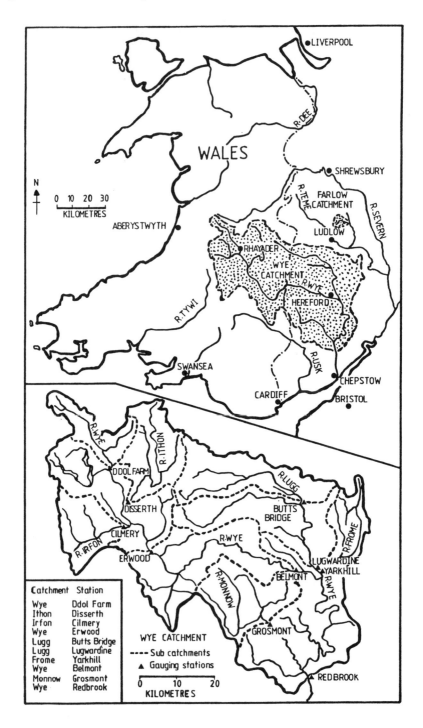

2. SOIL PARAMETERS

Considering that generally soils comprise the largest quantity of catchment sediment, pedological factors have not been extensively incorporated into suspended sediment yield equations. Therefore, the inclusion of a soil factor, rather than a geologic factor, in statistical models seems more appropriate. With the exception of the K factor in the USLE (Wischmeier and Mannering, 1969), the limited availability of detailed soil parameters has, in most cases, resulted in the use of selective terms such as soil particle size (Anderson and Wallis, 1965; Flaxman, 1972, 1974), soil aggregation indices (Anderson and Wallis, 1965) and soil infiltration capacities (Hindall, 1976). Furthermore, in an attempt to relate yields directly to sediment sources, surrogate parameters have been incorporated into other equations. Examples include total length of eroding banks (Striffler, 1965), contributing sediment area (Anderson, 1970; McPherson, 1975), soil erodibility (Larson and Hall, 1957), proneness to erosion (Jansen and Painter, 1974) and index of soil and transport conditions (Guyman, 1974).

Until recently the lack of soil data in Britain has meant that hydropedology has been under-used in catchment studies. In recent years, the Soil Survey of England and Wales has placed a stronger emphasis on the hydrological properties of soil, but mainly from a water yield rather than a sediment yield potential (Rudeforth and Thomasson, 1970; NERC, 1975; Farquharson et al., 1978). Using available water capacity (AWC) classes and mapping units proposed by Mackney and Burnham (1966), Mitchell (1979), in a study of the Farlow Catchment, Shropshire, found that the interpretation of the hydrological responses of soils depended on the mapping scale. In the same study, it was shown that high winter runoff potential (Hodgson, 1974, personal communication; Farquharson et al., 1978) correlated closely with the greatest density of streams and artificial drains (Mitchell, 1979).

Although Burnham (1964) produced a generalised map of the soils of Herefordshire, prior to 1983 only selected areas of the Wye Basin had been mapped at larger scales (Palmer, 1972; Hodgson and Palmer, 1971; Whitfield, 1971; Hodgson, 1972). Using the analysis from these and other detailed surveys, the Soil Survey produced, between 1979 and 1983, a national soil map of England and Wales published as six regional maps at a scale of 1:250,000 with accompanying bulletins. The majority of the Wye Catchment is covered by Bulletins No. 11 (Wales) (Rudeforth et al., 1984), No. 12 (Midland and Western England) (Ragg et al., 1984) and No. 14 (South-West England) (Findlay et al., 1984). Although some precision is lost, the survey covers the whole Wye catchment. The 1:250,000 Soil Association Map of England and Wales shows 56 soil associations in the Wye Catchment. The areal distribution of these rationalised soil series in each catchment can be used to estimate thickness of soil horizons, total soil depth, AWC, workability, texture and stoniness.

2.1 Soil Particle Size and Clay Ratio

Particle sizes have been derived from the profile description of the soil associations, according to the subdivisions in the Soil Survey Field Handbook (Hodgson, 1974). In each class, the average percentage of clay, silt and sand was calculated. If a soil association ranged across two classes, the range of values was adjusted accordingly to provide an estimated average particle size distribution. Therefore, the volume of clay, silt and sand in each horizon of each soil association can be estimated. This information, summed spatially on a catchment scale, provides the total volume of sand, silt and clay available for erosion (Table 1). These size distributions can be further used to calculate clay ratios (ratio of sand and silt to clay). These were found to have a direct relationship with erosion (Bouyoucos, 1935), although Bryan (1968) found the index less satisfactory. These apparent conflicting conclusions are partly dependent on the type of erosion; whereas soil detachment by raindrop impact increases with higher clay content (Rose, 1960), sheet and rill erosion respond directly to increased sand content. Using the percentage texture data for the soil associations of the Wye Catchment, clay ratios have been calculated to indicate areas of potential erosion.

2.2 Available Water Capacity

Available water capacity (the ability to hold a quantity of water between 0.05 and 15 bar suction in a metre of soil) for each soil association has been used to calculate average catchment AWC and the ratio of AWC to total volume of soil (AWCRAT) (Table 1). Field capacity data can be used to plan land drainage schemes, assess workability and trafficability of arable land and estimate potential grazing seasons. Dates for the return and end of field capacity can be adjusted for different soils to give estimates of machinery workdays in spring and autumn land work periods (Thomasson, 1982).

2.3 Poaching Risk

The grassland suitability classes based on the work of Harrod (1979) and Harrod and Thomasson (1980) assessed the balance between potential yield and risk of poaching (Figure 2). Restrictions similar to the trafficability by machinery apply to poaching by animals, therefore soils are classed on an increasing risk scale of 1 to 5. The yield category, involving soil water available to grass, is compared with maximum potential soil moisture deficit by four categories, based on a modified droughtiness class prepared by Thomasson (1979). The scheme identifies four grassland suitability classes, ranging from low to intense poaching risk. Using the generalised field capacity distribution and the land suitability for grassland, based on the soil associations, poaching risk can be mapped. This measure of grassland erosion potential can be examined in conjunction with stocking ratios, identifying areas at greatest risk.

3. AGRONOMIC PARAMETERS

Although soil factors have been used by some researchers, the effects of farming practice on sediment yield have been investigated more thoroughly. These include management practices (Chernyshev, 1972), land use cropping systems (Smith, 1958), compaction by tractors (Steinbrenner, 1955; Fullen, 1985) and compaction by animals (Alderfer and Robinson, 1947; Tanner and Mamanil, 1959; Federer et al., 1961; Lusby, 1970; Gifford and Hawkins, 1978; Mitchell, 1991).

Table 1. Average dependent and independent variables for the Wye catchments (1974 - 1980).

Variables	Ddol Farm WYE	Cilmery IRFON	Disserth ITHON	Erwood WYE	Belmont WYE	Butts Bridge LUGG	Lugwardine LUGG	Yarkhill FROME	Grosmont MONNOW	Redbrook WYE
DEPENDENT										
Suspended sediment yield (t.km^{-2})	17.07	34.59	38.98	28.50	35.25	35.25	26.79	12.02	36.14	32.73
INDEPENDENT										
Area (km^2)	174	244	358	1280	1900	371	886	144	357	4010
Pedological Particle size (A/O horizon):										
Clay %	22.19	21.96	23.66	22.22	19.74	19.60	19.33	25.19	21.23	19.63
Sand %	31.36	30.08	21.67	25.33	27.47	22.85	24.72	12.00	19.58	25.12
Silt %	40.66	41.31	46.69	45.32	46.44	50.47	49.71	61.92	56.98	49.91
Stones %	5.79	6.65	7.98	7.13	6.35	7.08	6.24	0.89	2.21	5.34
Clay Ratio a)	3.393	3.475	3.036	3.647	4.942	4.931	5.164	3.230	4.791	5.131
AWC mm b)	184	177	151	170	160	148	153	160	163	155
AWCRAT c)	0.226	0.234	0.168	0.216	0.204	0.172	0.178	0.237	0.230	0.200
Agronomic Arable/pasture ratio	0.038	0.300	0.059	0.044	0.091	0.327	0.379	0.509	0.248	0.209
Stocking Ratio d)	7.214	12.358	10.002	9.917	9.533	9.103	8.520	5.997	8.300	8.390

a) Clay ratio = $\dfrac{\% \, sand + \% \, silt}{\% \, clay}$

b) Available water capacity

c) AWCRAT = Ratio of AWC to total volume of soil

d) Stocking ratio = Total cattle and sheep per hectare of pasture

Figure 2. Influence of climate and soil moisture conditions on trafficability and poaching risk (from Harrod, 1979 and Harrod and Thomasson, 1980).

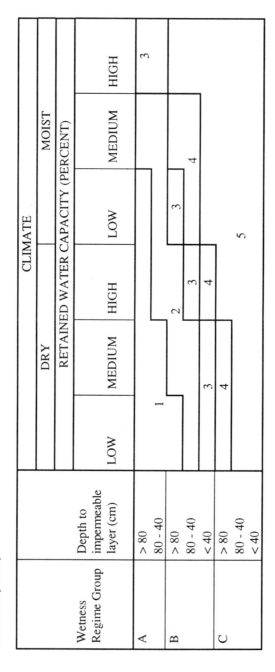

Trafficability:
1. Very high
2. High
3. Moderate
4. Low
5. Very low.

Poaching risk:
1. Very low
2. Low
3. Moderate
4. High
5. Very high

Retained water capacity classes:
High   > 45%
Medium 35 - 45%
Low   < 35%

Agronomic parameters were derived from the Ministry of Agriculture, Fisheries and Food's agricultural parish returns for 1977. The 318 parishes of the Wye Basin were sub-divided according to the catchment areas, with proportional allocation of data where parishes crossed inter-catchment boundaries. Despite some inaccuracies (Clark et al., 1983), the agricultural returns provided data for the calculation of arable/pasture ratios and stocking ratios (Table 1).

4. SUSPENDED SEDIMENT YIELDS OF CATCHMENTS OF THE WYE BASIN

The measurement of suspended sediment at selected stations in the Wye Basin has formed the basis of estimating catchment yields (Mitchell and Gerrard, 1987). Although various electronic devices have been used, rating curves constructed from point samples form the basis of estimating suspended sediment loads of rivers. As the movement of sediment in a river system is extremely complex, the precise measurement of load is subject to errors as reviewed by Walling (1977a; b) and Walling and Webb (1981). In order to minimise these errors, average suspended sediment yields were estimated using a multi-rating curve method, accounting for seasonal and stage tendency variations. These improvements in rating relationships are shown by the increase in correlation coefficients between general and winter rising-stage rating curves (Table 2). Annual yields for each catchment were generated using rating curves applied to a 15 minute discharge time base (Table 1). The yield of suspended sediments from the ten catchments can be compared with soil and agronomic parameters. The general relationships between the dependent and ten independent variables have been presented as a correlation matrix (Table 3).

5. DISCUSSION

5.1 Soil Particle Size

Although the range of proportions between the sub-catchments was narrow, the proportion of clay in the A/O horizon shows a distinct inverse relationship with suspended sediment yield, which accords with Hjulstrom (1935). The clay ratio for the A/O horizon indicates the importance of the availability of sand and silt sized materials. The spatial variation of the clay ratio (Figure 3) seems to have a significant influence on the suspended sediment yield, especially in the lowland catchments. Areas with the highest clay ratios (>8) are found in the lower parts of some catchments, in the vicinity of the gauging stations of Butts Bridge, Belmont and Redbrook and in the upper parts of the Monnow Catchment. The location of these potentially high sediment source areas in the vicinity of gauging stations can significantly influence estimated sediment delivery ratios (Walling, 1983). Availability of sediment between Hereford and Redbrook results in an unusual synchronisation of peak discharge with peak suspended sediment concentrations at Redbrook, the lowest gauging station.

Further insight into sediment dynamics can be made using rating curve constants and exponents (Muller and Forstner, 1968). Using the constants from sediment rating curves for rivers draining into the Severn Estuary, Brookes (1974) found good relationships both with catchment area and mean daily flow, but the exponents showed no relationships with any morphometric parameters. She found, instead, that the exponents correlated well with the mean size of bank material, which conforms with the principles of the erosional curve devised by Hjulstrom (1935).

Although the range in proportion of clay sized particles between catchments is small, the constants have a positive relationship ($r = 0.73$, significant at $p = 0.05$), while sand sized particles have an inverse relationship, with a poor correlation ($r = -0.64$, significant at $p = 0.05$). In contrast, the proportion of clay in the A/O horizon has a negative relationship with the exponents and a correlation coefficient of $r = -0.81$ (significant at $p = 0.01$) which accords with the suggestion that the exponent represents the intensity of erosion (Muller and Forstner, 1968). These tentative results are based on a small number of catchments, but with further results from basins with varying catchment characteristics, the interpretation and application of this approach may have some potential.

5.2 Available Water Capacity

Soil type and soil moisture conditions have a controlling influence on the yield of both water and sediment. Based on Soil Survey, an available water capacity map has been constructed for the Wye Catchment (Figure 4). Emphasising the spatial variation in the Wye Catchment, the greatest AWC (>250mm) occurs in both the peat soils of the upland plateaux and the alluvial soils of the lowland valleys. Theoretically, the areas with the lowest AWC and hence greater Hortonian overland flow, will tend to have increased erosion and greater sediment yield. Extensive areas with AWC less than 125mm are found in the Upper Lugg and Upper Ithon Catchments which, coupled with relief, must have a significant influence on sediment yield.

Although these upland catchments are in a partial rain-shadow, the lack of peat covering, as found in the more western, higher and wetter catchments, will have a distinct influence on the hydrological responses. Besides upland areas, large sections of middle and lower regions have low AWC, suggesting that high runoff in the lower section of catchments can supply large quantities of locally derived sediments, rather than this material being transported from upland areas. A significant proportion of the sediment yield at Redbrook may have been derived from the catchment area below Hereford, which helps to explain peak concentrations in suspended sediment coinciding with peak discharge, unlike other rivers draining into the Severn Estuary, with sediment concentrations lagging up to ten hours behind discharge peaks (Brookes, 1974). Although precision is lost in averaging AWC for whole catchments, a clear inverse relationship was found between AWCRAT (ratio of AWC to total volume of soil) and average suspended sediment yield.

222

Table 2. Constants, exponents and correlation coefficients for general and winter rising stage suspended sediment-discharge rating curves for the Wye catchments.

Catchment	Sampling Data		Suspended sediment-discharge rating curves					
			GENERAL			WINTER RISING STAGE		
	No. of samples	Max. discharge $(m^3.s^{-1})$	Constant	Exponent	r	Constant	Exponent	r
Wye at Ddol Farm	141	73	2.383	0.518	0.512	0.718	1.1013	0.683
Irfon at Cilmery	86	97	1.793	0.695	0.572	0.216	1.474	0.662
Ithon at Disserth	102	65	6.708	0.581	0.522	3.432	0.873	0.826
Wye at Erwood	144	457	8.987	0.496	0.606	0.445	0.963	0.739
Wye at Belmont	315	433	0.977	0.681	0.617	0.042	1.455	0.913
Lugg at Butts Bridge	123	29	5.271	0.894	0.738	1.972	1.611	0.860
Lugg at Lugwardine	138	52	1.833	1.046	0.829	0.760	1.505	0.934
Frome at Yarkhill	82	14	18.277	0.511	0.552	16.780	1.007	0.693
Monnow at Grosmont	94	133	4.090	0.836	0.768	2.906	1.057	0.885
Wye at Redbrook	299	374	0.436	0.885	0.726	0.074	1.320	0.890

Suspended sediment concentration = $a.Q^b$

r = correlation coefficient

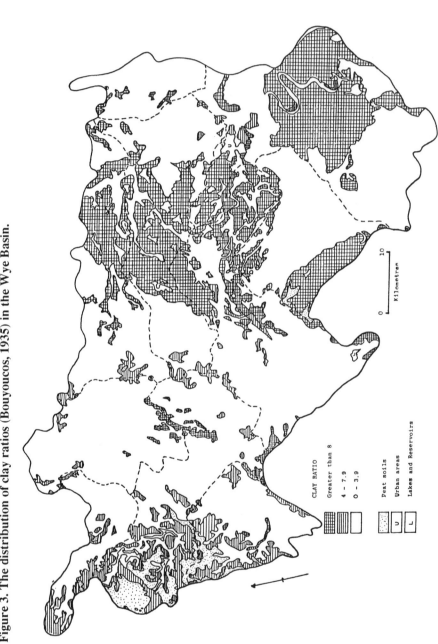

Figure 3. The distribution of clay ratios (Bouyoucos, 1935) in the Wye Basin.

CLAY RATIO

Greater than 8

4 - 7.9

0 - 3.9

Peat soils

Urban areas

Lakes and Reservoirs

0   10

Kilometres

Figure 4. Available water capacity between 0.05 and 15 bar suction in a metre of soil.

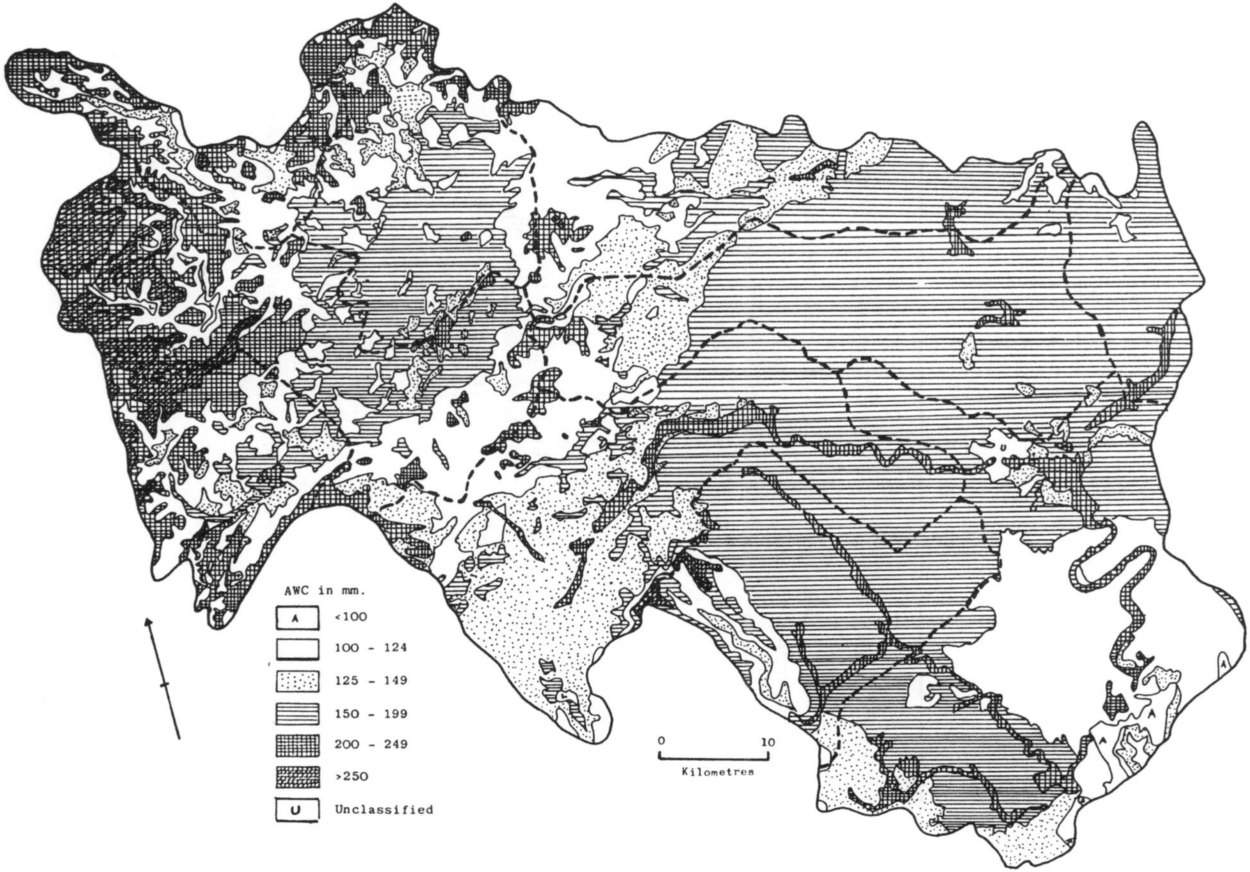

Table 3. Correlation matrix of dependent and independent variables of the Wye

Variables	1	2	3	4	5	6	7	8	9	10
1. Suspended sediment yield										
2. Area	0.229									
3. Clay %	-0.478	-0.510								
4. Sand %	0.250	0.190	-0.480							
5. Silt %	-0.249	-0.051	0.234	-0.936						
6. Stones %	0.525	0.124	-0.368	0.680	-0.809					
7. Clay ratio	0.369	0.537	-0.906	0.094	0.190	0.002				
8. AWC	-0.392	-0.243	0.286	0.530	-0.515	-0.051	-0.482			
9. AWCRAT	-0.490	-0.164	0.429	-0.039	0.122	-0.629	-0.345	0.749		
10. Arable/ pasture ratio	-0.399	-0.087	0.022	-0.760	0.854	-0.636	0.316	-0.549	-0.063	
11. Stocking ratio	0.719	0.010	-0.222	0.535	-0.621	0.681	-0.054	0.102	-0.166	-0.637

5.3 Stocking Ratio

As the Wye Basin is more noted as a stock farming region, the stocking ratios provide a more significant relationship with sediment yield than the arable/pasture ratio. The spatial distribution of the highest stocking ratios (Figure 5) emphasises the importance of the intermediate areas between the climatically harsh uplands and arable farming lowlands, forming a large part of the Irfon, Ithon, Monnow and Upper Lugg catchments. These densely stocked catchments, with moderately steep slopes are subjected to erosion due to close cropping of grass, poaching and river bank damage, resulting in high sediment yields (Thomas, 1964; Mitchell, 1979). Furthermore, if these areas of high stocking ratios possess soils of high poaching risk, erosion due to animals will be even greater. Using the categories proposed by Harrod (1979) and Harrod and Thomasson (1980) (Figure 2), a grassland poaching risk map of the Wye Catchment has been constructed (Figure 6). If this map is examined in conjunction with stocking ratios, the results help to explain the high sediment yield in particular catchments which hitherto appeared anomalous. In two parishes of the Farlow Catchment, the most significant feature in the changing land use was the large increase in stocking ratios since 1945, increasing from two to more than nine cattle and sheep per hectare of pasture. Erosion due to animals is becoming much more significant, especially in western Britain. Even in intermediate arable and mixed farming areas, large numbers of farm animals, especially sheep, are translocated during winter months, that is during the greatest 'erosion period'. This transhumance, particularly as practised in the Welsh Borderland, leads to increased erosion on even gentle slopes (Mitchell, 1991).

6. CONCLUSIONS

The availability of soil and agronomic data from the 1:250,000 Soil Association Map of England and Wales and MAFF's agricultural parish returns has meant that large basin surveys in Great Britain can use parameters which, hitherto, were restricted to small experimental catchments. The computation of parameters such as clay ratio, available water

Figure 5. Stocking ratios of the Wye catchment (cattle and sheep per hectare of pasture, including rough grazing).

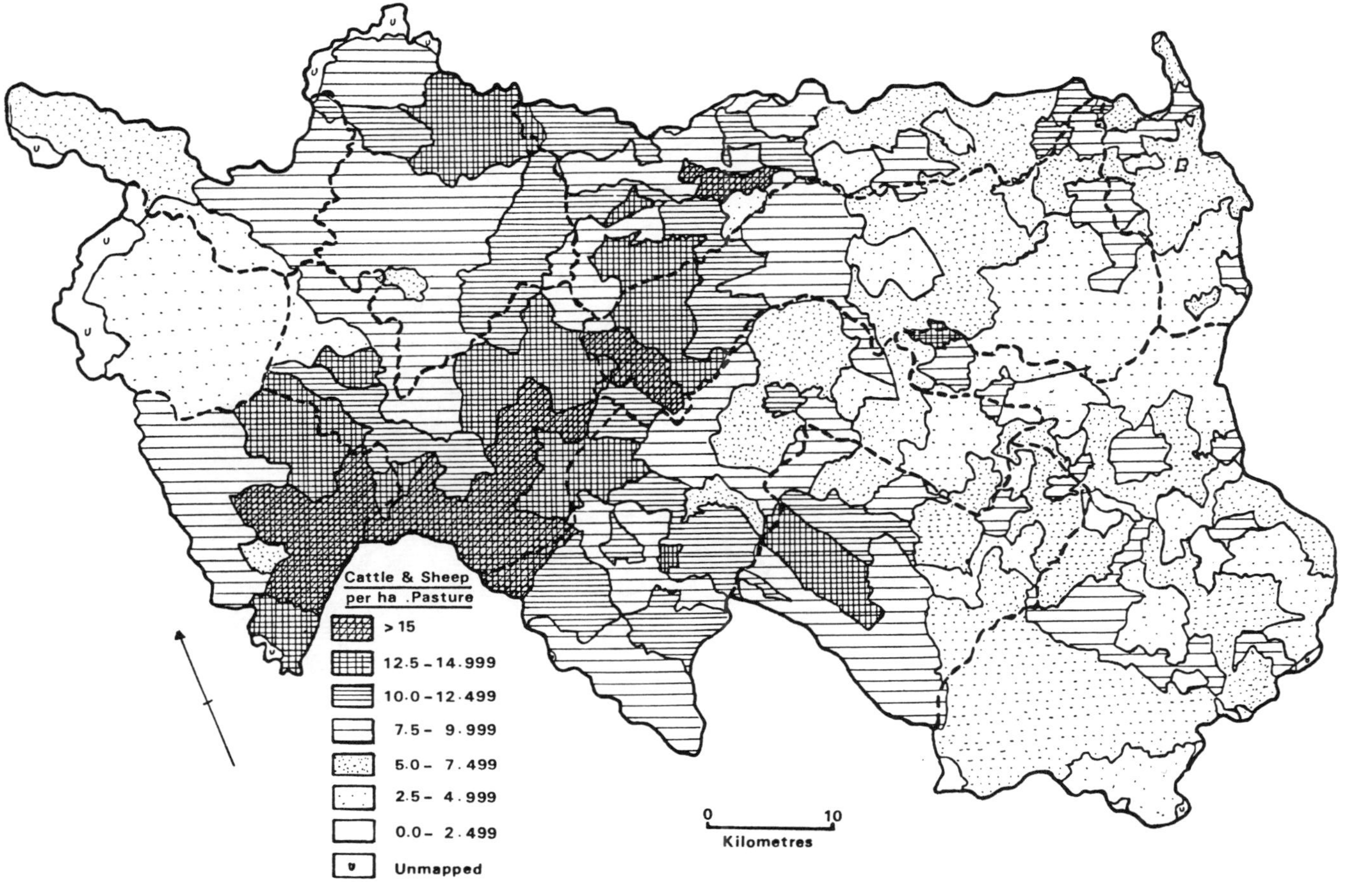

Cattle & Sheep
per ha .Pasture

>15
12.5 - 14.999
10.0 - 12.499
7.5 - 9.999
5.0 - 7.499
2.5 - 4.999
0.0 - 2.499
Unmapped

0   10
Kilometres

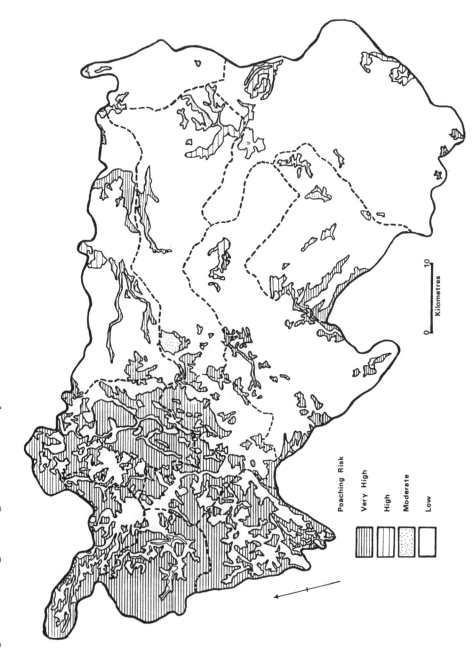

Figure 6. Poaching risk on grassland in the Wye catchment.

capacity, poaching risk, arable-pasture ratio and stocking ratio for the Wye Basin and nine sub-catchments provide useful insights to the origins of suspended sediment. Besides these parameters others, related to bank erosion and bed load, could be similarly derived. Further application of these parameters could form the basis of a national soil erosion potential map to complement the mean annual erosivity map of Great Britain (Morgan, 1980a; b).

ACKNOWLEDGEMENTS

The author is grateful to R. Hartnup of the Soil Survey and Land Research Centre, the Wye Division of the former Welsh Water Authority, the Ministry of Agriculture, Fisheries and Food and the Institute of Hydrology for providing data; the University of Wolverhampton for financial assistance and Dr. M.A. Fullen and Dr. A.J. Gerrard for advice and comments.

REFERENCES

ALDERFER, R.B. and ROBINSON, R.R. 1947. Runoff from pastures in relation to grazing intensity and soil compaction. Journal of American Society of Agronomy 39:948-958.

ANDERSON, H.W. 1970. Principal components analysis of watershed variables affecting suspended sediment discharge after a major flood. International Association of Hydrological Sciences 96:405-416.

ANDERSON, H.W. and WALLIS, J.R. 1965. Some interpretations of sediment sources and causes; Pacific Coast basins in Oregon and California. Proceedings Federal Inter-Agency Sedimentation Conference, USDA Misc. Publication 970:22-30.

BOUYOUCOS, G.J. 1935. The clay ratio as a criterion of susceptibility of soils to erosion. Journal American Society of Agronomy 27:738-741.

BROOKES, R.E. 1974. Suspended sediment and solute transport for rivers entering the Severn estuary. Unpublished PhD Thesis, University of Bristol.

BRYAN, R.B. 1968. The development, use and efficiency of indices of soil erodibility. Geoderma 2:5-26.

BURNHAM, C.P. 1964. The soils of Herefordshire. Transactions Woolhope Naturalist Field Club 38:27-35.

CHERNYSCHEV, Y.P. 1972. Hydrologic characteristics of soil erosion in the Central Chernozem Region. Soviet Hydrology selected papers 4:325-331.

CLARK, G., KNOWLES, D.J. and PHILLIPS, H.L. 1983. The accuracy of the Agricultural Census. Geography 68(2):115-26.

FARQUHARSON, F.A.K., MACKNEY, D., NEWSON, M.D. and THOMASSON, A.J. 1978. Estimation of runoff potential of river catchments from soil surveys. Soil Survey of England and Wales Special Survey No. 11. Harpenden.

FEDERER, C.A., TENPAS, C.B., SCHMIDT, D.R. and TANNER, C.B. 1961. Pasture soil compaction by animal traffic. Agronomy Journal 53:53-54.

FINDLAY, D.C., COLBORNE, G.J.N., COPE, D.W., HARROD, T.R., HOGAN, D.V. and STAINES, S.J. 1984. Soils and their use in South West England. Bulletin Soil Survey. Great Britain.

FLAXMAN, E.M. 1972. Predicting sediment yield in Western United States. Proceedings American Society Civil Engineers. Journal Hydraulic Division 98(HY12):2073-2085.

FLAXMAN, E.M. 1974. Predicting sediment yield in Western United States. In: Pacific Southwest Inter-Agency Committee, Report of Water Management Subcommittee, Erosion and Sediment Yield Methods.

FULLEN, M.A. 1985. Soil compaction, hydrological processes and soil erosion on loamy sands in East Shropshire. Soil and Tillage Research 6:17-29.

GIFFORD, G.F. and HAWKINS, R.H. 1978. Hydrologic impact of grazing on infiltration: a critical review. Water Resources Research 14:305-313.

GUYMAN, G.L. 1974. Regional sediment yield analysis of Alaska streams. American Society of Civil Engineers. Journal of Hydraulics Division 100(HY1):41-51.

HARROD, T.R. 1979. Soil suitability for grassland. In: Soil survey applications. Soil Survey Technical Monograph No. 13.

HARROD, T.R. and THOMASSON, A.J. 1980. Grassland suitability map of England and Wales 1:1M. Ordnance Survey, Southampton.

HINDALL, S.M. 1976. Prediction of sediment yields in Wisconsin streams. Proceedings 3rd Federal Inter-Agency Sedimentation Conference, 1.205-1.218.

HJULSTROM, F. 1935. Studies of the morphological activity of rivers as illustrated by the River Fyris. Bulletin Geological Institute, University of Uppsala 25:221-527.

HODGSON, J.M. 1972. Soils of the Ludlow district. Soil Survey of England and Wales. Harpenden.

HODGSON, J.M. 1974. Soil survey field handbook. Soil Survey Technical Monograph No. 5, Harpenden.

HODGSON, J.M. and PALMER, R.C. 1971. Soils in Herefordshire I. Sheet SO53 (Hereford South). Soil Survey Rec. No. 2, Soil Survey, Harpenden.

JANSEN, J.M.L. and PAINTER, R.B. 1974. Predicting sediment yield from climate and topography. Journal of Hydrology 21:317-380.

LARSON, F.H. and HALL, G.R. 1957. The role of sedimentation in watersheds. Proceedings American Society of Civil Engineers. Hydraulics Division. 83HY3:1-14.

LUSBY, G.C. 1970. Hydrologic and biotic effects of grazing versus non-grazing near Grand Junction, Colorado. US Geological Survey Professional Paper 700B.

MACKNEY, D. and BURNHAM, C.P. 1966. The soils of the Church Stretton district of Shropshire. Memoirs of the Soil Survey of Great Britain, Harpenden.

McPHERSON, H.J. 1975. Sediment yields from intermediate-sized stream basins in southern Alberta. Journal of Hydrology 25:243-257.

MITCHELL, D.J. 1979. Aspects of the hydrology and geomorphology of the Farlow Basin, Shropshire. Unpublished MSc Thesis, University of Birmingham.

MITCHELL, D.J. 1991. The use of vegetation and land use parameters in modelling catchment sediment yields. In: Thornes, J.B. (ed.). Vegetation and erosion. John Wiley, Chichester.

MITCHELL, D.J. and GERRARD, A.J. 1987. Morphological responses and sediment patterns. In: Gregory, K.J., Lewin, J. and Thornes, J.B. (eds). Palaeohydrology in practice: A river basin analysis. John Wiley, Chichester.

MORGAN, R.P.C. 1980a. Mapping soil erosion risk in England and Wales. In: De Boodt, M. and Gabriels, D. (eds). Assessment of erosion. Wiley, Chichester.

MORGAN, R.P.C. 1980b. Soil erosion and conservation in Britain. Progress in Physical Geography 4(1):24-47.

MULLER, G. and FORSTNER, U. 1968. General relationship between suspended sediment concentration and water discharge in the Alpenrhein and some other rivers. Nature 217:244-5.

NATURAL ENVIRONMENTAL RESEARCH COUNCIL 1975. Flood Studies Report, London, NERC.

PALMER, R.C. 1972. Soils in Herefordshire III. Sheet SO34 (Staunton-on-Wye). Soil Survey Rec. No. 11.

RAGG, J.M., BEARD, G.R., GEORGE, H., HEAVEN, F.W., HOLLIS, J.M., JONES, R.J.A., PALMER, R.C., REEVE, M.J., ROBSON, J.D. and WHITFIELD, W.A.D. 1984. Soils and their use in Midland and Western England. Bulletin. Soil Survey, G.B.

ROSE, C.W. 1960. Soil detachment caused by rainfall. Soil Science 89:1,28-35.

RUDEFORTH, C.C. and THOMASSON, A.J. 1970. Hydrological properties of soils in the River Dee Catchment. Soil Survey Special Survey No. 4, Harpenden.

RUDEFORTH, C.C., HARTNUP, R., LEA, J.W., THOMPSON, T.R.E. and WRIGHT, P.S. 1984. Soils and their use in Wales. Bulletin, Soil Survey, G.B.

SMITH, D.D. 1958. Factors affecting rainfall erosion and their evaluation. International Association of Scientific Hydrology. Pub. 43:97-107.

STEINBRENNER, E.C. 1955. The effect of repeated tractor trips on the physical properties of forest soils. Northwest Science 29:155-159.

STRIFFLER, W.D. 1965. Suspended sediment concentrations in a Michigan trout stream as related to watershed characteristics. In: Proceedings of the Federal Inter-Agency Sedimentation Conference (Agricultural Research Service Misc. Publication No. 970). US Department of Agriculture, Washington DC.

TANNER, C.B. and MAMARIL, C.P. 1959. Pasture soil compaction by animal traffic. Agronomy Journal 51:329-331.

THOMAS, T.M. 1964. Sheet erosion induced by sheep in the Plynlimon area, Mid-Wales. British Geomorphological Research Group Occasional Publication 2:11-14.

THOMASSON, A.J. 1979. Assessment of soil droughtiness. In: Jarvis, M.G., and Mackney, D. (eds). Soil survey applications. Soil Survey Technical Monograph 13:43-50.

THOMASSON, A.J. 1982. Soil and climatic aspects of workability and trafficability. The 9th Conference International Soil Tillage Research Organisation, Osijek, Yugoslavia.

WALLING, D.E. 1977a. Assessing the accuracy of suspended sediment rating curves for a small basin. Water Resources Research 13:531-538.

WALLING, D.E. 1977b. Limitations of the rating curve technique for estimating suspended sediment loads, with particular reference to British rivers. IAHS Publication 122:14-38.

WALLING, D.E. 1983. The sediment delivery problem. Journal of Hydrology 65:209-237.

WALLING, D.E. and WEBB, B.W. 1981. The reliability of suspended load data. Erosion and sediment transport measurement. Proceedings Florence Symposium. IAHS Publication 133:177-194.

WHITFIELD, W.A.D. 1971. Soils in Herefordshire I. Sheet SO52 (Ross-on-Wye west). Soil Survey Rec. No. 3.

WISCHMEIER, W.H. and MANNERING, J.V. 1969. Relation of soil properties to its erodibility. Soil Science Society of America Proceedings 33:131-137.

Chapter 23

IMAGING SPECTROSCOPY, GEOSTATISTICS AND SOIL EROSION MODELS

S.M. de Jong and H.Th. Riezebos
Department of Physical Geography, University of Utrecht,
PO Box 80.115, 3508 TC Utrecht, the Netherlands

Summary

In the last few decades a shift has taken place in erosion studies from lumped, empirical models towards more distributed and (quasi)-deterministic ones. Input maps for these models are generally derived from the conventional choropleth soil maps by digitising mapping units. However, the conventional choropleth map often comprises large lumped mapping units with a high, mostly unknown 'within-unit' variance. Hence, the accuracy of these maps is generally not sufficient to be applied to the latter type of models. Instead, accurate 'continuous maps' are required for each of the necessary model parameters.

This paper describes two possible methods to obtain continuous maps of some soil properties. One method follows a geostatistical approach, the other method uses imaging spectroscopy, a new and advanced remote sensing technique. Both methods are illustrated with examples from a study area in southern France.

It is concluded that the geostatistical approach yields accurate continuous maps, but the method is very laborious and expensive. Results of experiments with airborne imaging spectroscopy show that a limited number of soil properties can be identified by means of absorption features in the shortwave infrared part of the soil spectra. In combination with a-priori knowledge of the soil types in the region, these soil properties can be used to assess the spatial distribution of other relevant soil parameters. However, airborne imaging spectroscopy can only be applied successfully in areas lacking in dense vegetation cover. Furthermore, this new and advanced technique still suffers from a number of technical imperfections.

1. INTRODUCTION

Recent developments in computer technology have provided new tools to study and tackle environmental problems. Nowadays large amounts of data can be analysed and processed in a very short time. The computer has become a multi-purpose tool for studying and modelling our environment. Apart from computers, multi-spectral airborne and spaceborne images are now alsó at our disposal. Since the early 1970s satellites have scanned the earth and rendered digital images of several wavelengths ranging from the visible part of the spectrum to the middle infrared.

At the same time, developments in the field of erosion studies led to the design of new models with a large number of parameters to be processed and a large number of calculations to be performed. Examples of these models are ANSWERS (Beasley et al., 1980) and WEPP (Foster, 1987). These models are capable of simulating the erosion process in catchments and can be used to predict the effect of land management or land cover change upon erosion intensity. This new generation of erosion models is often linked to a Geographical Information System (GIS) thus simplifying the analysis, storage, processing and presentation of spatial data (Burrough, 1986; Raper, 1989). Examples of erosion modelling studies using a GIS can be found in De Roo et al. (1989), De Jong and Riezebos (1992) and De Jong and Riezebos (1988).

The conventional procedure of putting data into a Geographical Information System (GIS) is by digitising existing choropleth maps of physical attributes such as soils, geology and vegetation. The next step is to derive specific model input parameters (such as texture, organic matter content and hydraulic conductivity) from the mapping units, maintaining the original unit boundaries and ignoring 'within-unit' variance. If maps of all required properties are available in digital format, the erosion model can be run to predict annual soil losses, for example. Normally no attention is paid to the accuracy of the input maps and hence the accuracy of the resulting erosion risk map. Since the units of choropleth maps often have a high, mostly unknown 'within-unit' variance (Burrough and Heuvelink, 1992; Burrough, 1989, Burrough, 1986; Marsman and Gruijter, 1984; Beckett and Webster, 1971), it is not possible to assess the accuracy of the final product by means of error propagation.

The conclusion to be drawn is that erosion models are used to produce quantitative maps of predicted soil losses, yet rather inaccurate maps are generally used as input for these models.

More accurate input maps ('continuous maps') that take account of spatial variability are required to run such models properly. Two approaches can be followed to acquire such maps:

1. Geostatistics can be used to develop optimal sampling schemes for field surveys. Interpolation techniques are then applied to produce maps showing for each individual point estimated parameter values with a known error.
2. Remote sensing techniques can be used to map the spatial distribution of a number of soil properties, which can be used as indicators to assess values for other soil variables. A new instrument, the high resolution imaging spectrometer, has the potential to provide this kind of information.

Both techniques were applied to a study area in the Ardèche province in southern France.

2. THE STUDY AREA

The study area is located in the Bas-Vivarais, the southern part of the Ardèche province in France. The area is characterised by a Mediterranean climate with dry, hot summers and heavy rainstorms in spring and autumn. The soils are formed in Jurassic and Cretaceous carbonate rocks such as marls, calcareous marls and pure limestone (Bornand et al., 1977; BRGM, 1989).

Five soil types can be distinguished in the study area (Riezebos et al., 1990): Lithic Leptosol, Rendzic Leptosol, Calcaric Regosol, Vertic Cambisol and Eutric Regosol. The vulnerability of the five soil types to erosion varies with topographic position and with differences in organic matter content and texture. Common land degradation features in the area are rill and gully erosion and soil surface crusts. These erosion features occur mainly in the marlstone area or at the intermediate zones between limestone and marl.

The natural vegetation of the limestone plateaux consists of the so-called Mediterranean mixed forest, which is dominated by several drought resistant oak species. The marl areas and lithological transition zones are characterised by garrigue and rangelands. Garrigue is a degraded, open type of forest with many sclerophyllous shrubs (Di Castri et al., 1981). The rangelands and garrigues are extensively used for grazing. Agricultural land use comprises orchards, vineyards and some cereals.

3. THE GEOSTATISTICAL APPROACH

Geostatistics constitute a theory about the statistical behaviour of natural phenomena with a spatial variability (Davis, 1986). The spatial variation of any variable can be described with three major components (Burrough, 1986):

1. A deterministic component, e.g. a constant trend or constant average;
2. A stochastic component, which can be described by its statistical properties;
3. A random, spatially independent component.

The second and third component can be captured in the semivariance, a function between interval distance of sampling points and their variance.

$$\gamma^2 = \frac{1}{2}n \sum_{i=1}^{n} [z(x_i) - z(x_i + h)]^2$$

γ^2	: semivariance;
n	: number of sample pairs;
h	: sampling distance interval;
(x_i, x_i+h)	: pair of sample points at distance h

The semivariance can be plotted against the sampling distance in a so-called semivariogram. A semivariogram can be defined by three parameters: the nugget, the range and the sill. The range gives the maximal distance of spatial dependency, the nugget variance defines the very short distance variance and measuring errors, and the sill variance is comparable with the total variance in the data.

A geostatistical mapping approach may comprise the following steps: 1. The spatial variability of variables is assessed by nested analysis of variance; 2. The study area is sampled on a regular grid; the distance between the sampling points is based upon the results of step 1. Semi-variograms are calculated for each considered variable; and 3. Interpolation techniques such as Kriging are used to produce continuous maps of the study area. The accuracy of the final map can be assessed using the Kriging variance map.

Nested analysis of variance uses several levels of sampling distances (e.g. 4, 40, 400 and 1000m). An abrupt increase of variance between a specific sampling distance indicates a spatial structure in this range. The principles of this technique are described by Webster (1977) and Oliver and Webster (1986). The calculated distance of spatial variation serves as a guide value for step 2. The relevant variables are mapped on a regular grid, the distance of the grid points are determined in step 1. For each variable the semivariogram is calculated and the three components - nugget, sill and range are determined. Interpolation techniques such as Kriging (Isaaks and Srivastava, 1989) can now be applied to calculate a continuous map from the field point samples. Kriging is a weighted average interpolation method, which is based on the three components captured in the semivariogram. The weights that are assigned to respective sample points depend on the semivariances associated with the distances of these observations to the points to be interpolated. The resulting map of the interpolation has a minimal and known error.

A geostatistical mapping approach yields continuous input maps with a known variance. This variance makes it possible to assess the accuracy of the final product, e.g. the predicted soil loss map. The geostatistical approach was applied to a number of relevant erosion variables in the study area in the Bas-Vivarais (Riezebos, 1989; Van Beurden and Riezebos, 1988; De Jong and Riezebos, 1989).

Figures 1 (a and b) shows an example of the geostatistical approach: organic matter content of the soil is determined at a regular sampling distance of 150m. The upper map shows the Kriging map of organic matter content, the lower map shows the Kriging variance map belonging to the organic matter map. Table 1 shows the calculated range of values for a number of variables in the study area in the Bas-Vivarais. Sampling distances range from 15m to over 400m. Apparently, the spatial variation of erosion relevant parameters is very high. This implies that the distance between the individual sampling points must be small and the sampling distance differs per variable. For a small study area of 5 × 5km, for example, and a range distance of 100m, 50 × 50 = 2500 sampling points are required to calculate a continuous map using Kriging for one parameter. Disadvantages of the described geostatistical approach can be summarised as:

1. Two field surveys are needed. The first is required to perform the nested analysis of variance to assess the optimal sampling distance; the second field survey is required to map the study area according to a sampling grid based on the optimal distance.
2. The spatial variability of natural phenomena appears to be very high. Consequently, the range distance is small and the number of required field samples is not feasible in time or money.

Figure 1a. Kriging map of organic matter content.

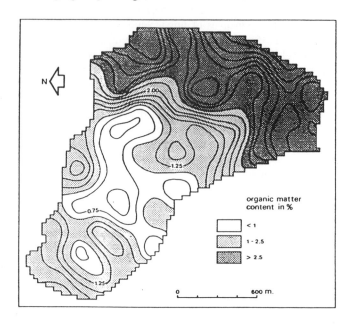

Figure 1b. Kriging variance map belonging to the organic matter map.

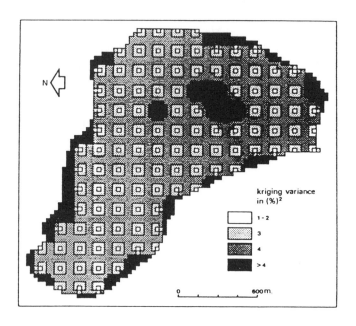

Although the geostatistical approach yields accurate maps of soil variables, the method does not seem feasible in view of the required efforts for field surveys, especially when labour costs are high. Further study of alternative methods to produce accurate input maps for models, which are less laborious, is a necessity.

Table 1. Some examples of calculated 'range values' for the study area in southern France.

Saturated hydraulic conductivity	150m
Percentage silt of top soil	250m
Base saturation	>400m
Porosity of top soil	250m
Bulk density	110m
Organic matter content of top soil	125m
Lime content of top soil	80m
Vegetative cover of a rangeland	25m

4. THE IMAGING SPECTROMETER

Several studies showed the usefulness of satellite images to produce accurate vegetation maps for areas in the Ardèche (Hill and Mégier, 1986; De Jong and Riezebos, 1991). The frequency of data acquisition, the large area covered by one scene and the digital format of the images make them ideal as input maps for distributed erosion and runoff models.

Regarding soil properties, the satellite data have so far proven less successful (Irons et al., 1989; Mulders, 1987). However, new and promising equipment is planned for the next generation of satellites. One of these new instruments is the imaging spectrometer. This instrument might improve the suitability of digital images for soil survey significantly. A spectrometer acquires contiguous reflectance information in many, narrow bands (Goetz et al., 1985). These contiguous spectra offer possibilities to identify surface minerals and soil constituents by means of absorption features in the spectra. Each absorption feature is typical for a specific constituent. A prerequisite to use this technique for soil mapping is that two conditions are met:

1. the soil is not covered by dense vegetation;
2. the soil types can be identified by one of the properties causing an absorption feature in the spectral curve.

In 1989, as an extension of the EISAC campaign, some experimental airborne spectroscopical images were acquired for a test site affected by erosion in the Ardèche in southern France. The properties of the GER-spectrometer are summarised in Table 2 and Figure 2. This project was initiated and financed by the Institute for Remote Sensing Applications (IRSA) of the Joint Research Centre (JRC) in Italy (Hill and Mégier, 1991; Hill, 1990).

Table 2. Technical specification of the 3 spectrometers of the GER-scanner.

Wavelength region	# Bands	Spectral Resolution
477-847nm	31	12.5nm
1440-1800nm	4	120nm
2005-2443nm	28	16nm

GIFOV at 3000m: appr. 10m
Scan angle: 90°

Figure 2. Typical spectral reflection curves for vegetation, soil and water and the position of the LANDSAT Thematic Mapper bands, the bands of the airborne spectrometer (GER) and the bands of the laboratory spectrometer.

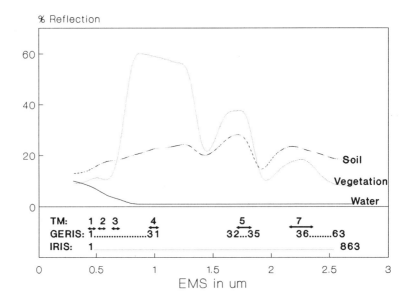

4.1. Laboratory Results

In order to interpret the spectroscopical images, a laboratory study was carried out in which samples of different soil types were taken to measure their spectral behaviour with an IRIS laboratory spectrometer (De Jong, 1991; 1992). The position of the spectral bands of the airborne spectrometer (GER), the laboratory spectrometer IRIS and of the Thematic Mapper satellite are shown in Figure 2.

Some laboratory results for pure limestone, a Vertic Cambisol and a Lithic Leptosol are shown in Figures 3 (a-c). The reflectance curves of both soils show absorption features at 2200nm, resulting from Al-OH minerals (Huntington et al., 1989), which are absent in the spectrum of limestone. The minerals in the present case are probably illite and kaolinite (Negendank et al., 1990). Furthermore, absorption features can be seen in Figures 3 (a - c) at 2350nm in the case of limestone and the Lithic Leptosol, due to absorption of carbonate (Hunt and Salisbury, 1971). Because of its low lime contents, the absorption feature is absent in the spectrum of the Vertic Cambisol (Fig. 3b). Some broad absorption features are recognised in the spectrum of the Vertic Cambisol in the wavelength region 1000 to 1100nm (De Jong, 1992). Absorption features at this position indicate the presence of iron (Hunt and Salisbury, 1970).

From these laboratory analyses, it is concluded that it is feasible to distinguish two soil types: Vertic Cambisols and the group of Leptosols based on the carbonates absorption features located in the range 2330-2350nm and based on iron in the range of 1000-1100nm.

4.2. Airborne Image Results

Figures 4 (a-c) shows results derived from the airborne image. Pixel-per-pixel comparison of spectra reveals a very noisy picture and shows no clear absorption features (Hill, 1990). Therefore, reflection values of bare surfaces are averaged for 10 pixels. The Lithic Leptosol shows an absorption feature at the position of carbonates (2350nm), which is absent in the Vertic Cambisol. Both soil types show absorption features at the 2200nm position (Al-OH minerals e.g. kaolinite and illite). The absorption features for iron (1000-1100nm) are outside the range of the GER-scanner and cannot be seen in Figure 4.

The limestone spectrum looks very noisy. Since the absorption features are not very pronounced the possibilities of mineral identification from airborne spectroscopical images seem rather limited so far. Two main causes can be identified, the non-purity of the pixel and technical restrictions of the equipment.

The spectral reflection curve of a pixel is an average response of the elements present in the pixel. Due to the 'within pixel variance' caused by numerous vegetation species, bare soil and rock outcrops, the spectral signature becomes indistinct. Several authors have discussed this problem of scale and sample size (Webster et al., 1989; Woodcock et al., 1988). This effect will play an important role in patchy Mediterranean landscapes and is part of a present study carried out in the Ardéche.

The technical problems of the GER spectrometer during the Ardéche flight are thought to be of great relevance and are described in detail by Hill (1990). Most important technical problems of the system are:

- band width deterioration up to 50nm causing smoothing or even extinction of the absorption features;
- moderate to bad signal-to-noise ratio ranging for the short wave infra-red range from 25:1 to about 10:1.

The poor signal-to-noise ratio makes it difficult to judge whether the images are suitable for recognition of certain soil properties.

Figure 3a. Laboratory spectra in SWIR of pure limestone.

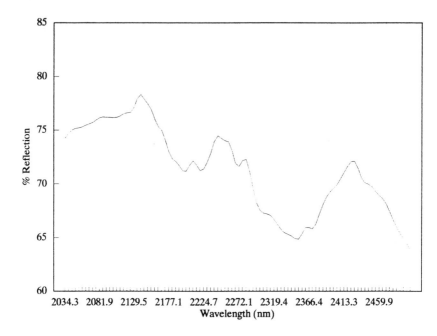

Figure 3b. Laboratory spectra in SWIR of a Vertic Cambisol.

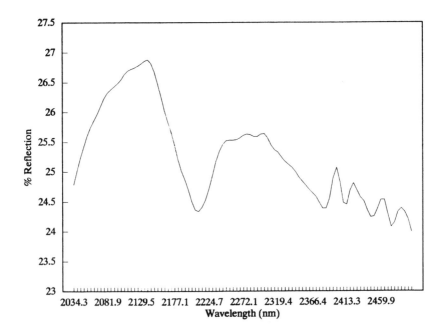

Figure 3c. Laboratory spectra in SWIR of a Lithic Leptosol.

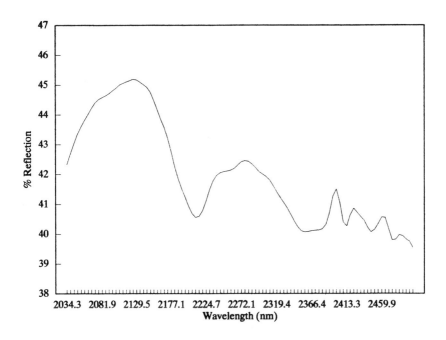

Figure 4a. GERIS spectra in SWIR of pure limestone.

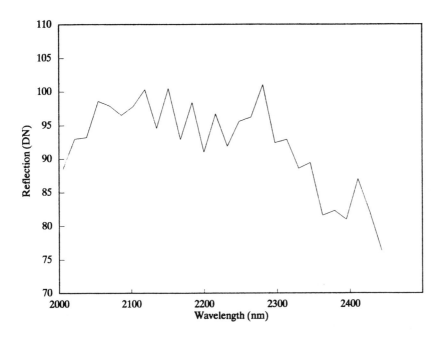

Figure 4b. GERIS spectra in SWIR of a Vertic Cambisol.

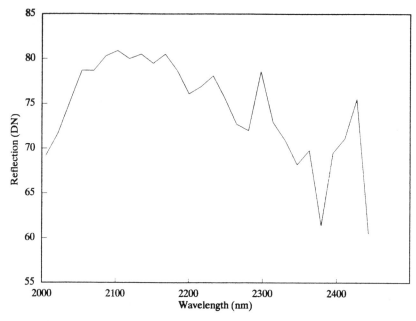

Figure 4c. GERIS spectra in SWIR of a Lithic Leptosol.

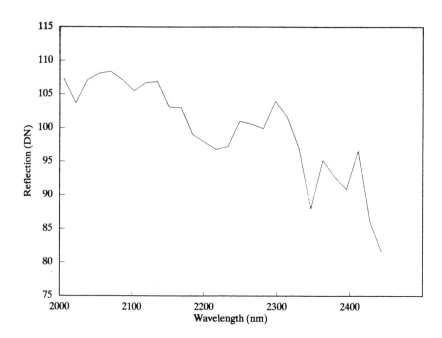

5. DISCUSSION AND CONCLUSIONS

Two methods are presented, which can be used to map the spatial distribution of soil variables. The geostatistical approach yields the most accurate maps with a known error but this method is very laborious and costs are high. The second method uses an imaging spectrometer. Absorption features theoretically give an indication of some soil constituents per pixel. Laboratory results are promising to detect differences of the soil contents of carbonate, iron and Al-OH minerals. However, the possibilities of the airborne images to identify a number of soil constituents proved to be less successful. The poor results of the airborne image may be caused to a large extent by technical limitations such as band width deterioration and a poor signal-to-noise ratio. Band width deterioration has a smoothing effect on absorption features. Moreover, a poor signal-to-noise ratio hinders recognition of the absorption feature. It is concluded that geostatistical survey yields the best results so far but that the imaging spectroscopy might be a useful alternative approach to map a number of soil properties, if certain technical limitations can be overcome.

ACKNOWLEDGEMENTS

The authors wish to thank the IRSA staff of the Joint Research Centre in Italy for their support and for putting the digital airborne images at our disposal for analysis. Especially, the efforts of Mr. J. Hill, Mr. J. Mégier and Mr. G. Maracci were indispensable.

REFERENCES

BEASLEY, D.B., HUGGINS, L.F. and MONKE, E.J. 1980. ANSWERS: a model for watershed planning. Transactions of the ASAE 23,4:938-944.

BECKETT, H.T. and WEBSTER, R. 1971. Soil variability: a review. Soils and Fertilizers 34.1

BORNAND, M., LEGROS, J.P. and MOINEREAU, J. 1977. Carte pédologique de France a moyenne echelle: Privas N-19. Centre National de Recherche Agronomiques, Versailles.

BRGM 1989. Carte Géologique de la France à 1:50,000. Nr. 888 Bessèges. Bureau de Recherche Géologiques et Minières, Orléans.

BURROUGH, P.A. 1986. Principles of geographical information systems for land resources assessment. Clarendon Press, Oxford.

BURROUGH, P.A. 1989. Fuzzy mathematical methods for soil survey and land evaluation. J. Soil Science 40:447-492.

BURROUGH, P.A. and HEUVELINK, G.B.M. 1992. The sensitivity of Boolean and continuous (fuzzy) logical modelling to uncertain data. Proc. 3rd Eur. Conf. GIS (EGIS '92), Munich, 23-27 March. pp. 1032-1041.

DAVIS, J.C. 1986. Statistics and data analysis in geology. John Wiley, New York.

DE JONG, S.M. 1991. Monitoring soil crusts using imaging spectroscopy in the Department Ardèche, France. Proc. IGARSS'91, 3-8 June, Helsinki. pp.1187-1189.

DE JONG, S.M. 1992. The analysis of spectroscopical data to map soil types and soil crusts of Mediterranean eroded soils. Soil Technology 5. In press.

DE JONG, S.M. and RIEZEBOS, H.TH. 1988. Erosiekartering met behulp van GIS: een Voorbeeld uit het Departement Ardèche, Frankrijk. Netherlands Geographical Studies 63, KNAG Utrecht. pp.45-53.

DE JONG, S.M. and RIEZEBOS, H.TH. 1989. Geostatistics in soil erosion studies. Proc. of Workshop on Integration of Remote Sensing and PC-Based Geographical Information Systems in the Management of European Marginal Land. 6-7 Sept. JRC Ispra, Italy. p.6.

DE JONG, S.M. and RIEZEBOS, H.TH. 1991. Use of a GIS database to improve multispectral image classification accuracy. Proc. 2nd Eur. Conf. GIS (EGIS'91). 2-5 April, Brussels. pp. 503-508.

DE JONG, S.M. and RIEZEBOS, H.TH. 1992. Assessment of erosion risk using multi-temporal remote sensing data and an empirical erosion model. Proc. 3rd Eur. Conf. GIS (EGIS'92). Munich, 23-27 March. pp. 893-900.

DE ROO, A.P.J., HAZELHOFF, L. and BURROUGH, P.A. 1989. Soil erosion modelling using 'ANSWERS' and Geographical Information Systems. Earth Surface Processes and Landforms 14:517-532.

DI CASTRI, F., GOODALL, D.W. and SPECHT, R.L. 1981. Ecosystems of the world 11: Mediterranean-type shrublands. Elsevier, Amsterdam.

FOSTER, G.R. 1987. User requirements. USDA-Water Erosion Prediction Project (WEPP). Draft 6.2:43 S; West Lafayette (Nat. Soil Erosion Lab.).

GOETZ, A.F.H., VANE, G., SOLOMON, J.E. and ROCK, B.N. 1985. Imaging spectrometry for earth remote sensing. Science 228:1147-1153.

HILL, J. 1990. Analysis of GER imaging spectrometer data acquired during the European Imaging Spectrometry Aircraft Campaign (EISAC'89). Proc. 10th EARSeL Symp. Toulouse, France. pp.35.

HILL, J. and MÉGIER, J. 1986. Rural land use inventory and mapping in the Ardèche Area. Improvement of automatic classification by multi-temporal analysis of TM data. Proc. EARSeL Symp., 25-28 June, Lyngby, Denmark. pp.75-85.

HILL, J. and MÉGIER, J. 1991. The use of imaging spectrometry in Mediterranean land degradation and soil erosion hazard assessments. Proc. 5th Int. Colloq. Physical Measurements and Signatures in Remote Sensing. Courchevel, France, 14-18 Jan. p.185-188.

HUNT, G.R. and SALISBURY, J.W. 1970. Visible and near infrared spectra of minerals and rocks: I. Silicate minerals. Modern Geology 1:283-300.

HUNT, G.R. and SALISBURY, J.W. 1971. Visible and near infrared spectra of minerals and rocks: II. Carbonates. Modern Geology 2:23-30.

HUNTINGTON, J.F., GREEN, A.A. and CRAIG, M.D. 1989. Identification, the goal beyond discrimination; the status of mineral and lithological identification from high resolution spectrometer data: examples and challenges. Proc. IGARSS'89 Symp. 10-14 July, Vancouver. p.6-11.

IRONS, J.R., WEISMILLER, R.A. and PETERS, G.W. 1989. Soil reflectance. In: G. Asrar (ed.). Theory and applications of optical remote sensing. John Wiley, New York. p.66-101.

ISAAKS, E.H. and SRIVASTAVA, R.M. 1989. An introduction to applied geostatistics. Oxford University Press, New York.

MARSMAN, B. and GRUIJTER, J.J. 1984. Dutch soil survey goes into quality control. In: Burrough,P.A. and Bie,S.W. (eds.). Soil information systems technology. Pudoc, Wageningen. pp. 127-34.

MULDERS, M.A. 1987. Remote sensing in soil science. Developments in Soil Science 15. Elsevier, Amsterdam.

NEGENDANK, J.F.W., BAUMANN, H., SIEMANN, H. and BRUCKENER, H.B. 1990. Geological mapping and mineralogical analysis of surface samples for soil erosion assessment with high resolution spectroscopy data (Ardèche). Final report study contr. 3790 89 08 ED ISP, University of Trier, Germany.

OLIVER, M.A. and WEBSTER, R. 1986. Combining nested linear sampling for determining the scale and form of spatial variation of regionalized variables. Geographical Analysis 18:227-242.

RAPER, J. (ed.). 1989. Three dimensional applications in Geographical Information Systems. Taylor & Francis, New York.

RIEZEBOS, H.TH. 1989. Application of nested analysis of variance in mapping procedures for land evaluation. Soil Use and Management 5.1:25-30.

RIEZEBOS, H.TH (ed.), DE JONG, S.M., VAN HEES, J.C. and HAEMERS, P.B.M. 1990. Physiographic and pedologic mapping for erosion hazard assessment (Ardèche Test Site). Final report, Contr. No. 3787-89-08 ED ISP NL, JRC-Ispra, Italy.

VAN BEURDEN, S.A.H.A. and RIEZEBOS, H.TH. 1988. The application of geostatistics in erosion hazard mapping. Soil Technology 1:349-364.

WEBSTER, R. 1977. Quantitative and numerical methods in soil classification and survey. Clarendon Press, Oxford.

WEBSTER, R., CURRAN, P.J. and MUNDEN, J.W. 1989. Spatial correlation in reflected radiation from the ground and its implications for sampling and mapping by ground-based radiometry. Remote Sensing of Environment 29:67-78.

WOODCOCK, C.E., STRAHLER, A.H. and JUPP, D.L.B. 1988. The use of variograms in remote sensing: II. Real digital images. Remote Sensing of Environment 25:349-379.

Chapter 24

VALIDATING THE ANSWERS SOIL EROSION MODEL USING [137]Cs

A.P.J. De Roo[1] and D.E. Walling[2]

[1]Department of Physical Geography, University of Utrecht, PO Box 80.115, 3508 TC Utrecht, the Netherlands

[2]Department of Geography, University of Exeter, Amory Building, Rennes Drive, Exeter, Devon, EX4 4RJ, UK

Summary

The ANSWERS model has been evaluated in several catchments in the loess area of Limburg (the Netherlands). To test the model under different conditions, it has been used in the Yendacott catchment, in Devon, UK. There, surface runoff and soil erosion have been measured and simulated. Field observation and hydrograph analysis confirm that saturation overland flow processes are of major importance. As a consequence, it is inevitable that the original ANSWERS model yields poor results, because the model is based on the Horton theory of storm runoff generation. The variable contributing area concept of Beven and Kirkby (1979) has been introduced into the model, using GIS software to calculate the ln(A/tanβ) parameter and the initial lambda value. The modified model results, with lambda calibrated for each storm, are improved, but the hydrograph still cannot be simulated closely. This may be because other important sources of water such as subsurface lateral flow caused by perched water tables or macro pores, are not represented, and groundwater flow is poorly simulated. The model is validated in a distributed way, using [137]Cs. Correlations of simulated soil loss with observed ^{137}Cs values are low, but statistically significant. This is not unexpected since it is difficult to simulate the hydrograph properly. It is concluded that quantitative spatial modelling of soil erosion still involves large percentage errors, especially when the model is used in areas other than those for which it was originally developed and tested. Although this type of modelling is useful for the estimation of spatial effects of soil conservation scenarios for example, there is a substantial risk of gratuitous application by users such as consultancy agencies.

1. INTRODUCTION

Surface runoff and soil erosion have been measured and simulated in a small catchment in Devon, UK. As a development of work undertaken in the Etzenrade catchment (De Roo, 1991; De Roo and Riezebos, 1992; De Roo et al., 1992) and the Catsop catchment (De Roo et al., 1989) in south Limburg (the Netherlands), the ANSWERS hydrological and soil erosion model (Beasley et al., 1980; Beasley and Huggins, 1982; Beasley et al., 1982) has been evaluated, tested and validated. The main reason why ANSWERS has been used in the Devon catchment is the evaluation of the model under different conditions.

2. THE ANSWERS MODEL

The ANSWERS model is an example of a distributed model linked to a GIS. The distributed parameter model <u>ANSWERS</u> (<u>A</u>real <u>N</u>onpoint <u>S</u>ource <u>W</u>atershed <u>E</u>nvironment <u>R</u>esponse <u>S</u>imulation), developed by Beasley et al. (Beasley et al., 1980; Beasley and Huggins, 1982), was used in this study for modelling surface runoff and soil erosion. The ANSWERS model is designed to simulate the hydrological behaviour of catchments having agriculture as their primary land use, during and immediately following a rainfall event. Its primary application is in planning and evaluating various strategies for controlling surface runoff and sediment transport from intensively cropped areas. The original version of the model has been fully integrated within the GIS so that data can be entered easily and the results can be displayed as maps and tables (De Roo et al., 1989).

A catchment to be modelled is assumed to be composed of square cells. Values of variables are defined for each element, namely slope, aspect, soil variables (porosity, moisture content, field capacity, infiltration capacity, USLE erodibility factor K), crop variables (coverage, interception capacity, USLE crop and management factors C and P) (Wischmeier and Smith, 1978), surface variables (roughness and surface retention) and channel variables (width and roughness). These values may be derived from digital maps or from interpolation of point data. From the above it is clear that several detailed maps of land attributes are needed to run the model. These maps can be stored and edited in a GIS.

2.1 Conceptual structure of the ANSWERS model

A rainfall event can be simulated with increments of one second up to several minutes, depending on the gridsize. The continuity equation is used to establish the composite response of the single elements. The output of up-slope elements becomes the input of downslope elements. Several physically-based mathematical relationships are used to describe interception, infiltration, surface retention, drainage, overland flow, channel flow, subsurface flow, detachment by rainfall and/or overland flow and sediment-transport by overland flow (interrill erosion).

After rainfall begins, some is intercepted by the vegetation canopy (with coverage PER) until such time as the interception storage potential (PIT) is met. When the rainfall rate exceeds the interception rate, infiltration begins. Infiltration rates are influenced by the initial soil moisture content (ASM), total porosity (TP), soil moisture content at field capacity (FP), the infiltration rate at saturation (FC), the initial infiltration rate (FC+A) and the infiltration control zone depth (DF). Since the infiltration rate decreases in an exponential manner as the soil water storage increases, a point may be reached when the rainfall rate exceeds the combined infiltration and interception rates. When this occurs, water begins to accumulate on the surface in micro-depressions (retention storage), defined by the roughness parameters RC and HU. Once surface retention exceeds the capacity of the micro-depressions, overland flow begins (influenced by Manning's n, slope and aspect). A steady state infiltration rate (FC) may be reached if the duration and the intensity of the rainfall event are sufficiently large. When rainfall ceases, infiltration continues until depression storage water is no longer available.

Both soil detachment and transport can be caused by either raindrop impact (DTR) or overland flow (DTO). Whether or not a detached soil particle moves depends upon the sediment load in the flow and its capacity for sediment transport (TC). When water and sediment reach an element with a channel, they are transported to the catchment outlet. Sedimentation within a channel occurs when the transport capacity has been exceeded.

3. VALIDATING DISTRIBUTED HYDROLOGICAL AND SOIL EROSION MODELS

Distributed hydrological and soil erosion models can be validated by comparing measured and simulated values of discharge and sediment concentration at the catchment outlet. Also, overland flow patterns can be observed in the field and compared to the simulated patterns. Furthermore, soil erosion patterns can be evaluated using [137]Cs (Walling et al., 1986; De Roo, 1991; Quine and Walling, 1991). Thus, the distributed model can be evaluated in a distributed way.

Tracers such as ^{137}Cs can be used in soil erosion research to identify sources of sediments on a broad, catchment scale. The sampling of a drainage basin can be undertaken within a few days, without inconvenience to farmers, and the results represent long-term average soil erosion rates. Caesium-137, a semi-'natural' radionuclide, remains in the soil for a substantial time and is easily detectable. The Cs-labelled sediment behaves within the natural environment as natural materials would do.

Several techniques exist to examine errors in rainfall-runoff models. A number of indices have been developed to compare model-generated hydrographs with measured hydrographs. McCuen and Snyder (1975) developed a goodness of fit index. They adjusted the commonly used Pearson product-moment correlation coefficient by using a ratio of the standard deviations of the two hydrographs being compared. This was done due to the lack of sensitivity of the Pearson coefficient to differences in the size of the two hydrographs.

Nash and Sutcliffe (1970) introduced another measure for the goodness of fit, the coefficient of efficiency. The efficiency of a model is defined by R^2:

$$R^2 = \frac{F_0{}^2 - F^2}{F_0{}^2} \tag{1}$$

$$F_0{}^2 = \sum (q - q^*)^2 \tag{2}$$

$$F^2 = \sum (q' - q)^2 \tag{3}$$

where: q = measured discharge
           q^* = average measured discharge
           q' = simulated discharge

This coefficient was also used for comparison of measured and simulated hydrographs. If the coefficient of efficiency is negative, the model results are worse than the case where the average measured discharge was used as an estimate. Aitken (1973) proposed several other indices and tests, such as the sign test, which were used here, but are not discussed.

4. DESCRIPTION OF THE AREA

The Yendacott catchment, situated about 10km north of the city of Exeter (Devon, UK), has a drainage area of 147ha. The catchment is a part of the Jackmoor Brook basin, that has a drainage area of 930ha (Peart and Walling 1986, Loughran et al. 1987). Topography is generally subdued with slope angles rarely exceeding 5°. The catchment is developed on sandstones, breccias and conglomerates of Permian age and the soils are dominated by gleyed brown earths of the Rixdale series, with some brown earths of the Shaldon series occurring around the western margins (Clayden, 1971). Some small pockets of soils belonging to the Cutton series, developed on coarse alluvium and colluvium, also occur in the valley bottoms. Soil texture is mainly clay loam and clay (Table 1).

Table 1:   Soil texture of soil samples from the Yendacott catchment.

	No. of samples	% clay	% silt	% sand
clay	8	44.3	20.5	35.2
clay loam	16	33.8	27.8	38.4
sandy clay loam	4	28.4	23.4	48.3
loam	3	26.7	32.6	40.7
Total	31	35.1	25.8	39.1

Approximately 14% of the area is covered by mixed deciduous woodland, largely in the valley bottoms (Table 2). Over the remainder of the basin, cultivation is dominated by cereal crops (wheat and barley) grown in rotation with ley grass, fodder beans, potatoes and swedes.

Table 2:   Land use in the Yendacott catchment (percentage of total area).

	1986/87	1987/88	1988/89
Woodland	13.73	13.73	13.73
Grassland	15.18	16.10	12.43
Farms	0.27	0.27	0.27
Wheat	39.43	22.16	44.36
Barley	16.78	36.03	25.54
Oats	4.05	11.69	3.67
Swedes	5.03	-	-
Peas	5.52	-	-

The individual fields, which are frequently bounded by hedges and banks, are characteristically around 12ha in size. As a result of the unfavourable drainage characteristics of the soils of the Rixdale series, sub-surface tile drainage and ditches have been introduced into several parts of the basin. Mean annual precipitation is estimated at 800mm. whilst mean annual runoff from the basin is approximately 350mm. More than 50% of the mean annual precipitation falls during the months of November to March when many of the fields have limited vegetation cover.

5. CATCHMENT HYDROLOGICAL RESPONSE

Precipitation was measured at Orchard Close (Thorverton), 1500m outside the catchment, with a tipping bucket rain gauge A second, less reliable autographic rain gauge at Ratcliffe Farm (Thorverton), 500 m outside the catchment, was used only when the Orchard Close rain gauge was not operating. Discharge was measured at the gauging station at Rixenford Lane using a trapezoidal flume and a stilling well with a recorder. Sediment loadings were measured using a calibrated turbidity meter. The discharge and sediment loadings data were combined, yielding soil loss estimates from the catchment. The hydrological data for between autumn 1986 and spring 1991 were collated.

From Table 3 it is concluded that for this set of storms, the total amount of rainfall can predict the total discharge ($r=0.619$) and the total soil loss ($r=0.659$). Total discharge can be predicted using a multiple regression including total rainfall ($\beta=0.769$, $p=0.000$, significant) and initial soil moisture content ($\beta=0.309$, $p=0.059$, significant). The cumulative kinetic energy since the start of the growing season is not significant in this equation ($\beta=0.218$, $p=0.168$). Peak discharge can be predicted using single regression of the initial soil moisture content ($\beta=0.332$, $p=0.038$, significant). Peak discharge could not be predicted significantly using a multiple regression including total rainfall ($\beta=-0.299$, $p=0.704$, not significant), cumulative kinetic energy ($\beta=0.256$, $p=0.111$, not significant) and rainfall kinetic energy ($\beta=1.044$, $p=0.204$, not significant).

Table 3: Correlations between the basic hydrological data for the Yendacott catchment.

	RAIN	KEWISCHM	KEWCUM	ASM	OBSM3	OBSLS	OBSKG	OBS-TIMETPK
RAIN	1.000							
KEWISCHM	0.981	1.000						
KEWCUM	-0.256	-0.314	1.000					
ASM	-0.307	-0.340	0.107	1.000				
OBSM	0.619	0.592	0.054	0.096	1.000			
OBSLS	0.557	0.557	0.041	0.096	0.865	1.000		
OBSKG	0.659	0.651	-0.118	-0.082	0.815	0.688	1.000	
OBS-TIMETPK	0.777	0.737	-0.275	-0.138	0.579	0.452	0.673	1.000

RAIN = Rainfall at Orchard Close (mm); KEWISCHM = Kinetic Energy (Wischmeier method) ($J.m^{-2}$); KEWCUM = Cumulative Kinetic Energy ($J.m^{-2}$) since start of growing season; ASM = Initial soil moisture content for cereal crops (%); OBSM3 = Observed Discharge (m^3); OBSLS = Observed Peak Discharge ($l.s^{-1}$); OBSTIMETPK = Observed Time To Peak Discharge (min); OBSKG = Observed Soil Loss (kg).

6. VALIDATION OF THE ORIGINAL ANSWERS MODEL

The Yendacott catchment was modelled by constructing elements of 20m × 20m. GIS processing (digitizing and rastering) was undertaken using the GENAMAP GIS (formerly called DELTAMAP) (GENASYS, 1990). A Digital Elevation Model (DEM) was constructed by digitizing the contour map, which was rastered to cells of 10*10 using block kriging. Maps of slope and aspect were derived from the DEM using the PC-RASTER GIS software, developed in Utrecht. The land use maps were digitized and rastered. Soil texture data were used to calculate porosity, saturated conductivity and the USLE K-factor, using pedo-transfer functions (Cosby et al., 1984; Wischmeier and Smith, 1978). Although using textural data to calculate the soil hydrological variables can lead to large errors, this approach was used because no other data were available. These soil data, the land use, management and channel descriptions were entered. Channels and management practices were also digitized and rastered. The original ANSWERS model yields very poor results. The main reason for this is that the model only simulates Horton overland flow. Field observation and hydrograph analysis confirm that saturation overland flow processes are of major importance in the Yendacott catchment. As a consequence, it is inevitable that the model yields poor results.

7. INTRODUCING THE VARIABLE CONTRIBUTING AREA CONCEPT INTO ANSWERS

Catchment runoff may occur in at least four major ways (Beven and Kirkby, 1979):
-        Horton overland flow: rainfall intensity exceeds infiltration capacity;
-        Partial area conceptual model of runoff (Hewlett, 1961): rainfall intensities exceed infiltration or storage capacity within a variable area of near-saturated soils;
-        Rain falling on stream channels and completely saturated soils;
-        'Subsurface storm flow' (flow within the soil) and 'return flow' (subsurface flow emerging to flow over the surface).

The last three processes contribute to saturation overland flow, which is often dealt with using 'variable contributing areas' (Gregory and Walling, 1973; Beven and Kirkby, 1979). Other common associate terms are: partial contributing areas (Weyman, 1975), minimum contributing areas (Weyman, 1975), variable source areas (Gregory and Walling, 1973; Hewlett, 1982; Ward and Robinson, 1990), source areas (Betson and Marius, 1969) and partial areas (Dunne and Black, 1970a; b). In agricultural catchments, tile drains may also contribute to catchment runoff. Field observation and hydrograph analysis confirmed that the variable contributing area concept is applicable in the Yendacott catchment.

Saturation overland flow from 'contributing areas' can be simply represented as:

$$q_{of} = i * A_c \qquad\qquad (4)$$

with:

q_{of} = overland flow ($m^3 h^{-1}$)
A_c = the saturated area (m^2)
i = instantaneous rainfall intensity ($m\ h^{-1}$)

In the distributed model, the infiltration rate is set to zero in 'contributing' elements (pixels, gridcells). Beven and Kirkby (1979) introduced a method to determine the 'contributing area' or the saturated area, in which the saturated area A_C is that for which:

$$\ln \frac{a}{\tan \beta} \geq \frac{S_T - S_3}{m} + \lambda \qquad (5)$$

with:
a   = the area drained (m ^{2});
$\tan \beta$ = slope gradient (-);
S_T = maximum storage value (mm), consisting of S_1(max) + S_2(max);
S_1 = surface interception and depression store (mm);
S_2 = near surface infiltration store (mm);
S_3 = subsurface store (mm) (negative value represents a moisture deficit);
m   = a calibration constant (mm);
lambda = a constant, related to A, a and $\tan \beta$ (-);
A   = the total area of the catchment (m^2).

The factors a (the area drained) and $\tan \beta$ (slope gradient) can be derived from Digital Elevation Models (DEM's), e.g. using the WATERSHED Tools, a GIS package developed by Van Deursen (Van Deursen and Kwadijk, 1990). Using neighbourhood algorithms (Marks et al. 1984; Jenson and Domingue, 1988), slope, aspect and curvature can be calculated, and drainage patterns and catchments can be extracted. For each pixel the number of 'upstream elements' is calculated, which equals the area drained when multiplied by the area of one pixel. Using a raster GIS program, the maps of a and $\tan \beta$ are used to calculate the $\ln(a/\tan \beta)$ map. An example is given in Figure 1.

Beven and Kirkby (1979) proposed that the lambda parameter can be approximated by:

$$\lambda = \frac{1}{A} * \sum \ln \frac{a}{\tan \beta} * \Delta A \qquad (6)$$

with:
A   = the total area of the catchment (m ^{2})
ΔA = some elemental area of constant a/$\tan \beta$ in the sub-basin

The subsurface storage S_3 varies during a rainfall event, thus explaining the term 'variable contributing area'. The parameter m of the exponential subsurface store is estimated by making discharge measurements at the basin outlet, thus by calibration.

Figure 1. The ln(a/tanß) parameter for the Yendacott catchment.

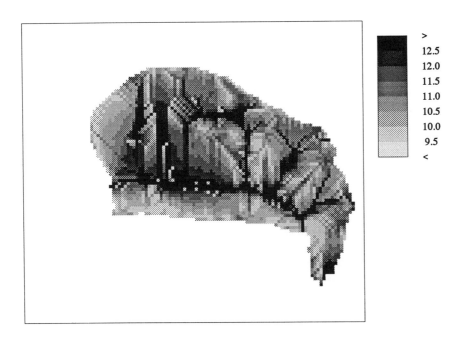

>	
12.5	
12.0	
11.5	
11.0	
10.5	
10.0	
9.5	
<	

This concept of variable contributing area has been incorporated into ANSWERS. The map of ln(a/tanβ) (the area drained/slope gradient), the lambda parameter and the m parameter are used as input. In ANSWERS, saturation overland flow is produced when:

$$\ln \frac{a}{\tan \beta} \geq \frac{PIV}{m} + \lambda \qquad (7)$$

where:

PIV = the potential infiltration volume (mm), defined as:

$$PIV = DF * TP * (1 - ASM) + DR - F \qquad (8)$$

with:

DF   = infiltration control zone depth (mm);
TP   = total porosity (%)
ASM = initial soil moisture content (% saturation);
DR   = cumulative drainage to subsurface store (mm);
F   = cumulative infiltration (mm);

8. VALIDATION OF THE MODIFIED MODEL

The simulation results of the modified ANSWERS model, including the variable contributing area concept, are summarized in Table 4.

Table 4:   **Correlations between measured data and simulated results with the modified ANSWERS model (not calibrated) for the Yendacott catchment.**

	OBSM3	OBSLS	OBSKG	OBSTIMETPK
SIMM3	0.616	0.567	0.657	0.754
SIMLS	0.317	0.378	0.369	0.304
SIMKG	0.514	0.570	0.632	0.685
SIMTIMETPK	0.661	0.555	0.800	0.865

OBSM3	=	Observed Discharge (m ^{3})
SIMM3	=	Simulated Discharge (m ^{3})
OBSLS	=	Observed Peak Discharge (l.s^{-1})
SIMLS	=	Simulated Peak Discharge (l.s^{-1})
OBSTIMETPK	=	Observed Time To Peak Discharge (min)
SIMTIMETPK	=	Simulated Time To Peak Discharge (min)
OBSKG	=	Observed Soil Loss (kg)
SIMKG	=	Simulated Soil Loss (kg)

Comparing these results with Table 3 leads to the conclusion that it is better to use a regression equation using total rainfall to predict total discharge, peak discharge or soil loss than to use the modified ANSWERS model. Another reason for not being satisfied with the model results is that the average McCuen correlation coefficient is very low. For the Yendacott catchment, the average Pearson coefficient was 0.575, but the average McCuen coefficient was 0.277, which is very low.

It was expected that calibration of lambda would improve the simulations and increase the McCuen coefficients. The method of Beven and Kirkby (1979) for the calculation of lambda is useful as a first approximation.

Figure 2 shows the correlation between simulated and observed discharge for all storms. Figure 3 shows an example hydrograph with observed and simulated discharge for the Yendacott catchment.   Table 5 summarizes the correlations between observations and simulations. Table 6 summarizes the simulation results for the individual rainstorms. Table 5 shows that it is indeed possible to improve the simulation of the hydrograph when calibrating the model solely on lambda. The results seem promising, but a careful look at Table 6 shows that the McCuen coefficients used for the comparison between observed and simulated hydrograph are low in many cases.

Table 5. Correlations between measured data and simulated results with the modified ANSWERS model (calibrated with lambda) for the Yendacott catchment.

	OBSM3	OBSLS	OBSKG	OBSTIMETPK	LAMBDA
SIMM3	0.864	0.883	0.836	0.641	-0.467
SIMLS	0.840	0.973	0.716	0.470	-0.540
SIMKG	0.846	0.944	0.809	0.573	-0.563
SIMTIMETPK	0.590	0.469	0.745	0.926	-0.115

OBSM3	=	Observed Discharge (m ^{3})
SIMM3	=	Simulated Discharge (m ^{3})
OBSLS	=	Observed Peak Discharge (l.s^{-1})
SIMLS	=	Simulated Peak Discharge (l.s^{-1})
OBSTIMETPK =		Observed Time To Peak Discharge (min)
SIMTIMETPK =		Simulated Time To Peak Discharge (min)
OBSKG	=	Observed Soil Loss (kg)
SIMKG	=	Simulated Soil Loss (kg)

A comparison of Tables 4 and 5 demonstrates the improvement of the ANSWERS simulations obtained by calibration on lambda. Not only the simulation of peak discharge, but also the estimates of total discharge, total soil loss and the time to peak discharge improve. Also, the median McCuen coefficient and the Nash coefficient have increased significantly. A comparison of Tables 5 and 3 leads to the conclusion that the simulation results obtained with the modified ANSWERS model are now better than the prediction of discharge using a multiple regression including total rainfall and initial soil moisture content, as explained above. Apart from other advantages, such as the spatial pattern of soil loss or gain and overland flow that can be obtained, and the possibility to evaluate soil conservation scenarios (De Roo et al., 1989), spatial simulation modelling using the modified ANSWERS model seems to yield better results than simple regression analysis.

Thus, introducing the variable contributing area concept improves the model simulations of discharge and soil loss. However, calibration on the lambda parameter for each storm seems necessary and this is undesirable. The best fitting lambda can be predicted using a multiple regression involving initial soil moisture content, the cumulative kinetic energy since the start of the growing season, and the total rainfall during a storm. Thus, the data set has been split in two parts, and the equation to estimate the best fitting lambda derived from calibrating the first data set (eight storms in 1986-1987) has been used to estimate lambda for the second (validation) data set (24 storms 1988-1991).

Figure 2. Correlation between observed and simulated catchment discharge for all storms (calibrated with lambda).

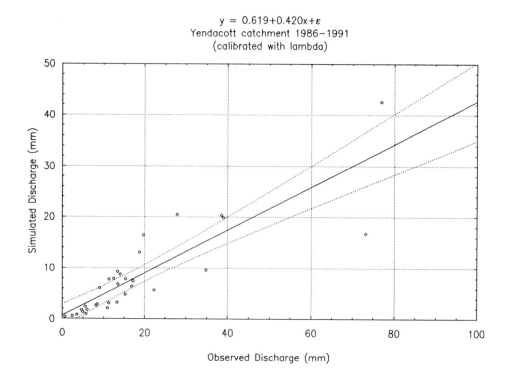

$$y = 0.619+0.420x+\varepsilon$$
Yendacott catchment 1986–1991
(calibrated with lambda)

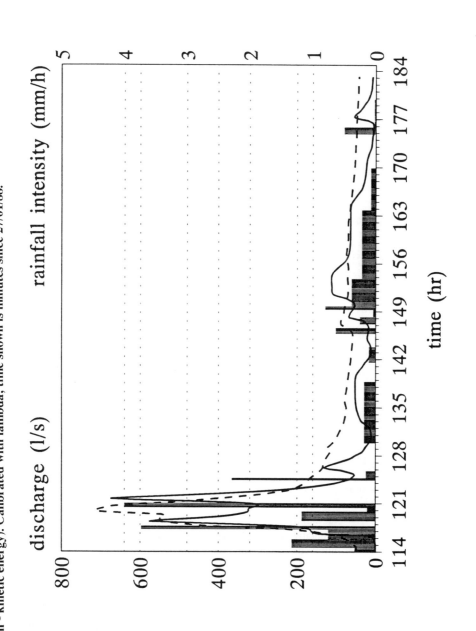

Figure 3. Observed (dashed line) and simulated (solid line) catchment discharge for the event on 31/01/88 (22.9 mm rainfall, 209 J.m⁻¹ kinetic energy). Calibrated with lambda; time shown is minutes since 27/01/88.

Table 6. **Simulation results for the Yendacott catchment**

Date	Observed discharge (m³)	Simulated discharge (m³)	Observed soil loss (kg)	Simulated soil loss (kg)	Lambda value calibrated	Pearson correlation coefficient	McCuen's coefficient	Nash efficiency coefficient
861019	924	541	324	858	12.5	0.6049	0.4765	-0.376
861113	40910	29992	92326	202409	10.1	0.7594	0.6427	0.557
870402	113136	62490	-1	236788	9.7	0.7602	0.5342	0.461
870406	22495	7171	-1	237641	0.5	0.8397	0.5789	-0.389
871027	13186	8968	25421	341051	1.1	0.4867	0.3879	-0.400
871108	3547	839	3404	1383	12.5	0.8201	0.7666	-0.432
871111	19848	10111	22412	491671	0.1	0.6521	0.4248	0.128
871115	16708	11419	-1	66579	9.5	0.6093	0.5070	-0.037
880127	56678	29782	79820	131052	10.0	0.5730	0.3831	0.161
880131	27612	19161	-1	93857	9.1	0.8349	0.7195	0.629
880203	11968	3862	-1	11323	10.7	0.6500	0.5516	-1.940
880207	19530	4817	16144	184121	0.4	0.7811	0.7037	-0.174
880317	18299	11598	6653	229781	0.9	0.6250	0.4434	-0.497
880320	12414	4220	4185	8610	10.9	0.8117	0.6962	-0.036
880323	88442	661	2577	3750	11.6	0.7378	0.7130	-1.062
881008	6911	2512	5770	7642	12.3	0.7598	0.5155	0.355
881011	7293	1915	2592	5572	12.2	0.5845	0.4491	-0.170
890223	57339	29081	96172	111035	10.5	0.3924	0.2931	-0.111
890320	5136	1127	-1	1610	12.0	0.8046	0.6312	0.140
891213	19677	13714	19595	53573	10.9	0.2691	0.2640	-0.500
891215	25158	10984	9349	42087	10.8	-0.1990	-0.1748	-2.431
891218	22515	11492	11153	59764	10.3	0.4753	0.4063	-0.118
891219	107614	24575	109403	154428	9.5	0.2076	0.0813	-0.268
891224	24852	9346	8205	52874	9.9	0.5898	0.5478	-0.097
900127	28938	24114	13451	136997	9.0	0.6262	0.5961	0.260
900129	20625	12895	6907	7522	07.0	0.8140	0.7338	0.517
900130	16553	4531	4572	22599	10.1	0.8780	0.8642	-0.837
900201	51075	14077	8467	80290	9.0	-0.1674	-0.0949	-0.867
910108	8041	3524	15842	115511	0.9	0.2085	0.1949	-3.155
910109	32761	8335	2653	45220	9.7	0.5856	0.5003	-1.219
910118	16117	3133	850	9400	11.2	0.4863	0.4695	-1.287
910222	8413	1438	2980	6098	10.5	0.4554	0.2930	-8.128

The best fitting lambda equation (R=0.881) is:

$$\lambda = 16.6 - 0.0585*ASM - 0.0283*RAIN - 0.000548*KEWCUM \qquad (9)$$

with RAIN = Rainfall (mm);
KEWCUM = Cumulative Kinetic Energy (J.m⁻²) since start of growing season
ASM = Initial soil moisture content for cereal crops (%)

This equation has been used to estimate lambda, and all storms were simulated again to obtain a validation set. It is no surprise that the simulation results have deteriorated (Tables 7 and 8). However, the results are still better than the results before calibration (Table 7). Even with the improvement obtained by incorporating the 'variable contributing area' concept, the hydrograph still cannot be simulated properly. In this catchment, both subsurface lateral flow, caused by perched water tables or macro pores, producing little or no soil erosion, and saturation overland flow (as observed in the field), producing substantial amounts of soil erosion, contribute significantly to the hydrograph. However, the former is not simulated in the modified ANSWERS model. Thus, an important source of water is not taken into account.

Table 7. Correlations between measured data and simulated results with the modified ANSWERS model (lambda estimated using equation 9) for the Yendacott catchment.

	OBSM3	OBSLS	OBSKG	OBSTIMETPK	LAMBDA
SIMM3	0.752	0.615	0.633	0.717	-0.461
SIMLS	0.576	0.502	0.350	0.401	-0.526
SIMKG	0.728	0.589	0.610	0.718	-0.520
SIMTIMETPK	0.656	0.543	0.803	0.870	-0.166

Table 8. Summary of the goodness of fit indices for hydrographs for the Yendacott catchment and the most important correlation coefficients between observed and simulated data, without and with calibration on lambda, and for the validation set.

	before calibration		after calibration		validation	
	median	(s.dev.)	median	(s.dev.)	median	(s.dev.)
Pearson's correlation	0.637	(0.269)	0.617	(0.268)	0.616	(0.276)
McCuen's coefficient	0.246	(0.198)	0.504	(0.236)	0.287	(0.227)
Coefficient of Efficiency	-0.499	(12.898)	-0.221	(1.606)	-0.918	(10.964)
corr. s/o Total discharge	0.616		0.864		0.752	
corr. s/o Peak discharge	0.378		0.973		0.502	
corr. s/o Soil loss	0.632		0.809		0.610	
corr. s/o Time to peak disch.	0.865		0.926		0.870	

Furthermore, the Nash coefficient of efficiency is a better goodness of fit index for hydrographs than the McCuen and Pearson coefficients, that are highly correlated. The Nash coeffient of efficiency in combination with the McCuen coefficient gives the best results.

9. THE DISTRIBUTION OF ^{137}Cs IN THE YENDACOTT CATCHMENT

Walling et al. (1986), Loughran et al. (1987) and Walling and Bradley (1988) have reported information regarding the distribution of ^{137}Cs in the Yendacott catchment. Walling and Bradley (1988) investigated a cultivated part of the catchment with average slope angles of 3° (5%). 162 point data were interpolated to a map using block kriging. A substantial part of this area (68.6%) can be classified as stable in terms of soil erosion, with a ^{137}Cs content between 2 and 3 kBqm^{-2}. 13.0% of the area has a ^{137}Cs content higher than 3 kBqm^{-2}, and can be classified as depositional areas. 18.3% of the area has a ^{137}Cs content lower than 2 kBqm^{-2}, and can be classified as erosion areas.

10. COMPARING ^{137}Cs DATA WITH SIMULATED SOIL EROSION RATES

The ^{137}Cs data were used in two ways to validate the erosion model. The first approach is to compare the caesium inventory for each individual sampling point with the soil erosion estimate provided by the model. The second approach is to interpolate the point caesium data to a map, and compare this map with the soil erosion maps simulated by the model. A map showing total soil erosion between autumn 1986 and spring 1991, produced by adding all the simulations with the modified ANSWERS model, is shown in Figure 4. Correction of the ^{137}Cs content for ploughing and variable atmospheric input were not made. The results are shown in Table 9.

In general, the correlations between simulated soil loss and observed soil loss are low. The correlations between the individual point values of ^{137}Cs and the simulated soil loss are not significant. However, due to the large number of observations, the correlations between the individual rastercells of the interpolated ^{137}Cs map and the cells of the simulated soil loss map higher than 0.05 are significant, although they are low (Figure 5).

Figure 4. Total soil loss between autumn 1986 and spring 1991, simulated with the modified ANSWERS model.

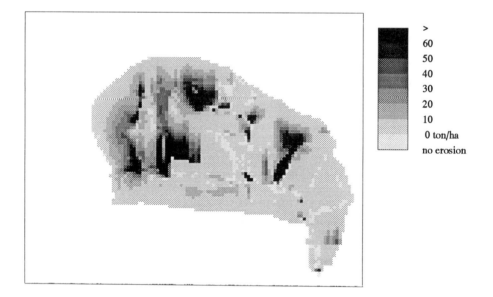

	>
	60
	50
	40
	30
	20
	10
	0 ton/ha
	no erosion

Table 9. Correlation of the measured ^{137}Cs in the soil and the simulated soil erosion rates in the Yendacott catchment: total annual values and values for some major storms.

	^{137}Cs(block)		^{137}Cs(point)	
	calibration	validation	calibration	validation
number of observations	1288	1288	162	162
Soil loss 1986-1991	0.152	-0.012	0.079	-0.010
Total erosion 1986	0.150	0.121	0.083	0.081
Total erosion 1987	0.153	0.146	0.077	0.083
Total erosion 1988	0.097	0.081	0.079	-0.050
Total erosion 1989	0.135	-0.001	0.080	-0.090
Total erosion 1990	0.154	0.171	0.073	0.137
Total erosion 1991	0.107	0.149	0.083	0.079
Storm 861113	0.150	0.121	0.083	0.082
Storm 870402	0.177	0.070	0.074	-0.050
Storm 880127	0.132	0.129	0.081	0.072
Storm 880131	0.075	0.078	0.071	0.071
Storm 890223	0.165	0.179	0.075	0.074
Storm 891219	0.158	0.193	0.077	0.055
Storm 900127	0.154	0.154	0.074	0.074
Storm 900201	0.081	0.080	0.069	0.069

Figure 5. Correlation between soil gain (autumn 1986 - spring 1991) simulated with the modified ANSWERS model (lambda calibrated) and the interpolated ^{137}Cs values.

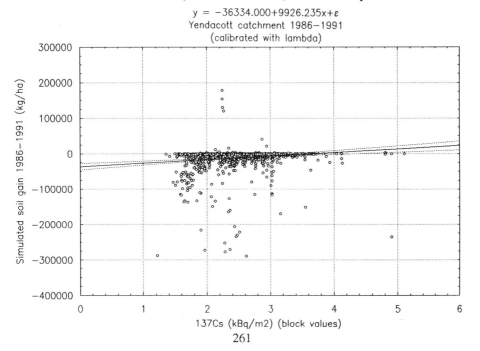

$$y = -36334.000 + 9926.235x + \varepsilon$$
Yendacott catchment 1986−1991
(calibrated with lambda)

11. CONCLUSIONS

Based on the results obtained in the Yendacott catchment and the Dutch experimental catchments, a new model will be developed. Together with the University of Amsterdam and the Winand Staring Centre in Wageningen (the Netherlands), a model is being developed that includes subsurface flow and saturation overland flow. This model will be used to predict discharge and soil loss in catchments.

ACKNOWLEDGEMENTS

The authors wish to thank the following people at the University of Exeter: Keith Garrett for his introduction to the Yendacott catchment and for providing basic data, Dave Hogan for information about soils and Tim Quine for his useful remarks and his hospitality during the stay of the first author in Exeter.

REFERENCES

AITKEN, A.P. 1973. Assessing systematic errors in rainfall-runoff models. Journal of Hydrology, Vol.20, 121-136.

BEASLEY, D.B. and HUGGINS, L.F. 1982. ANSWERS - Users manual. Purdue University, West Lafayette, IN., 54 p.

BEASLEY, D.B., HUGGINS, L.F. and MONKE, E.J. 1980. ANSWERS: A Model for Watershed Planning. Transactions of the ASAE, Vol.23, No.4, 938-944.

BEASLEY, D.B., HUGGINS, L.F. and MONKE, E.J. 1982. Modelling sediment yields from agricultural watersheds. Journal of Soil and Water Conservation, Vol.37 No.2, 113-117.

BETSON, R.P. and MARIUS, J.B. 1969. Source areas of storm runoff. Water Resources Research, Vol. 5, No. 3, 574-582.

BEVEN, K.J. and KIRKBY, M.J. 1979. A physically based, variable contributing area model of basin hydrology. Hydrological Sciences Bulletin, Vol.24, No.1, 43-69.

CLAYDEN, B. 1971. Soils of the Exeter district. Memoirs of the Soil Survey of Great Britain, England and Wales. Harpenden, Herts, U.K.

COSBY, B.J., HORNBERGER, G.M., CLAPP, R.B. and GINN, T.R. 1984. A statistical exploration of the relationships of soil moisture characteristics to the physical properties of soils. Water Resources Research, Vol.20, 682-690.

DE ROO, A.P.J. 1991. The use of ^{137}Cs as a tracer in an erosion study in South Limburg, The Netherlands and the influence of Chernobyl fallout. Hydrological Processes, Vol.5, 215-227.

DE ROO, A.P.J. and RIEZEBOS, H.TH. 1992. Infiltration experiments on loess soils and their implications for modelling surface runoff and soil erosion. CATENA, Vol.19., 221-239.

DE ROO, A.P.J., HAZELHOFF, L. and BURROUGH, P.A. 1989. Soil erosion modelling using ANSWERS and Geographical Information Systems. Earth Surface Processes and Landforms, Vol.14, 517-533.

DE ROO, A.P.J., HAZELHOFF, L. and HEUVELINK, G.B.M. 1992. The use of Monte Carlo simulations to estimate the effects of spatial variability of infiltration on the output of a distributed hydrological and erosion model. Hydrological Processes, Vol.6, 127-143.

DUNNE, T. and BLACK, R.D. 1970a. An experimental investigation of runoff production in permeable soils. Water Resources Research, Vol.6, No.2, 478-490.

DUNNE, T. and BLACK, R.D. 1970b. Partial area contributions to storm runoff in a small New England watershed. Water Resources Research, Vol.6, No.5, 1296-1311.

GENASYS 1990. GENAMAP Reference Manual ver.4.2. Genasys Inc., Fort Collins CO, USA.

GREGORY, K.J. and WALLING, D.E. 1973. Drainage basin form and process; a geomorphological approach. Edward Arnold Publishers Ltd., London. pp. 458.

HEWLETT, J.D. 1961. Watershed Management. In: Report for 1961 Southeastern Forest Experiment Station, US Forest Service, Ashville, N.C.

HEWLETT, J.D. 1982. Principles of forest hydrology. The University of Georgia Press, Athens. pp. 183.

JENSON, S. and DOMINGUE, J. 1988. Extracting topographic structure from digital elevation data for geographic information system analysis. Photogrammetric Engineering and Remote Sensing, 1593-1600.

LOUGHRAN, R.J., CAMPBELL, B.L. and WALLING, D.E. 1987. Soil erosion and sedimentation indicated by caesium[137]: Jackmoor Brook catchment, Devon, England. CATENA, Vol. 14, 201-212.

MARKS, D., DOZIER, J. and FREW, J. 1984. Automated basin delineation from digital elevation data. Geo-processing, Vol.2, 299-311.

MCCUEN, R.H. and SNYDER, W.M. 1975. A proposed index for comparing hydrographs. Water Resources Research, Vol.11, No.6, 1021-1024.

NASH, J.E. and SUTCLIFFE, J.V. 1970. River flow forecasting through conceptual models, 1. A discussion of principles. Journal of Hydrology, Vol.10, 282-290.

PEART, M.R and WALLING, D.E. 1986. Fingerprinting sediment source: the example of a drainage basin in Devon, UK. IAHS Publ. no. 159, 41-55.

QUINE, T.A. and WALLING, D.E. 1991. Rates of soil erosion on arable fields in Britain: quantitative data from caesium-137 measurements. Soil use and management, Vol.7, 169-176.

VAN DEURSEN, W.P.A. and KWADIJK, J. 1990. Using the WATERSHED tools for modelling the Rhine catchment. In: Harts, J.J., Ottens, H.F.L. and Scholten, H.J. (eds). Proceedings of EGIS '90: First European Conference on Geographical Information Systems, Amsterdam. pp. 254-262.

WALLING, D.E. and BRADLEY, S.B. 1988. The use of caesium-137 measurements to investigate sediment delivery from cultivated areas in Devon, UK. In: Sediment budgets. Proceedings of the Porto Alegre Symposium. IAHS Publ. no. 174, 325-335.

WALLING, D.E., BRADLEY, S.B. and WILKINSON, C.J. 1986. A Caesium-137 budget approach to the investigation of sediment delivery from a small agricultural drainage basin in Devon, UK. IAHS Publ. no. 159, 423-435.

WARD, R.C. and ROBINSON, M. 1990. Principles of hydrology. Third edition. McGraw-Hill Book Company, London. p. 365.

WEYMAN, D.R. 1975. Runoff processes and streamflow modelling. Oxford University Press. p. 54.

WISCHMEIER, W.H. and SMITH, D.D. 1978. Predicting rainfall erosion losses - a guide to conservation planning. U.S. Department of Agriculture, Agricultural Handbook No. 537. p.58.

Chapter 25

A SPLASH DELIVERY RATIO TO CHARACTERIZE
SOIL EROSION EVENTS

F.J.P.M. Kwaad

Laboratory of Physical Geography and Soil Science, University of Amsterdam,
Nieuwe Prinsengracht 130, 1018 VZ Amsterdam, the Netherlands

Summary

In 1986, a plot study was set up on sloping, cultivated loess soils in south Limbourg (the Netherlands) to evaluate the effects of various cropping systems of fodder maize on runoff, erosion and crop yield. From November 1988 until March 1990 splash erosion was also measured on the experimental plots. In this paper, splash erosion is compared with total soil loss. For this purpose a 'splash delivery ratio' is used. This is the ratio of total soil loss from a plot (or field) to the amount of soil detached by rain splash. Four conditions are distinguished:

(a) delivery ratio = 0; no runoff, only detachment by drop impact;
(b) delivery ratio <1; less soil is removed from the plot in runoff than is detached by drop impact; the transport capacity of runoff is exceeded by the amount of detached soil; transport-limited erosion;
(c) delivery ratio = 1; soil loss in the runoff equals the quantity of soil detached by drop impact; more soil could be transported by runoff than is in fact transported; sediment concentration of the runoff is lower than the maximum possible concentration; detachment-limited erosion;
(d) delivery ratio >1; more soil is removed from the plot in runoff than is detached by drop impact (e.g. rill erosion).

During the period of measurement only the first two conditions have been observed. A delivery ratio of nil occurred in the months with no runoff, as was the case in all summer months of 1989. All other months had delivery ratios from 0 to 47.3%. However, two subgroups could be distinguished: (b1) cropping systems with tillage in autumn and a winter or spring cover crop, which did not reach higher delivery ratio values than 8.4%, and (b2) cropping systems with spring tillage only and no winter cover crop, which reached delivery ratio values up to 47.3%. Condition (d) was not observed during the period of study, nor were rills formed on any of the plots.

1. INTRODUCTION

In 1986 a plot study was set up at the Experimental Farm Wijnandsrade to evaluate the effects of three cropping systems of fodder maize on runoff, erosion and crop yield on sloping loess soils in south Limbourg (Kwaad, 1991, this volume; Kwaad and Van Mulligen,

1991). From November 1988 to March 1990 splash erosion was also measured on the experimental plots.

Soil erosion is generally seen as a two phase process, consisting of the detachment of individual soil particles, and their subsequent transport by agents such as running water and wind (Morgan, R.P.C., 1986). Since the work of Ellison (1947) and Young and Wiersma (1973), raindrop impact is considered the most important agent of soil particle detachment. All physically-based mathematical models of soil erosion make use of the distinction between detachment and transport by raindrops and runoff, starting with the seminal paper of Meyer and Wischmeier (1969). According to Morgan, R.P.C. (1986), the severity of erosion depends on the quantity of material supplied by detachment and the capacity of the eroding agents to transport it. Where the agents have the capacity to transport more material than is supplied by detachment, the erosion is described as *detachment-limited*. Where more material is supplied than can be transported, the erosion is *transport-limited*. The recognition of which factor, detachment or transport, is limiting is important because conservation measures are mostly needed when erosion is detachment-limited. This constitutes an inherently dangerous condition, as an excess transport capacity of runoff is implicit, which could eventually lead to erosion by running water (rill and gully erosion).

In theoretical discussions and modelling of soil erosion, intensive use is made of the concepts of detachment and transport, and detachment-limited and transport-limited erosion. However, not very many field data are available on the relative significance in quantitative terms of detachment and transport under varying conditions. Exceptions are Bollinne (1982), Morgan, C. (1986), and Govers and Poesen (1988).

During the plot study at Wijnandsrade total soil loss and erosion by splash were measured in order to obtain on the relative significance of the two processes. To evaluate the importance of splash erosion vs. total soil loss, a 'splash delivery ratio' is used. A 'sediment delivery ratio' was introduced by Roehl (1962) as the percentage relationship between the sediment yield at a specific measuring point in a watershed and the gross, or total erosion occurring in the watershed upstream from that point. Govers and Poesen (1988) calculated a delivery ratio as 'the ratio between the mass of sediment effectively removed from the interrill plot by interrill wash to the mass of sediment detached by drop impact'. The concept is used here as the proportion of splash eroded material that is lost from a plot (or field) in runoff. Soil loss in runoff and splash erosion are expressed in $g.m^{-2}.time-unit^{-1}$ (Bollinne, 1982; Poesen and Torri, 1988).

2. MATERIALS AND METHODS

Twelve plots with three replications of three cropping systems of fodder maize and permanently bare soil were arranged in a randomised block design on a field of 6% slope. Cropping systems of maize were:

(a) conservation cropping system A; maize with winter rye as the winter cover crop and tillage only in the autumn;
(b) conservation cropping system B; maize with summer barley as the early-spring cover crop and tillage in the autumn and early spring;
(c) conventional system C; maize, with stubble in winter and spring tillage; and

265

(d) permanently bare fallow C'; tillage as for system C   (Geelen, 1987).

Direction of tillage and sowing was up and down slope. Naturally occurring runoff was collected from part of the plots, which measured 22m in length and 1.80m wide (Mutchler et al., 1988; Brakensiek et al., 1979). The plots were laid out in October 1985. Maize was grown continuously from 1986 to 1989. Systematic measurements of runoff were carried out from May 1987 until March 1990. The interval between visits was four weeks. In November 1988, Bollinne splash collectors, with a diameter of 7.5cm and a surface area of $44.2cm^2$ (Bollinne, 1975) were installed on the plots (one per plot). To prevent overland flow from entering the cups, the rim of the cups was kept about 3mm above the surrounding soil surface, which was rather rough on some cropping systems. Splash material was collected at 4-weekly intervals until March 1990. Collected amounts are only 0.65 of the real mass of detached soil, due to underestimation of the real mass by using a cup of 7.5cm diameter (Poesen and Torri, 1988). So, collected amounts were multiplied by a correction factor of 1.538.

Soils at the experimental site are truncated gleyic luvisols in loess with less than 0.80m of overlying colluvial loess. Average particle size distribution of the plough layer of the plots was 13% clay, 81% silt and 6% sand. Organic matter content was 1.8%.

Average yearly precipitation at the site is $750mm.yr^{-1}$, with rain in all seasons. High intensity rainfall is restricted to April-October (Levert, 1954). A 30-minute intensity of $24mm.h^{-1}$ is exceeded once a year (Buishand and Velds, 1980). Average annual rainfall erosivity (Wischmeier and Smith, 1978) is 75 (metric units) according to Bergsma (1980).

Figure 1. Monthly precipitation.

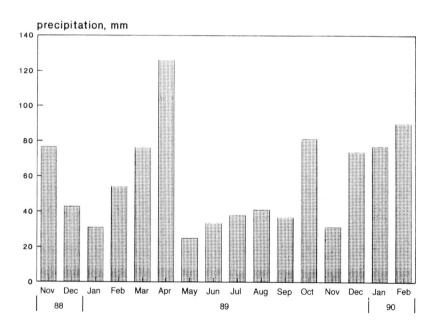

3. RESULTS

3.1. Seasonal differences

In Figure 1 monthly rainfall is given for the period of the splash measurements. Amounts ranged from 24mm to 126mm per month. From Figure 2 it appears that most runoff occurred in winter 1988/89 and early spring 1989. Hardly any runoff occurred in summer 1989. A similar trend can be observed in the runoff percentages. Soil loss in runoff is shown in Figure 3. Again, highest amounts occurred in winter 1988/89 and spring 1989, and also in winter 1989/90. Splash erosion is shown in Figure 4. Here, three main periods with relatively high rates of erosion (spring 1989, summer 1989 and winter 1989/90) can be seen. The trend follows more or less the trend in rainfall amounts (Figure 1). Delivery ratios are given in Figure 5. The highest values clearly occur in winter. It is remarkable that the delivery ratio did not exceed, or reach, 100%. Sediment concentration of runoff in months when runoff amounts >0.250mm is given in Figure 6. Low sediment concentrations seem to coincide in some months with high values of the delivery ratio, and vice versa (e.g. March 1989). No rills were formed on any of the plots during the course of the four year study.

Figure 2. Runoff amounts.

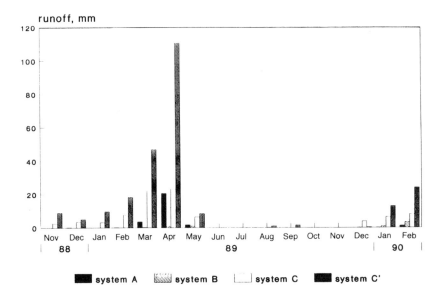

3.2. Differences between cropping systems

Differences between cropping systems are discussed in detail by Kwaad (this volume). System A was characterised by a fine and smooth winter rye seedbed in November which slaked during the winter. The winter rye was low until April. At the time of maize drilling the soil surface had become strongly slaked, even and compact. No tractor wheelings were present. Runoff was low in winter. System B had a rough surface after autumn tillage with coarse clods and a high random microrelief. Some slaking occurred in winter, but surface storage capacity remained high. Also, the surface roughness remained relatively high after summer barley drilling. Runoff was low in winter. During maize drilling, shallow tractor wheelings formed. Systems C and C' had a fine and smooth seedbed with pronounced tractor wheelings after maize drilling in early May. Rapid slaking occurred in spring and continued in summer on system C'. After the maize harvest, an even, completely slaked, compact soil surface was present, with deep tractor wheelings. Runoff was high in winter.

Figure 3. Soil loss in runoff.

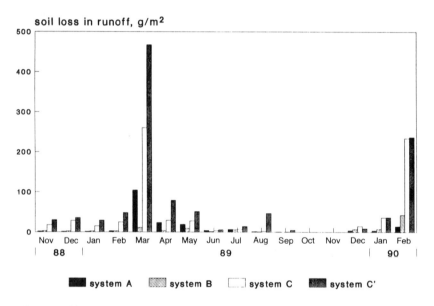

Regarding runoff volume, runoff percentage, soil loss in runoff and delivery ratio, cropping systems can be ranked: C' > C > A and B. Regarding splash erosion, no such clear ranking of cropping systems can be given. The order of cropping systems changed from month to month. In 7 out of 16 months, splash erosion rate was highest on cropping system B.

4. DISCUSSION

If Figures 4 and 5 are compared, the delivery ratio shows a clearer and simpler trend than splash erosion. It is not quite clear why the order of the splash erosion rates for the different cropping systems changed from month to month. Cropping system B, which had the highest splash erosion rates in 7 out of 16 months, had the highest random microrelief of the four systems. This implies a higher exposed surface area per m^2 than the other systems, plus a possibility of splash creep (Moeyersons and De Ploey, 1976) down the sides of micro-elevations.

Figure 4. Splash erosion.

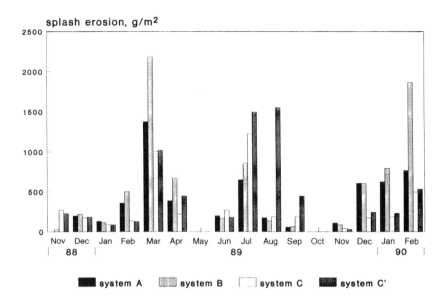

Figure 5. Delivery ratios.

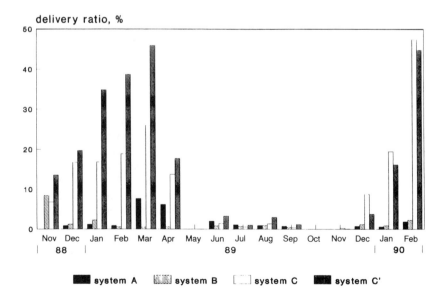

269

Figure 6. Sediment concentrations for monthly runoff amounts > 0.250mm.

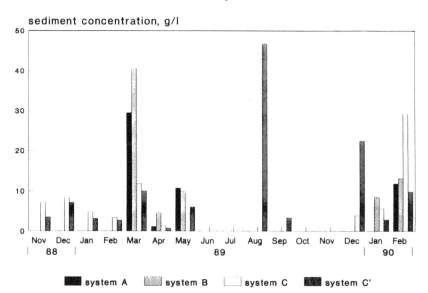

In theory, the following subdivision of events, based on the splash delivery ratio, can be inferred:

(a) delivery ratio = 0; no runoff, only detachment by drop impact;
(b) delivery ratio <1; less soil is removed from the plot in runoff than is detached by drop impact; transport capacity of the runoff is exceeded by the amount of detached soil; transport-limited erosion;
(c) delivery ratio = 1; soil loss in the runoff equals the quantity of soil detached by drop impact; more soil could be transported by runoff than is in fact transported; sediment concentration of the runoff is lower than the maximum possible concentration; runoff is not loaded to capacity; detachment-limited erosion;
(d) delivery ratio >1; more soil is removed from the plot in runoff than is detached by drop impact (e.g. rill erosion).

In fact, during the period of measurement only the first two conditions were observed. Delivery ratios ranged from 0 to 47.3%. Two subgroups can clearly be distinguished: (b1) systems A and B, which did not reach delivery ratio values greater than 8.4%, and (b2) systems C and C', which reached delivery ratio values up to 47.3%. Subgroup b1 clearly had transport-limited conditions with relatively high sediment concentrations of runoff, whereas subgroup b2 went some way toward detachment-limited conditions in winter. The difference between the two subgroups can be attributed to the difference in runoff regime. On systems A and B runoff was strongly reduced compared to systems C and C', especially in winter, due to the difference in soil surface state. So, on systems A and B insufficient runoff occurred to remove the available mass of detached soil particles.

The question arises as to why not all soil particles that were detached by drop impact, were removed from the plots of systems C and C' in winter, as runoff volumes were high and yet sediment concentrations of the runoff were relatively low for these plots (Figure 6).

Sediment concentrations of runoff as high as 82g.l[-1] have been measured during earlier years of the plot study. Runoff could have transported more material than it in fact did. The following explanations are tentatively put forward as to why the delivery ratio did not reach 100%:

(a) Transport capacity of runoff has been too low in terms of particle size. Velocity of overland flow was too low to move detached particles of all sizes, including aggregates. Indeed, part of the eroded soil material in both the storage tanks of the plots and in the splash collectors consisted of soil aggregates and small clods of millimetres in size.
(b) Runoff was generated only on part of the plot surface (e.g. in tractor wheelings) and/or only part of the plot surface contributed to the runoff that was collected in the storage tanks, so that some of the soil particles detached by drop impact did not come into contact with running water or did not reach the storage tanks.
(c) By working on a monthly basis, the effects of rainfall events with and without runoff are combined. Individual events may have had higher delivery ratios than are calculated from the monthly figures.
(d) Splash erosion rates on cropping systems C and C' may have been overestimated (and delivery ratios underestimated). Lower splash rates may have occurred locally (e.g. in tractor wheelings) than near the splash cups, which were placed between the wheelings.

5. CONCLUSIONS

It seems that during the period of study, detachment-limited erosion did not occur on the plots. This would imply that no excess transport capacity of runoff was available, and that there was no risk of erosion by runoff itself. Indeed, at no time during the four years of the study did rills form on any of the plots (Bryan, 1987).

ACKNOWLEDGEMENTS

Financial support was received from the Foundation Experimental Farm Wijnandsrade, the Landscape Engineering Service at Roermond and the Provincial Government of Limbourg. The plots were located on the property of the Experimental Farm Wijnandsrade, and all agricultural operations were carried out by the Experimental Farm with Mr. Kerckhoffs as acting manager. The cropping systems of maize were devised primarily by Ing. P. Geelen, research worker at the Experimental Farm.

REFERENCES

BERGSMA, E. 1980. Provisional rain erosivity map of the Netherlands. In: de Boodt, M. and Gabriels, D. (eds). Assessment of erosion. Wiley, Chichester.

BOLLINNE, A. 1975. La mesure de l'intensité du splash sur sol limoneus. Pédologie 25:199-210.

BOLLINNE, A. 1982. Etude et prévision de l'érosion des sols limoneux cultivés en moyenne Belgique. Thesis Liège.

BRAKENSIEK, D.L., OSBORN, H.B. and RAWLS, W.J. (eds). 1979. Field manual for research in agricultural hydrology. US Department of Agriculture. Agriculture Handbook No. 224.

BRYAN, R.B. 1987. Processes and significance of rill development. Catena Supplement 8:1-15.

BUISHAND, T.A. and VELDS, C.A. 1980. Neerslag en verdamping. K.N.M.I., de Bilt.

ELLISON, W.D. 1947. Soil erosion studies - Parts 1 and 2. Agricultural Engineering 28:145-146,197-201.

GEELEN, P. 1987. Bestrijding van watererosie. Jaarverslag over 1986 van de Proefboerderij Wijnandsrade

GOVERS, G. and POESEN, J. 1988. Assessment of the interrill and rill contributions to total soil loss from an upland field plot. Geomorphology 1, 343-354.

KWAAD, F.J.P.M. 1991. Summer and winter regimes of runoff generation and soil erosion on cultivated loess soils (The Netherlands). Earth Surface Processes and Landforms 16: 653-662.

KWAAD, F.J.P.M. 1994. Cropping systems of silo maize to reduce erosion on cultivated loess soils. In: Rickson, R.J. (ed.). Conserving Soil Resources: European Perspectives. CAB International, Wallingford, Oxfordshire.

KWAAD, F.J.P.M. and VAN MULLIGEN, E.J. 1991. Cropping system effects of maize on infiltration, runoff and erosion on loess soils in South Limbourg (The Netherlands): a comparison of two rainfall events. Soil Technology 4:281-295.

LEVERT, C. 1954. Regens, een statitische studie. Staatsdrukkerij, Den Hag.

MEYER, L.D. and WISCHMEIER, W.H. 1969. Mathematical simulation of the process of soil erosion by water. Transactions of the American Society of Agricultural Engineers 12:754-758,762.

MOEYERSONS, J. and DE PLOEY, J. 1976. Quantitative data on splash erosion, simulated on unvegetated slopes. Zeitschrift für Geomorphologie, Supplement 25:120-131.

MORGAN, C. 1986. The relative significance of splash, rainwash and wash as processes of soil erosion. Zeitschrift für Geomorphologie 30:329-337.

MORGAN, R.P.C. 1986. Soil erosion and conservation. Longman, Harlow.

MUTCHLER, C.K., MURPHREE, C.E. and McGREGOR, K.C. 1988. Laboratory and field plots for soil erosion studies. In: Lal, R. (ed.). Soil erosion research methods. Soil and Water Conservation Society, Ankeny.

POESEN, J. and TORRI, D. 1988. The effect of cup size on splash detachment and transport measurements. Part 1. Field measurements. Catena Supplement 12:113-126.

ROEHL, J.W. 1962. Sediment source areas, delivery ratios and influencing morphological factors. Int. Assoc. Scient. Hydrol. Publ. 59:202-213.

YOUNG, R.A. and WIERSMA, J.L. 1973. The role of rainfall impact in soil detachment and transport. Water Resources Research 9:1629-1636.

WISCHMEIER, W.H. and SMITH, D.D. 1978. Predicting rainfall erosion losses. A guide to conservation planning. US Department of Agriculture. Handbook No. 537.

Chapter 26

MODELLING THE EFFECTS OF CLIMATE CHANGE ON RUNOFF AND EROSION IN CENTRAL SOUTHERN NORWAY

P. Botterweg

JORDFORSK, Centre for Soil and Environmental Research,

N-1432 Ås, Norway

Summary

The expected climate changes caused by the enhanced greenhouse effect of CO_2 and NO_x gases has initiated several studies to evaluate the effects on global and local ecosystems. The study presents an evaluation of the effect of climate change on runoff and erosion with the SOIL/CREAMS-model. Runoff and erosion from a field in central southern Norway have been simulated for a 19 year period, and compared with measured, expected and extreme weather data. Expected and extreme weather data are generated by increasing daily measured values of temperature and precipitation. The simulations showed that for the whole period, runoff and erosion do not change significantly, but the response to climate changes varies significantly between different years. On average, a decrease of runoff and erosion is found in winter (including snowmelt), but an increase during the growing season and autumn was found. Maximum daily values of runoff and erosion increase up to 200%, together with an increase in the number of events. Climate change may cause an increased loss of nutrients during the growing season.

1. INTRODUCTION

Since the late 1970s both scientists and politicians have become more aware and anxious about the long term effect of increasing CO_2 and NO_x concentration in the atmosphere. The enhanced greenhouse effect caused by the changes in the atmosphere may change the climate on earth. These changes may in some way affect ecological systems. International research programmes have been initiated to quantify the expected climate and environmental changes (IGBP, 1988). The Norwegian Air Research Institute (NILU, 1990) expects an increase in mean temperature and in the amount of precipitation in Norway. It is also expected that precipitation will come more in the form of intense storms, rather than long-lasting precipitation with low intensity. No changes in mean wind direction or velocity are expected. NILU has used two different scenarios, based on General Circulation Models, resulting in an expected, most likely climate change and an extreme, less likely climate change, respectively (Table 1). However, it has to be realised that magnitude or even direction of the estimates are uncertain and are still under discussion among meteorologists. That discussion lies outside the scope of this paper, and the climate changes as estimated by NILU are accepted.

Climate change may cause changes in agriculture, both in regard to the crops grown and management practices. As an obvious consequence, the level of non-point source pollution, including soil erosion, will probably increase. Erosion of arable land is a severe problem in southern parts of Norway. This study investigates the effect of climate change on runoff and erosion using the SOIL/CREAMS model. The study was requested by a committee that analysed the consequences of climate change for Norway.

Table 1. Expected and extreme (in parentheses) climate changes for the coastal and continental region of Norway (NILU, 1990).

	Coastal Region	Continental Region
TEMPERATURE		
Winter (Dec-Feb)	+3.0 (+3.5)°C	+3.5 (+5.0)°C
Summer (June-Aug)	+1.5 (+2.5)°C	+2.0(+3.0)°C
PRECIPITATION (mm)		
Winter (Dec-Feb)	+15 (+15)%	+10 (+15)%
Spring (Mar-May)	+10 (+15)%	+10 (+15)%
Summer (June-Aug)	+5 (+20)%	+5 (+20)%
Autumn (Sept-Nov)	+5 (+15)%	+5 (+15)%

2. THE EFFECT OF CLIMATE CHANGE ON RUNOFF AND EROSION

The effect of climate change on erosion can be evaluated using a mathematical description of the erosion processes as given by Foster et al. (1990). Total erosion from an area is determined by interrill erosion, rill erosion and sediment transport capacity. Semi-mathematical descriptions of these processes are given in equations 1-3.

$$IE = f(EI, R, K, S, M) \tag{1}$$
$$RE = f(R, K, L, M) \tag{2}$$
$$TC = f(R, L, S) \tag{3}$$

where

IE = interrill erosion $\qquad$ R = runoff (mm)
RE = rill erosion $\qquad$ K = soil erodibility
TC = transport capacity $\qquad$ L = slope length (m)
EI = Wischmeier's rainfall erosivity $\qquad$ S = slope (%)
M = management factor

How will these factors be affected by climate changes?

2.1 Wischmeier's rainfall erosivity (EI)

Rainfall erosivity can be estimated from daily precipitation as defined by Lombardi (1979):

$$EI = a \cdot P^b \tag{4}$$

where P = daily precipitation, and a, b are coefficients. EI is an estimation of the energy available for splash erosion. It is therefore reasonable to accept that the coefficients a and b depend on precipitation characteristics, especially intensity. Regression analyses from approximately 2700 USA data points for daily precipitation (inch/day), results in a = 7.897 and b = 1.5. For Norway the coefficients are unknown, but a formula as found by Lombardi (1979) can be expected to yield reasonable results. Thus, an increase of daily precipitation (P) will increase EI directly. Styczen and Nielsen (1989) showed that splash erosion increases proportionally with rainfall intensity, and an increase of one or both coefficients is therefore likely.

2.2 Runoff (R)

Runoff is the difference between precipitation (P) and infiltration (I):

$$R = P - I \qquad\qquad\qquad\qquad (5)$$

The expected change in precipitation has been quantified (Table 1), but how will infiltration change when both temperature and precipitation increases? The following main effects of increased temperature on infiltration may apply for areas that have a stable winter:

1. increased evapotranspiration during the whole year;
2. increased number of events with rain on frozen soil;
3. more freeze-thaw cycles in the top-soil;
4. less snow accumulation.

Higher precipitation intensity and precipitation amount causes additional effects:

5. more sealing of the top-soil;
6. higher water content in the soil.

Without modelling the infiltration process on a daily basis it is difficult to estimate the combined effect of points 1-6, which separately have different effects on infiltration, e.g. increased evapotranspiration reduces the soil water content and increases infiltration capacity (Hillel, 1980). Less snow accumulation reduces the amount of melt water in spring and reduces runoff. Lundin (1990) found that alternating frost/thaw reduces infiltration. Thus, more events with snowmelt and/or rain on frozen soil will increase runoff.

2.3 Soil erodibility factor (K)

In the laboratory or field soil shear strength (which is positively correlated with erodibility) can be measured, but a direct measurement of soil erodibility is not possible. Wischmeier et al. (1971) constructed a soil erodibility nomograph for farm land which showed that K is positively correlated with the percentage of silt plus percentage of very fine sand, and negatively correlated with the content of organic matter in the top-soil. The expected rise of temperature will stimulate the biological activity in the top-soil, and the organic matter content in Norwegian soils may decrease, leading to a higher erodibility. The effect of alternating frost/thaw on soil erodibility is not clear (Benoit et al., 1990). However, thawing soil has a higher K-value than soil under other conditions (Kok, 1989). Therefore, an increase of K during the winter may be expected with a higher mean temperature.

2.4 Terrain factors slope (S), and slope length (L)

Of course, topography is not directly affected by climate changes, but there may be an indirect impact. From an agricultural point of view the climate in Norway will become more suitable than today. Agricultural production per ha may increase, and the area used for production may expand to areas at higher altitude and latitude. On average for the whole country, the increase of productive land may change S and L.

2.5 Management factor (M)

The management factor used here includes soil tillage and crop selection. The expected warmer climate means that the area usable for production of small grains (e.g. wheat and barley) will expand further to the north. At the same time other crops, demanding a warmer climate (e.g. sugar beet and maize) may be introduced in the southern part of Norway. The use of winter wheat will also increase. These changes in agriculture may lead to a higher percentage of cultivated fields unprotected by plant cover during the autumn or spring. A summary of expected impacts on erosion is given in Table 2.

Table 2. Summary of theoretically expected changes in erosion, caused by climate changes.

EROSION = f(EI, R, K, S, L, M)		(1-3)
Expected change/effect on erosion		
++	EI	= rainfall erosivity
??	R	= runoff
++	K	= soil erodibility
0(+)	L	= slope length
0(+)	S	= slope
0(+)	M	= management

++ = increase; ?? = unknown; 0(+) = no effect on single field, but positive effect for the whole country.

3. MODEL SIMULATION

To evaluate effects of climate change on runoff and erosion the SOIL/CREAMS model (Botterweg, 1990) has been used to simulate a 19-year period under different climate regimes.

3.1 The model

The SOIL/CREAMS model is a combination of the physically based hydrology model SOIL (Jansson, 1990; 1991) and the erosion part of the CREAMS model (Knisel, 1980). SOIL simulates water and heat dynamics in a layered soil profile covered with vegetation. Water flow is assumed to be laminar and solved with Richards equation for unsaturated flows.

Heat flow in the model is the sum of conduction and convection. Compartments for snow, intercepted water and surface ponding are included to account for processes at the upper soil boundary. Different types of lower boundary conditions can be specified, including ground water flow. Weather input variables for the model are daily values for temperature, precipitation, wind velocity, relative humidity and cloud cover.

SOIL has been shown to simulate hydrology satisfactorily for a wide range of soil types and vegetation covers in different climate zones. Jansson (1991) presents a list of references with applications of SOIL. User selected output variables can be given on a daily basis. The model is calibrated with both surface runoff and drain runoff measurements (Botterweg, 1990).

The erosion part of the CREAMS model is a modification of the USLE as proposed by Foster et al. (1981). Morgan et al. (1987) found that CREAMS simulates yearly amounts of runoff quite well, but the accuracy of daily values is poor. CREAMS does not have satisfactory routines for handling snow cover and snow melt. For these reasons, Botterweg (1990) substituted the hydrology part with the SOIL model. The erosion part is not physically based but has the fundamental characteristics of the USLE with its shortcomings (Morgan, 1986; Svetlosanov and Knisel, 1982). In this study, field parameters were kept constant and in the input file only daily values for EI and runoff differ for the three climatic scenarios. The main interest in this study are changes in soil loss, not absolute values, and CREAMS's erosion part was therefore accepted as reliable. The erosion part has been calibrated with measured soil loss.

3.2 The climate

Three climate regimes (measured, expected and extreme climate) have been used to simulate runoff and soil loss from an arable field. Measured weather data from 1970-1988 were taken from a weather station situated 30km west of the site. Based on Table 1 data, for each day, expected and extreme temperature has been calculated by increasing daily mean temperature with a date-dependent value (Figure 1). In the same way daily precipitation has been increased by a date-dependent percentage (Figure 1). Today, meteorologists are not able to give other estimates for climate change, and changes in precipitation pattern and intensity could not be included in this study.

3.3 The field

The arable field, 0.9ha in size, is situated 30km north of Oslo in the continental climate zone. The soil has a clay content of about 50%. The 155m long slope of the field varies between 4 and 8%. For each year the same agricultural management, common for the area, is assumed. Small grains (wheat and barley) are the main crops. The seed bed is prepared and sown in the second half of April. The crop is harvested in the last week of August and the field is ploughed in the second half of October.

3.4 Simulation

Runoff and soil loss from the field have been simulated with the measured or control weather data, expected weather data and extreme weather data respectively for a period of 19 years. Each climate ran both with and without frost-limited infiltration. This was realised by a switch in the SOIL model, that gives an infiltration rate of zero when soil temperature falls below zero degrees. This method is a simplification. Thunholm and Lundin (1990) found a more complicated relation between ice content and infiltration. In this study only the combined effect of increased temperature and increased precipitation could be evaluated. Possible changes of precipitation intensity, K, L, S or M have not been taken into account.

Figure 1. Daily increase in temperature (°C) and precipitation (%) for expected and extreme climate change (NILU, 1990).

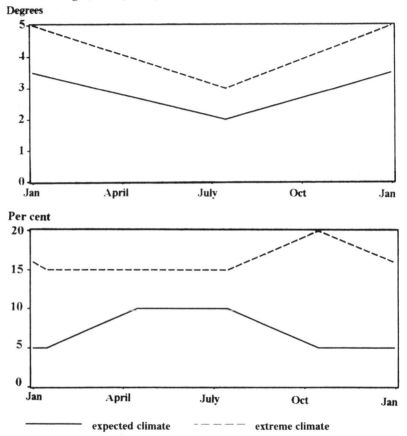

4. RESULTS

The results presented are based on daily values for runoff and soil loss over the period 1970-1988. Based on the daily output, the number of days with runoff and days with erosion have been calculated. Each year has been divided into three seasons, winter (January-April), spring/summer (May-September) and autumn (October-December). Winter includes the period of snowmelt. Spring/summer represents the growing season and autumn is the period with uncovered soil after harvesting, harrowing and/or ploughing.

The total amount of runoff for the 19-year period decreased for the expected climate change, but increased in the case of the extreme climate change (Fig. 2). When infiltration is limited by frost, the total number of days with runoff increased by less than 10% under the changed climates. Without frost-limited infiltration, a slight reduction in the number of days with runoff was found.

For soil loss the changes were different (Fig. 2). On the one hand, the number of days with soil loss increased independently of the effect of frost on infiltration. On the other hand, the total amount of soil loss decreased more than runoff, and no increase in soil loss was measured for the extreme climate change.

Thus, looking to total values for runoff and soil loss, climate change did not have a statistical effect. However, the response in the different years to climate changes varied considerably. In some years runoff and soil loss increased with up to 40%, while in other years a reduction of the same magnitude was found. The year 1973 gave an extreme relative change in soil loss, especially during the autumn (Table 3).

Table 3. Simulated, total soil loss, during three seasons in 1973, for control climate and expected and extreme changed climate.

		Soil Loss (kg.ha^{-1})		
Season (1973)	Frost Limited Infiltration	Control Climate	Expected Climate	Extreme Climate
Winter	OFF	1.1	15.7	276.9
Spring/summer	-	27.8	92.3	194.8
Autumn	-	1.1	845.6	191.7
SUM	-	30.0	953.6	663.4
Winter	ON	293.2	438.8	775.3
Spring/summer	-	124.1	164.6	192.0
Autumn	-	19.3	1474.6	511.1
SUM	-	436.6	2078.0	1478.4

Figure 2. Number of days with runoff and total runoff (mm), and the number of days with soil loss and total soil loss (t.ha^{-1}), 1970-1988 for 1) control; 2) expected and 3) extreme climate. (A = frost-limited infiltration option OFF; B = frost-limited infiltration option ON).

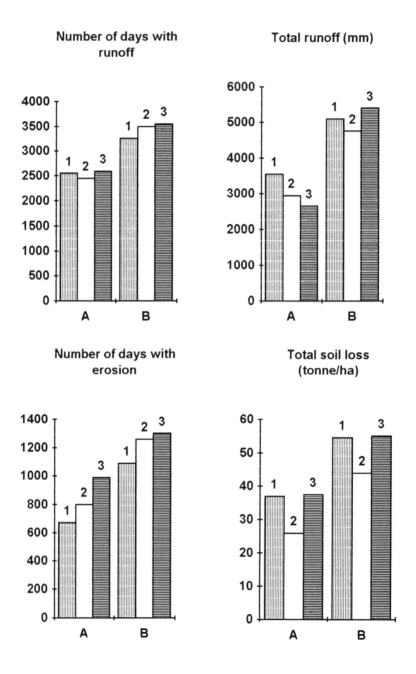

4.1 Runoff

During the winter the number of days with runoff decreased where frost had no effect on infiltration (Table 4). With frost-limited infiltration the number of days with runoff increased by about 15%. During summer and autumn, days with runoff increased slightly. Total runoff per season decreased during the winter independent of the effect of frost on infiltration. During the growing season, runoff increased by 15 and 40% respectively for expected and extreme climate change. In autumn, runoff increased by 25 and 61% respectively. A percentile distribution of daily runoff during the autumn shows that the number of days with high runoff values increases (Table 5). Maximum, measurable, daily runoff increases from 19mm to nearly 35mm.

Table 4. Total runoff (R) and number of days (N) with runoff per season for the period 1970-1988 for control, expected and extreme climate, with and without frost limited infiltration.

Season	Frost-limited infiltration	Climate					
		Control		Expected		Extreme	
		R (mm)	N	R (mm)	N	R (mm)	N
Winter	OFF	2092	1020	1207	752	1436	874
Spring/summer	-	639	752	736	759	895	850
Autumn	-	816	808	1023	941	1313	904
Winter	ON	2926	1284	2479	1457	2538	1479
Spring/summer	-	640	831	736	832	894	852
Autumn	-	1475	1138	1556	1179	1957	1188

Table 5. Percentile values and maximum runoff (mm.day^{-1}) during the autumn for three weather scenarios. (N=total number of events).

Climate	Frost-limited infiltration	Percentile				Max	N
		75	90	95	99		
Control	OFF	0.6	3.1	6.0	12.4	19.0	808
Expected	-	0.8	3.3	6.0	13.6	20.7	941
Extreme	-	1.2	4.8	7.2	15.8	34.7	904
Control	ON	1.3	4.2	6.5	11.6	18.5	1138
Expected	-	1.3	4.1	6.5	12.9	21.0	1179
Extreme	-	1.7	5.2	7.4	16.2	34.7	1188

4.2 Erosion

Independent of the effect of frost on infiltration, the total number of days with erosion increased for all seasons (Table 6). The total amount of soil loss decreased in winter for both expected and extreme climate changes (Table 6). During the growing season and autumn soil loss increased by as much as 60%. A percentile distribution showed that the reduction of erosion in winter is equally distributed over all events, independent of the amount of soil loss per event. During the autumn, extreme events increased in both number and in the amount of soil loss per event (Table 7). Thus, both detachment and transport capacity has increased. The maximum daily soil loss increases from 650 to nearly 1400 kg.ha^{-1}. A summary of the results is presented in Table 8.

Table 6. Total soil loss (SL) and number of days (N) with soil loss per season for the period 1970-1988 for control, expected and extreme climate, with and without frost limited infiltration.

| Season | Frost-limited infiltration | Climate | | | | | |
| | | Control | | Expected | | Extreme | |
		SL t.ha^{-1}	N	SL t.ha^{-1}	N	SL t.ha^{-1}	N
Winter	OFF	26.1	260	11.2	262	15.2	329
Spring/summer	-	1.4	193	1.9	219	3.1	332
Autumn	-	10.2	219	13.0	307	29.7	321
Winter	ON	33.8	394	22.0	494	24.7	509
Spring/summer	-	1.9	279	2.4	302	3.1	332
Autumn	-	18.9	403	19.6	449	28.0	460

Table 7. Percentiles values and maximum soil loss (kg.ha^{-1}.day^{-1}) during autumn for three weather scenarios. (N=total number of events).

| Climate | Frost-limited infiltration | Percentile | | | | Max | N |
		75	90	95	99		
Control	OFF	41	148	249	481	651	219
Expected	-	26	126	262	593	728	307
Extreme	-	55	161	309	570	1392	321
Control	ON	45	143	231	513	627	403
Expected	-	42	134	217	479	695	449
Extreme	-	56	166	299	602	1393	460

5. DISCUSSION

This preliminary effort to estimate the effect of climate changes on runoff and erosion shows: (i) an analysis based on the mean of 30 years' weather data did hide the variation in response that exists, and (ii) this would have smoothed extreme events. Runoff and erosion are typically the result of one single storm, and are affected by the weather conditions prior to the storm. Extreme events contribute considerably to total runoff and erosion, and these events are therefore important to incorporate in an evaluation. This study shows that the effect of climate change on runoff and erosion has to be analysed on an event basis, including intensity and other storm factors which were not considered here. It has to be realised that using predictive, simulation models always includes extra uncertainty because of the lack of validation.

Table 8. Summary of relative changes (%) in runoff and erosion for a clay soil in central southern Norway, as caused by expected and extreme climate change, for a 19-year period. (N=number of days with runoff/erosion; T=total runoff/erosion; MAX=maximum runoff/erosion per day).

| Season | Frost-limited infiltration | RUNOFF | | | | | |
| | | Expected climate change | | | Extreme climate change | | |
		N	T	Max	N	T	Max
Winter	OFF	-26	-42	-23	-14	-31	-22
Spring/summer	-	1	15`	-7	13	40	7
Autumn	-	16	25	9	12	61	83
Winter	ON	13	-15	-11	15	-13	-15
Spring/summer	-	0	19	25	3	40	23
Autumn	-	4	5	14	4	33	86

| Season | Frost-limited infiltration | EROSION | | | | | |
| | | Expected climate change | | | Extreme climate change | | |
		N	T	Max	N	T	Max
Winter	OFF	0	-57	-18	26	-41	-23
Spring/summer	-	13	35	28	40	125	39
Autumn	-	72	27	12	47	193	214
Winter	ON	25	-35	-27	29	-28	-25
Spring/summer	-	8	30	27	19	64	47
Autumn	-	11	4	11	14	48	222

The study shows that over a long period, annual runoff and erosion change insignificantly. However, significant differences in response between years and between seasons were found. During winter, runoff and erosion decrease; during the growing season and autumn, runoff and erosion increase. It has to be noted that the winter results are less certain than the results for the other seasons.

The effects of frost and thaw processes on infiltration/runoff are shown to be a weak point in SOIL (Botterweg, 1990), and an improvement on this point is necessary. It is recommended that Lundin's (1989) analysis of frost impact on infiltration be incorporated in the model. Another source of error is the estimate of a soil erodibility factor for thawing soil. CREAMS operates with a constant value for the whole year, which is unrealistic for soils that are frozen for part of the year (Kok, 1989). The number of days with runoff and erosion increases for the winter season, but the total amounts of runoff and soil loss decreases. Therefore, the model is expected to be sensitive to changes in hydrological conductivity and erodibility in periods with alternating frost and thaw.

Oygarden (1989) showed that the field simulated in this study had the highest concentration of nutrients in runoff during the growing season. Increased runoff during that period will then lead to a significantly increased loss of nutrients. Growing row-crops (like maize and sugar beet) may cause increased runoff, too, as compared to growing small grains. Even higher nutrient losses can be expected. In summer the effect of lost nutrients will be more severe, compared with colder periods. This study showed that to reduce erosion from agricultural areas, measures that have an effect during autumn have to be given highest priority. This study clearly indicates that a closer co-operation with meteorologists is necessary. Besides data for mean changes of temperature, precipitation and other climatic factors, an analysis of the frequency distribution of extreme events is needed to evaluate the consequences of climate change.

REFERENCES

BENOIT, G.R., YOUNG, R.A. and WILTS, A. 1990. Runoff and erosion during simulated rainfall on frozen field plots with different depths of surface thaw and level of erodibility. In: Cooley, K.R. (ed.). Proceedings, International Symposium on Frozen Soil Impacts on Agricultural, Range and Forest Lands. March 21-22, 1990. Spokane, Washington. CRREL Special Report 90-1. pp.135-144.

BOTTERWEG, P.F. 1990. The effect of frozen soil on erosion - a model approach. In: Cooley, K.R. (ed.). Proceedings, International Symposium on Frozen Soil Impacts on Agricultural, Range and Forest Lands. March 21-22, 1990. Spokane, Washington. CRREL Special Report 90-1.

FOSTER, G.R., LANE, L.J., NOWLIN, J.D., LAFLEN, J.M. and YOUNG, R.A. 1981. Estimating erosion and sediment yield on field-sized areas. Transactions of the ASAE 23:1253-1262.

FOSTER, G.R., LANE, L.J., NOWLIN, J.D., LAFLEN, J.M. and YOUNG, R.A. 1990. A model to estimate sediment yield from field-sized areas: development of the model. In: Knisel, W.G. (ed.). CREAMS; A field scale model for chemicals, runoff and erosion from agricultural management systems. SDA, Conservation Research Report No. 26:36-64.

HILLEL, D. 1980. Fundamentals of soil physics. Academic Press, London.

IGBP 1988. The International Geosphere-Biosphere Program: A study of global change. Report No. 4. A plan for action. IGBP, Stockholm.

JANSSON, P. 1990. SOIL water and heat model. Technical description. Department of Soil Sciences, Agricultural University of Sweden, Uppsala.

JANSSON, P. 1991. SOIL model, User's manual. Department of Soil Sciences, Agricultural University of Sweden, Communications 91:7, Uppsala.

KNISEL, W.G. 1980. CREAMS; A field scale model for chemicals, runoff and erosion from agricultural management systems. SDA, Conservation Research Report No. 26.

KOK, H. 1989. Freeze-thaw induced variability of soil shear strength. PhD thesis, University of Idaho.

LOMBARDI, F. 1979. Universal Soil Loss Equation (USLE), runoff erosivity factor, slope length exponent, and slope steepness exponent for individual storms. PhD thesis, Purdue University, W. Lafayette, Ind.

LUNDIN, L.-C. 1989. Water and heat flows in frozen soils. Basic theory and operational modelling. Comprehensive summaries of Uppsala Dissertations from the Faculty of Science 186, Uppsala.

LUNDIN, L.-C. 1990. Simulating the freezing and thawing of arable land in Sweden. In: Cooley, K.R. (ed.). Proceedings, International Symposium on Frozen soil impacts in agricultural, range and forest lands, 21-22 March, 1990. Spokane, Washington. CRREL Special Report 90-1. pp.87-98.

MORGAN, R.P.C. 1986. Soil erosion and conservation. Longman. Harlow.

MORGAN, R.P.C., MORGAN, D.D.V. and FINNEY, H.J. 1987. Predicting hillslope runoff and erosion in the Silsoe area of Bedfordshire, England using the CREAMS model. In: Pla Sentis, I. (ed.). Soil conservation and productivity. Proceedings of the IV ISCO Conference. Sociedad Venezolana de la Ciencia del Suelo, Maracay, Venezuela.

NILU. 1990. Drivhuseffekten og klimautviklingen. Bidrag til den interdepartementale klimautredningen (with English summary). Norwegian Institute for Air Research, NILU OR 21/90, Oslo, 1990.

OYGARDEN, L. 1989. Handlingsplan mot landbruksforurensninger. Rapport nr. 6. Utproving av tiltak mot araelavrenning i Akershus, GEFO, Ås.

STYCZEN, M. and NIELSEN, S.A. 1989. A view of soil erosion theory, process-research and model building: possible interactions and future developments. Quaderni di Scienza del Suolo II:27-47

SVETLOSANOV, V. and KNISEL, W.G. (eds). 1982. European and United States case studies in application of the CREAMS model. IIASA Collaborative Proceedings series, CP-82-S11, Luxembourg.

THUNHOLM, B. and LUNDIN, L.-C. 1990. Infiltration into a seasonally frozen clay soil. In: Cooley, K.R. (ed.). Proceedings, International Symposium on Frozen Soil Impacts on Agricultural, Range and Forest Lands. 21-22 March, 1990. Spokane, Washington. CRREL Special Report 90-1:156-160.

WISCHMEIER, W.H., JOHNSON, C.B. and CROSS, B.V. 1971. A soil erodibility nomograph for farmland and construction sites. Journal of Soil and Water Conservation 26:189-193.

Chapter 27

THE EUROPEAN SOIL EROSION MODEL: AN UPDATE ON ITS STRUCTURE AND RESEARCH BASE

R.P.C. Morgan
Silsoe College, Cranfield University, Silsoe,
Bedfordshire, MK45 4DT, UK

Summary

The European Soil Erosion Model (EUROSEM) is a process-based erosion prediction model designed to predict erosion in individual events and to evaluate soil protection measures. Early versions of the model, described at international conferences in Addis Ababa, Salt Lake City and Dehra Dun, are up-dated and the present version of the model is described.

The model uses a mass balance equation to compute sediment transport, erosion and deposition over the land surface. The model simulates the volume of rainfall reaching the ground surface as direct throughfall, leaf drainage and stemflow. The rate of detachment of soil particles by raindrop impact is computed as a function of the energy of the direct throughfall and leaf drainage, the detachability of the soil and the depth of surface water. The detachment of soil particles by runoff is determined as a function of the difference between transport capacity and existing sediment concentration in the flow, simultaneous deposition of sediment from the flow and the cohesion of the soil. Transport capacity is modelled as a function of unit stream power (product of velocity and slope). Compared with other similar models, EUROSEM simulates tillage and crop cover (vegetation) effects in a dynamic way and accounts for soil protection measures by describing the soil, microtopographic and vegetation conditions associated with each practice.

EUROSEM must be linked to a hydrological model capable of predicting and routing runoff. The present version is linked to KINEROS, a kinematic runoff model.

1. INTRODUCTION

The development of policies in Europe to control erosion is, at present, hindered by the lack of a satisfactory system for assessing the problem. Assessment methods based on scoring systems for rainfall, soils, slope and land use (Auerswald and Schmidt, 1986; Rubio, 1988; Giordano et al., 1991) provide data on the spatial distribution of erosion risk but limited information on erosion rates which cannot be validated. Also, they do not provide the necessary data for designing soil conservation measures or evaluating their effects. These deficiencies can only be overcome by combining erosion risk assessments with predictions from erosion models.

The European Soil Erosion Model (EUROSEM) is being developed by a team of scientists from ten European countries as a tool for assessing erosion risk and evaluating soil

286

protection measures (Chisci and Morgan, 1988). The model uses a process-based approach to predict erosion for individual storms from fields and small catchments. Although this European effort parallels similar work in the USA on CREAMS (Foster et al., 1981), ANSWERS (Beasley et al., 1980) and WEPP (Nearing et al., 1989), there are important differences between EUROSEM and these other models. First, EUROSEM operates for single events and is not a continuous simulation model, which means that it requires input data on the starting conditions for each storm. Second, EUROSEM provides hydrographs and sedigraphs as routine output for each storm as well as totals of runoff and soil loss. Third, EUROSEM uses a more explicit modelling of (1) the transfer of splashed particles on the interrill areas to rills; (2) the role of the plant cover; (3) simultaneous deposition and erosion of material by flow; and (4) changes in soil surface conditions during a storm. Early versions of the model were described by Morgan et al. (1990; 1992a). This paper describes Version 2.4, as applied at the scale of a field or single slope segment and, therefore, provides an up-date on the previous papers. Together, the papers illustrate the evolution of EUROSEM over time.

2. STRUCTURE AND OPERATIONAL PROCEDURE

A flow chart of the model is presented in Figure 1. The model has a modular structure with each part being developed in as much detail as the existing level of knowledge allows. This structure will enable improvements to be made as new research improves our understanding of erosion processes and the way they can be described. Additional modules are planned to deal with snowmelt, frozen soils, crusting, cracking soils and stoniness.

EUROSEM is designed to operate for successive one-minute time steps within a storm. The basic inputs are the length and width of the individual fields or slope segments to which the model is being applied, and the rainfall depths for successive time periods in the storm within which rainfall intensity is more or less uniform. These and other input requirements are listed in Table 1. The operating equations are set out in Table 2.

EUROSEM deals in turn with the interception of rainfall by the plant cover; the volume and kinetic energy of the rainfall reaching the ground surface as direct throughfall and leaf drainage; the volume of stemflow; the volume of surface depression storage; the detachment of soil particles by raindrop impact; detachment and deposition of soil particles by runoff; and the transport capacity of the runoff.

The model computes soil loss as a sediment discharge, defined as the product of the volume of runoff and the sediment concentration in the flow, to give a volume of sediment passing a given point in a given time. The computation is based on the dynamic mass balance equation (Table 2; equation 1) (Bennett, 1974; Kirkby, 1980; Woolhiser et al., 1990), illustrated in Figure 2. The term, e, in this equation is defined (equation 2) as the summation of the rate of soil particle detachment by raindrop impact and the net rate of soil particle detachment by flow. The latter represents the relative balance between the rates of particle detachment and simultaneous sediment deposition.

Figure 1. Flow chart of the European Soil Erosion Model (EUROSEM).

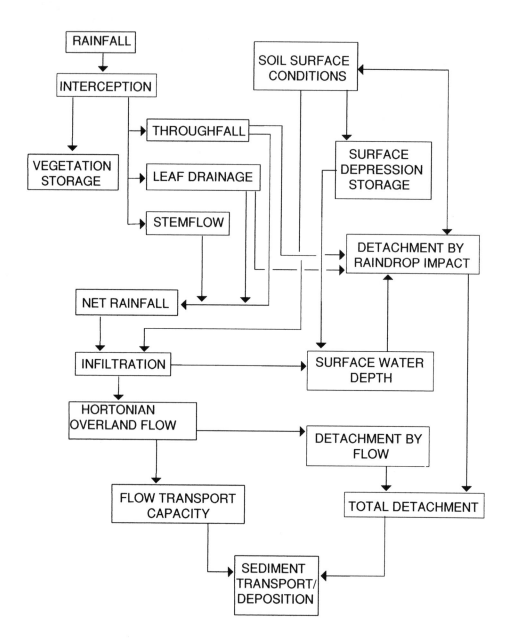

Table 1. Input requirements for EUROSEM.

RAINFALL	
Rainfall depth (mm) for each time step in the storm	(R)
Duration of the storm (min)	
SOIL	
Detachability index (g.J)	(k)
Settling velocity of the soil particles (m.s^{-1})	(v_s)
Cohesion (kPa) of the soil at saturation	(COH)
Median particle size (µm)	(D50)
Manning's n for soil particle roughness	(n)
Saturated hydraulic conductivity	(k_s)
TOPOGRAPHY	
Width of the slope segment (m)	(W)
Length of the slope segment (m)	(L)
Roughness ratio	(RR)
Slope steepness (m.m)	(S)
Manning's n for microtopographic roughness	(n_t)
Number of concentrated flow paths per m width	(DEPNO)
Average depth of concentrated flow paths (m)	(RILLD)
Average width of concentrated flow paths (m)	(RILLW)
Side slope of concentrated flow paths (rills) (1:X)	(Z)
VEGETATION	
Percentage vegetation cover	(COV)
Maximum depth of interception storage (mm)	(IC_{max})
Average acute angle of plant stems to ground (°)	(PA)
Plant shape factor (grasses/others)	(SHAPE)
Percentage basal area of plant stems	(PBASE)
Average height of the plant canopy (m)	(PH)
Manning's n for plant-induced roughness	(n_v)

Note: other input parameters are as specified for the KINEROS model (Woolhiser et al., 1990). A full list of the input requirements is contained in the User Guide (Morgan et al., 1991).

EUROSEM has been designed so that it can be linked to hydrological models from which values of surface runoff can be generated. EUROSEM is, at present, attached to the KINEROS model (Woolhiser et al., 1990) which is an event oriented, physically-based distributed model that uses a kinematic wave equation, solved by a finite difference technique, to model the movement of runoff and sediment over the land surface. It should be feasible to attach EUROSEM to other single-event, distributed hydrological models and work is in progress to link it with the Système Hydrologique Européen (SHE) model (Danish Hydraulic Institute, 1985).

Table 2. Operating equations for EUROSEM.

MASS BALANCE

$$\partial(AC) / \partial t + \partial (QC) / \partial x - e (x, t) = q_s (x, t) \qquad (1)$$
$$e = DET + DF \qquad (2)$$

RAINFALL INTERCEPTION

$$IC = R . COV \qquad (3)$$
$$DT = R - IC \qquad (4)$$
$$IC_{store} = IC_{max} (1 - \exp R_{cum}/IC_{max}) \qquad (5)$$
$$SF = 0.5 \, TIF (\cos PA . \sin^{-2} PA) \qquad (6)$$
for grasses, and
$$SF = 0.5 \, TIF (\cos PA) \qquad (7)$$
for other plant species.
$$LD = TIF - SF \qquad (8)$$

SOIL SURFACE CONDITION

$$DS = 0.15 \{(1 - RR)/DEPNO\} \qquad (9)$$

RUNOFF GENERATION

$$k_{sveg} = k_s . 1/(1 - PBASE) \qquad (10)$$
$$Q = a \, h^m \qquad (11)$$
$$a = (S)^{0.5}/n \qquad (12)$$
$$m = 5/3 \qquad (13)$$

SOIL DETACHMENT BY RAINDROP IMPACT

$$DET = k . KE^{1.0} . e^{-2h} \qquad (14)$$
$$KE(DT) = 8.95 + (8.44 . \log I) \qquad (15)$$
$$KE(LD) = (15.8 . PH^{0.5}) - 5.87 \qquad (16)$$

SOIL DETACHMENT BY FLOW

$$DF = y \, w \, v_s (TC - C) \qquad (17)$$
$$y = u_{gmin}/u_{gcrit} \qquad (18)$$
$$u_{gcrit} = 0.89 + 0.56 \, COH \qquad (19)$$

TRANSPORT CAPACITY OF FLOW

$$TC = 0.063 (Su - 0.4)^{0.56} \qquad D50=50 \ \mu m \qquad (20)$$
$$TC = 0.038 (Su - 0.4)^{0.75} \qquad D50=100 \ \mu m \qquad (21)$$
$$TC = 0.033 (Su - 0.4)^{0.77} \qquad D50=120 \ \mu m \qquad (22)$$
$$TC = 0.027 (Su - 0.4)^{0.82} \qquad D50=150 \ \mu m \qquad (23)$$
$$TC = 0.022 (Su - 0.4)^{0.89} \qquad D50=200 \ \mu m \qquad (24)$$
$$TC = 0.017 (Su - 0.4)^{0.96} \qquad D50=250 \ \mu m \qquad (25)$$

SEDIMENT DEPOSITION

$$DEP = w \, v_s (TC - C) \qquad (26)$$

Table 2 continued.

UNIFIED RILL MODEL

$$\frac{dQ}{dh} = a\left\{ P(m+1)r^m \frac{dr}{dh} + r^{m+1}\frac{dP}{dh}\right\} \tag{27}$$

$$Y = \frac{1.05W_{ir}(h/d)^j}{1+(h/d)^j} + zh \tag{28}$$

$$DR = DET . q / v \tag{29}$$

where C = sediment concentration (m ^{3}.m^{-3})

A = cross sectional area of the flow (m ^{2})

Q = discharge (m ^{3}.s^{-1})

q_s = external input or extraction of sediment per unit length of flow (m ^{3}.s^{-1} per m)

e = net detachment rate or rate of erosion of the bed per unit length of flow (m ^{3}.s^{-1} per cm)

x = horizontal distance (m)

t = time

DET = rate of soil particle detachment by raindrop impact (m ^{3}.s^{-1})

DF = net rate of soil particle detachment by flow (m ^{3}.s^{-1})

IC = depth of rainfall intercepted by the vegetation cover (mm)

DT = depth of direct throughfall (mm)

IC_{store} = depth of rainfall stored on the vegetationcover (mm)

R_{cum} = cumulative depth of rainfall (mm) since start of storm

SF = depth of stemflow (mm)

TIF = depth of temporarily intercepted throughfall (mm)

LD = depth of leaf drainage (mm)

DS = depth of surface depression storage (mm)

DEPNO = the number of depressions (individual low points) along the transect of roughness measurement

k_{sveg} = the saturated hydraulic conductivity of the soil with the vegetation (m.h^{-1})

h = average water depth (m)

KE = kinetic energy of the rainfall (J.m^{-2}.mm^{-1})

I = rainfall intensity (mm.h^{-1})

w = flow width (m)

u_{gmin} = the minimum value required for the critical grain shear velocity to detach soil particles (cm.s^{-1})

u_{gcrit} = the critical grain shear velocity for rill initiation (cm.s^{-1})

P = wetted perimeter (m)

r = hydraulic radius (m)

Y = equivalent horizontal distance between base of rill sidewall and intersection of water surface on the sidewall (see Figure 3)

Table 2 continued.

w_{ir} = width between the centre line of the rill and the centre line of the interrill area (m)
d = depth between the base of the rill and the average height of the interrill surface (m)
j = an exponent
DR = rate of supply to rills of soil particles detached by raindrop on the interrill area (g.m^{-2})
q = discharge per unit length of flow (m ^{3}.s^{-1} per m)

For other notation, see Table 1.

Figure 2. Representation of the mass balance equation.

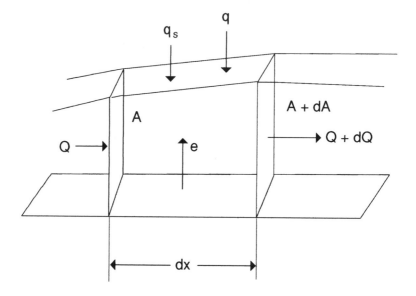

3. RAINFALL INTERCEPTION

From the rainfall depth for each time period in the storm, the model calculates the rainfall intensity and rainfall volume (i.e. depth × area). Account is also kept of the cumulative rainfall depth during the storm.

On reaching the canopy of the vegetation, the rainfall is divided into two parts (equation 3). These are that reaching the soil surface as direct throughfall (equation 4), falling either on open ground or passing through gaps in the canopy, and that which strikes the vegetation cover. A proportion of the intercepted rainfall is stored on the leaves and branches of the vegetation. This is termed the interception store (equation 5; Merriam, 1973). The remainder reaches the ground surface as either stemflow (equations 6 and 7 for grasses and other vegetation types respectively, adapted from van Elewijck, 1989) or leaf drainage (equation 8) which together comprise temporarily intercepted throughfall. The total rainfall reaching the ground is the sum of the contributions from direct throughfall, leaf drainage and stemflow.

Figure 3. Representation of the hydraulics of flow in the unified rill scheme described by equation 26 (after Smith, personal communication, 1992).

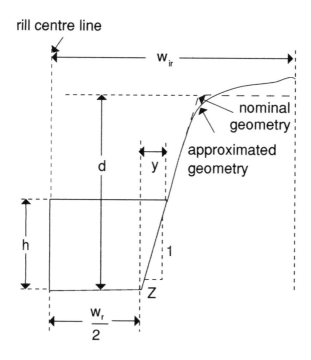

4. SOIL SURFACE CONDITION

The roughness of the soil surface, including roughness brought about by tillage, determines the volume of water that can be held on the surface as depression storage. The roughness of the soil surface is expressed by a roughness ratio (RR) defined as the ratio of the straight line distance between two points on the ground to the actual distance measured over all the microtopographic irregularities. The average depth of depression storage is estimated from measurements of roughness made in the downslope direction using equation 9. This equation assumes that only 15% of the overall roughness of the soil surface is effective surface depression storage. Studies by Martin (1980) showed that depression storage varied from 10% of the area on a loamy sand surface with a local micro-relief of 1cm, to 16% on a surface with a local micro-relief of 10cm. If these studies are typical, the chosen figure of 15% may well be an overestimate. It is hoped that trials with EUROSEM will confirm whether this is the case and that future research will enable an improved modelling procedure to be used.

5. RUNOFF GENERATION

Runoff is generated as a flow depth using the KINEROS model (Woolhiser et al., 1990), after modifications to deal with rainfall interception and surface depression storage, as outlined above. KINEROS simulates surface runoff as infiltration-excess overland flow using the Smith and Parlange (1978) procedure to model infiltration. The values of the saturated hydraulic conductivity used in KINEROS are adjusted for EUROSEM, using equation 10 (Holtan, 1961), to allow for the effect of vegetation on infiltration rates.

When the net rainfall intensity at the ground surface exceeds the infiltration rate and surface depression storage is satisfied, the excess comprises surface runoff. In the KINEROS model, runoff on a slope plane is viewed as a one-dimensional surface flux in which discharge is related to the depth of flow by equation 11. The values for exponents a and m in this equation are determined, as in KINEROS, from Manning's hydraulic resistance equation (equations 12 and 13), assuming turbulent flow.

6. SOIL DETACHMENT BY RAINDROP IMPACT

Soil detachment by raindrop impact is considered for direct throughfall and leaf drainage. The ability of stemflow to detach soil particles is ignored in this part of the model. It is taken account of indirectly, however, through the contribution of stemflow to surface runoff and hence to soil detachment by surface flow. Soil detachment is related to the kinetic energy of the rain reaching the ground surface, the surface water depth and the detachability of the soil (equation 14). The detachment rates are converted from weights into mass, based on a particle specific gravity of 2.65 $Mg.m^{-3}$.

The energy of the rainfall reaching the ground surface as direct throughfall is assumed to be the same as that of the natural rainfall. It is estimated as a function of rainfall intensity from equation 15, derived by Brandt (1989), assuming that the raindrop size distribution follows that described by Marshall and Palmer (1948). The energy of the leaf drainage is estimated from equation 16 developed experimentally by Brandt (1990), except that when the average height of the plant canopy is less than 14cm, the kinetic energy of the leaf drainage is set to zero to avoid negative energy values. The total kinetic energy of the rainfall can be calculated by multiplying the energies obtained from equations 15 and 16 by the respective depths of direct throughfall and leaf drainage received, and summing the two values.

Values for soil detachability are related to soil texture and are taken from graphs and tables presented in a review of soil detachment experiments by Poesen (1985), and corrected according to the procedure proposed by Poesen and Torri (1988). Although Torri et al. (1987) show that the value of the exponent relating detachment to surface water depth also depends on soil texture, insufficient experimental work is available to define the relationship over a wide range of soils. A working value of 2.0 is therefore proposed as representative of a range of values between 0.9 and 3.1.

7. SOIL DETACHMENT BY RUNOFF

Soil detachment by runoff is modelled in terms of a generalised erosion-deposition theory proposed by Roger Smith of the USDA, Fort Collins, Colorado (personal communication,

1992). This assumes that the transport capacity of the runoff (TC) reflects a balance between the two continuous counteracting processes of erosion and deposition. It implies that the ability of flowing water to erode its bed is independent of the amount of material it carries and is only a function of the energy expended by the flow, particularly the shear between the water and the bed, and the turbulent energy in the water. The implication seems entirely reasonable in the light of work by Rauws and Govers (1988) which shows that sediment detachment by overland flow is related to the grain shear velocity of the flow; and studies by Govers (1987) which indicate that the initiation of soil particle movement is associated with turbulent perturbations within the flow.

According to the generalised theory, the transport capacity represents the sediment concentration at which the erosive ability of the flow and accompanying deposition are in balance. From this condition, a general equation (equation 17) can be derived in which soil particle detachment by flow is modelled as a function of the difference between the sediment concentration in the flow at transport capacity and the actual sediment concentration, any balancing simultaneous deposition and the resistance of the soil to detachment. The resistance of the soil is described (equation 18) by the ratio of the minimum critical grain shear velocity required to detach soil particles (assumed here to be 1cm.s^{-1}) and the critical grain shear velocity actually required, obtained (equation 19; Rauws and Govers, 1988) as a function of the cohesion of the soil.

8. TRANSPORT CAPACITY OF THE FLOW

The capacity of runoff to transport detached soil particles is modelled as a function of unit stream power based on the work of Govers (1990) which showed that the transporting capacity of overland flow could be predicted from simple hydraulic parameters. Equations 20 to 25, derived experimentally by Govers (1990) are used to model the transport capacity of the flow as a function of the median particle size of the surface soil.

9. SEDIMENT DEPOSITION

When the sediment concentration in the flow exceeds the transport capacity, e (equations 1 and 2) becomes negative and represents a net deposition rate. This situation arises when rates of soil detachment by raindrop impact are very high or when the model is being run over a series of slope segments, and runoff and sediment are routed from one segment to another of lower gradient. Since, effectively, excess sediment concentration will be deposited at a rate dependent upon the settling velocity of the particles (equation 26), there may be short time periods and short distances along the slope plane over which sediment will continue to be transported in excess of transport capacity until the pick-up rate and transport capacity come into equilibrium. Although, as pointed out by Kirkby (1980), this approach to modelling the interaction between erosion and deposition has not been exhaustively tested, it has the advantage of smoothing out the processes over time and space. An alternative approach, allowing all the excess material to be dumped immediately (Meyer and Wischmeier, 1969) causes large discontinuities in erosion and deposition rates to occur along a slope plane.

10. CALCULATION OF SOIL EROSION

Equation (1) is solved numerically to obtain the net rate of erosion (e) and the sediment discharge (product QC) for each time step and for between 5 and 15 points or nodes along a slope segment. Details of the procedure and the routing equations are contained in the KINEROS manual (Woolhiser et al., 1990).

When EUROSEM is applied to a relatively smooth slope plane (i.e. one without any rills or plough furrows) the model simulates interrill erosion with a high proportion of the soil surface covered by shallow overland flow. When rills or other defined channels exist on the slope plane, the model places the runoff into the rills which then fill with increasing runoff depth until they overflow onto the surrounding surface. Under these conditions, discharge cannot be related to flow depth, using the rating curve defined by equation 11, because depth is no longer equivalent to the hydraulic radius. Instead, a new rating curve is used (equation 27; Smith, personal communication), in which the relationship between discharge and flow depth is dependent upon the wetted perimeter and the hydraulic radius as defined by the unified profile for the rill side wall and neighbouring interrill surface shown in Figure 3. The model contains continuous functions of changes in wetted perimeter and hydraulic radius with flow depth in relation to $Y*$, where $Y* = Y/w_{ir}$ (equation 28).

When flow does not overtop the rills, a constraint has to be placed on the rate at which soil particles detached by raindrop impact on the interrill area are supplied to the rills. This is achieved using equation 29 (Styczen and Nielsen, 1989; Smith, personal communication) which represents the balanced (transport capacity) sediment concentration of the interrill flow inputs to the rills. Effectively, this means that interrill erosion is modelled as a combined process of detachment of soil particles by raindrop impact and their conveyance by flow; detachment by flow is assumed to be zero and any particles detached by raindrops in excess of the transport capacity of the flow are deposited on the interrill areas.

11. MODEL OPERATION

The version of EUROSEM described here is designed to simulate erosion on a single slope plane or segment; it will thus predict erosion from single fields. Other versions will enable erosion to be simulated over a series of segments arranged in a downslope sequence, or arranged to form a catchment of hillslopes and channels. The model will predict the total runoff and erosion in an individual event, the hydrograph and sediment graph during the event, and the location of areas of erosion (sediment sources) and deposition (sediment sinks). From the output of the model, it is possible to infer whether the main source of sediment is from splash erosion or from detachment of soil particles by interrill or rill flow. The model also simulates changes in the micro-topography of the soil surface and the width and depth of the rills as they occur during the storm in response to erosion and deposition. Armouring and crusting of the soil surface are not considered in this version of the model but it is intended that sub-routines will be included in later versions to deal with these processes. Additional sub-routines are also planned for snowmelt and effects of stoniness.

EUROSEM differs from most other process-based soil erosion models in its treatment of the effects of soil, soil surface state and vegetation. Soil erodibility is represented as a dynamic property through the use of soil cohesion and a detachability index. Soil surface state is accounted for through changes in surface roughness. Vegetation is modelled through its effects on the volume and energy of the rain reaching the ground surface, infiltration, roughness imparted to the flow, and reinforcement of soil cohesion by the root system. Soil conservation measures can therefore be simulated by describing the soil, microtopography and vegetation conditions associated with each practice.

EUROSEM will be made available on a 3.5" 720 k diskette which will run on any IBM-compatible computer with 640 k of random access memory using the MS-DOS operating system. A User Guide (Morgan et al., 1991) describes the input data required, how that might be obtained, and how to create the necessary input and output files to run the model. A documentation manual (Morgan et al., 1992b) describes the background to the model in more detail. The results of a validation exercise on the current version of the model and some examples on how the model may be used are given elsewhere in this volume (Quinton, this volume; Rickson, this volume).

ACKNOWLEDGEMENTS

Financial support from Directorate-General XII of the Commission of the European Communities under the Fourth Environment Programme (Research Grant EV4VI*1591) and the STEP Programme (Research Grant PL 900247) is acknowledged. The enthusiastic support, constructive criticisms and assistance of Dr Roger Smith (USDA Fort Collins, Colorado) are greatly appreciated.

Scientists who have been involved with the EUROSEM programme are: R.P.C.Morgan and R.J.Rickson (Project Coordinators), J.Albaladejo Montoro, V.Andreu, H.R.Bork, P.Botterweg, V.Castillo, J.A.Catt, G.Chisci, B.Diekkrüger, W.Everaert, G.Govers, B.Hasholt, A.J.Johnston, E.Klaghofer, Y.Le Bissonnais, G.Monnier, T.Panini, J.W.A.Poesen, J.N.Quinton, J.L.Rubio, V.Sardo, P.Strauss, M.E.Styczen, D.Torri, G.Tsakiris, M.Vauclin, H.Vereeken and R.Webster.

REFERENCES

AUERSWALD, K. and SCHMIDT, F. 1986. Atlas der Erosionsgefährdung in Bayern. Karten zum flächenhaften Bodenabtrag durch Regen. GLA-Fachberichte B1B, Bayerisches Geologisches Landesamt, München.

BEASLEY, D.B., HUGGINS, L.F. and MONKE, E.J. 1980. ANSWERS: A model for watershed planning. Transactions of the American Society of Agricultural Engineers 23, 938-944.

BENNETT, J.P. 1974. Concepts of mathematical modelling of sediment yield. Water Resources Research 10, 485-492.

BRANDT, C.J. 1989. The size distribution of throughfall drops under vegetation canopies. Catena 16, 507-524.

BRANDT, C.J. 1990. Simulation of the size distribution and erosivity of raindrops and throughfall drops. Earth Surface Processes and Landforms 15, 687-698.

CHISCI, G. and MORGAN, R.P.C. 1988. Modelling soil erosion by water: why and how. In: Morgan, R.P.C. and Rickson, R.J. (eds). Erosion assessment and modelling. Commission of European Communities Report No. EUR 10860 EN, 121-146.

DANISH HYDRAULIC INSTITUTE. 1985. Introduction to the SHE. Hørsholm.

FOSTER, G.R., LANE, L.J., NOWLIN, J.D., LAFLEN, J.M. and YOUNG, R.A. 1981. Estimating erosion and sediment yield on field-sized areas. Transactions of the American Society of Agricultural Engineers 24, 1253-1263.

GIORDANO, A., BONFLIS, P., BRIGGS, D.J., MENEZES DE SEQUIERA, E., ROQUERO DE LABURU, C. and YASSOGLOU, N. 1991. The methodological approach to soil erosion and important land resource evaluation of the European Community. Soil Technology 4, 65-77.

GOVERS, G. 1987. Initiation of motion in overland flow. Sedimentology 34, 1157-1164.

GOVERS, G. 1990. Empirical relationships on the transporting capacity of overland flow. International Association of Hydrological Sciences Publication 189, 45-63.

HOLTAN, H.N. 1961. A concept for infiltration estimates in watershed engineering. USDA Agricultural Research Service ARS-41-51.

KIRKBY, M.J. 1980. Modelling water erosion processes. In: Kirkby, M.J. and Morgan, R.P.C. (eds). Soil erosion. John Wiley, Chichester. pp. 183-216.

MARSHALL, I.S. and PALMER, W.M. 1948. The distribution of raindrops with size. Journal of Meteorology 5, 165-166.

MARTIN, L. 1980. An assessment of soil roughness parameters using stereophotography. In: De Boodt, M. and Gabriels, D. (eds). Assessment of erosion. John Wiley, Chichester. pp. 237-248.

MERRIAM, R.A. 1973. Fog drip from artificial leaves in a fog wind tunnel. Water Resources Research 9, 1591-1598.

MEYER, L.D. and WISCHMEIER, W.H. 1969. Mathematical simulation of the process of soil erosion by water. Transactions of the American Society of Agricultural Engineers 12, 754-758, 762.

MORGAN, R.P.C., QUINTON, J.N. and RICKSON, R.J. 1990. Structure of the soil erosion prediction model for the European Community. In: Proceedings, International symposium on water erosion, sedimentation and resource conservation. Central Soil and Water Conservation Research and Training Institute, Dehra Dun. pp. 49-59.

MORGAN, R.P.C., QUINTON, J.N. and RICKSON, R.J. 1991. EUROSEM: A user guide. Silsoe College, Silsoe.

MORGAN, R.P.C., QUINTON, J.N. and RICKSON, R.J. 1992a. Soil erosion prediction model for the European Community. In: Hurni, H. and Tato, K. (eds.). Erosion, conservation and small-scale farming. Proc. VIth ISCO Conference, Addis Abeba, Ethiopia. Geographica Bernesia. ISCO.WASWC.

MORGAN, R.P.C., QUINTON, J.N. and RICKSON, R.J. 1992b. EUROSEM: Documentation manual. Silsoe College, Silsoe.

NEARING, M.A., FOSTER, G.R., LANE, L.J. and FINCKNER, S.C. 1989. A process-based soil erosion model for USDA-Water Erosion Prediction Project Technology. Transactions of the American Society of Agricultural Engineers 32, 1587-1593.

POESEN, J. 1985. An improved splash transport model. Zeitschrift für Geomorphologie 29, 193-211.

POESEN, J. and TORRI, D. 1988. The effect of cup size on splash detachment and transport measurements. Part I: Field measurements. Catena Supplement 12, 113-126.

QUINTON, J.N. 1994. Validation of physically-based erosion models. In: Rickson, R.J. (ed.). Conserving Soil Resources: European Perspectives. CAB International. Wallingford, Oxfordshire.

RAUWS, G. and GOVERS, G. 1988. Hydraulic and soil mechanical aspects of rill generation on agricultural soils. Journal of Soil Science 39, 111-124.

RICKSON, R.J. 1994. Potential applications of the European Soil Erosion Model (EUROSEM) for evaluating soil conservation measures. In: Rickson, R.J. (ed.). Conserving Soil Resources: European Perspectives. CAB International. Wallingford, Oxfordshire.

RUBIO, J.L. 1988. Erosion risk mapping in areas of Valencia Province (Spain). In: Morgan, R.P.C. and Rickson, R.J. (eds). Erosion assessment and modelling. Commission of European Communities Report No. EUR 10860 EN, pp. 25-39.

SMITH, R.E. and PARLANGE, J.Y. 1978. A parameter-efficient hydrologic infiltration model. Water Resources Research 14, 533-538.

STYCZEN, M. and NIELSEN, S.A. 1989. A view of soil erosion theory, process-research and model building: possible interactions and future developments. Quaderni di Scienza del Suolo 2, 27-45.

TORRI, D., SFALANGA, M. and DEL SETTE, M. 1987. Splash detachment: runoff depth and soil cohesion. Catena 14, 149-155.

VAN ELEWIJCK, L. 1989. Influence of leaf and branch slope on stemflow amount. Paper presented to British Geomorphological Research Group Symposium on Vegetation and Geomorphology, Bristol.

WOOLHISER, D.A., SMITH, R.E. AND GOODRICH, D.C. 1990. KINEROS: A kinematic runoff and erosion model: documentation and user manual. USDA Agricultural Research Service ARS-77.

Chapter 28

VALIDATION OF PHYSICALLY BASED EROSION MODELS, WITH PARTICULAR REFERENCE TO EUROSEM

J.N. Quinton

Silsoe College, Cranfield University, Silsoe,
Bedfordshire, MK45 4DT, UK

Summary

Validation of physically-based erosion models is needed so that they can be used with confidence. However, there are problems associated with model validation, often due to the different perceptions of model validation by model users. Two groups of users can be recognised: Field personnel and researchers. These groups have different requirements from model validation. Field personnel are interested in predictive models and researchers are interested in models for understanding. Validating for prediction involves comparing model output with observed values, and checking the error. Models for understanding can be viewed as a scientific hypothesis: they can never be shown to be true, but they may be corroborated. Failed model validation can be due the validation data set, imperfect measurements, or the model. If it is the model it is unlikely that the whole model is wrong. To highlight the problem area the model should be tested in sections.

The European soil erosion model (EUROSEM) was tested on a data set from the Woburn erosion experiment. As a predictive model it is able to predict soil loss to within 1 t.ha^{-1} in 90% of simulations, and runoff to within 0.8mm in 50% of simulations. Observed and predicted sedigraphs show good agreement. The EUROSEM hydrograph is sharper than that which was observed.

1. INTRODUCTION

In recent years much time, energy and resources have been expended in the development of physically based erosion models, e.g. CREAMS (Knisel, 1980); WEPP, (Lane and Nearing, 1989); KINEROS, (Woolhiser et al., 1990); EUROSEM, (Morgan et al., 1992a). Physically based erosion models are those which use mathematical expressions to simulate erosion processes. The underlying philosophy behind this approach is that, given a high degree of understanding of the processes and how they respond to stresses, the system's response to any set of stresses can be defined or predetermined, even if the magnitude of the new stresses falls outside the range of historically observed stresses (Konikow and Patten, 1985).

To date, validation of these models has been limited to a small number of comparisons between observed and predicted data, (Foster et al.,1981), with little analysis of the model's performance and no recommendations to users.

In other scientific disciplines, such as ecology, modellers have begun to realise that if they want their models to be used they must provide adequate predictions (Caswell, 1976; Bunnell, 1989). It is only recently that physically based erosion modellers have begun to view validation with the same degree of importance (De Roo and Walling, this volume).This paper seeks to explore the philosophy behind the validation of erosion models and to illustrate that philosophy using EUROSEM (Morgan et al., 1992a).

2. PHILOSOPHY OF VALIDATION

One of the chief reasons for validating models is to prevent their disuse. If a model is to be used, the user must have confidence in it. This can be built in two ways: by applying it to situations similar to those on which it will be used, and showing that it performs adequately; and by demonstrating that the model is based on sound science.

Models are representations of reality. The construction of a model is a design problem, in which the modeller searches for agreement between the model and a set of demands placed upon it. The model cannot be validated unless these demands and the environment in which it is to be used have been specified. The demands placed on the model will depend on the model user; different users will require different information from a validation. Table 1 taken from Marcot et al. (1983) illustrates the number of different criteria for model validation.

Broadly speaking, for most physically based erosion models two sets of users can be identified:

(i) Field personnel and policy makers who will have little concern for the internal workings of the model. They will primarily be interested in the predictive ability of the model

(ii) Researchers, who may be involved in the model's development or interested in using the model. In either case, they are interested in gaining insight into the workings of the model (Caswell, 1976; Bunnell, 1989).

It becomes clear that the two sets of users have different needs from models, and they will therefore require different information from a model validation. The following two sections will discuss validating for prediction and for understanding.

2.1 Validating predictive models

Predictive models are designed primarily to provide accurate quantitative predictions of the behaviour of a system. There is no need to understand its mechanics. The concern is to evaluate the models accuracy and the range of conditions over which it is useful (Caswell, 1976).

For a predictive model the validation process is theoretically straight-forward. Limits of the permissible error are set, the model run, and the difference between output and the real world is computed and compared with the permitted error. The model can then be accepted or rejected.

Table 1. Criteria for Validation (adapted from Marcot et al., 1983).

Criterion	Explanation	References
Precision	Number of significant figures in a prediction or simulation.	Hall and Day 1977
Generality	Capability of a model to represent a broad range of similar systems.	Levins 1966, Walters 1971
Realism	Accounting for relevant variables and relations.	Levins 1966, Walters 1971
Accuracy	How well a simulation simulates reality.	Hall and Day 1977
Robustness	Conclusions that are not particularly sensitive to model structure.	Hall and Day 1977
Validity	A model's capability to produce all empirically correct predictions.	Mankin et al., 1977, Gass 1977
Usefulness	If at least some of the model's predictions are correct.	Mankin et al., 1977, Gass 1977, Schrank and Holt 1967
Reliability or Adequacy	The fraction of the model's predictions that are empirically correct.	Mankin et al., 1977, Leggett and Williams 1981
Resolution or Depth	The number of parameters of a system which the model attempts to mimic.	Bledsoe and Jamieson 1969, Farmer et al., 1982
Wholeness or Breadth	The number of processes and interactions reflected in the model.	Holling 1966, Farmer et al., 1982
Heurism	The degree to which the model usefully furthers empirical and theoretical investigation.	Marcot et al., 1983
Adaptability	Potential for further development and application.	Marcot et al., 1983 Schellenberger 1974
Availability	Existence of other, simpler, validated models that perform the same function.	Schrank and Holt 1967
Face Validity	Model credibility	Gass 1977
Sensitivity	Model variables and parameters matching real-world counterparts.	Gass 1977, Van Horn 1969
Hypothesis Validity	The realism with which the subsystem models interact.	Gass 1977
Technical and operational validity	Identification and validity/importance of all divergence in model assumptions.	Schellenberger 1974

When deciding on the criteria for rejection the intended use of the model should always be considered. A 30% error may be perfectly adequate for one user, but totally unacceptable for another. If the model is not being tested for a particular application no statement can be made as to whether or not it is validated - the error can only be quantified. Decisions whether or not to use the model should be left to the user.

A further complication in the validation process is whether the model should be calibrated before testing. Calibration is commonly used during hydrological model testing (DHI, 1991; Klemes, 1982) but as pointed out by Beven (1989) any model with sufficient degrees of freedom can be made to fit a hydrograph. Therefore if the model must be calibrated, its physically based nature may be compromised, if not lost entirely.

It can be argued that to calibrate a physically based model by altering inputs within the limits of variation encountered in the field is permissible, and that by doing so the model's physical basis remains intact. This is particularly true when some variables included in the model are difficult or impossible to determine accurately in the field. In EUROSEM's case this might include Manning's n and the maximum interception storage.

If the model user calibrates the model before application, the validation could justifiably include a calibration stage. This is not just testing the model, but also how well the operator can perform calibration.

Validation of physically based models can also be viewed not only as a test of the model but also as a test of the user. The model's performance will depend on the user's experience of the model and his perception of the model application

2.2 Validating for understanding

Models developed to increase understanding require a different approach. The concern is not with the accuracy of our prediction, but the truth of process description.

Trying to infer the veracity of a physically based erosion model from observations is an exercise in induction - attaining the truth from observations of a particular phenomenon (Popper, 1959; Medawar, 1968; Caswell, 1976). Induction was discounted as a suitable method for scientific research during the latter part of the nineteenth century, and has been replaced by the hypothetico-deductive method (Medawar, 1968).

Thus, a physically based model can be viewed as a scientific hypothesis: a statement which the scientist believes to be true, but, as Kant (1885) states, 'hypotheses always remain hypotheses i.e. suppositions to the full reality of what we can never obtain.' In other words we can never show that a hypothesis is true, as a hypothesis can never be a complete description of reality.

Although we are unable to validate a model designed to increase understanding, we are able to refute or corroborate it (Popper, 1963; Caswell, 1976; Bunnell, 1989). Each test that the model passes will increase confidence in the model, but it may fail the next test and be refuted.

303

Models are not always constructed with a single purpose in mind. Different groups involved in the construction of the model may have different goals for the model. The goals themselves may be equally valid but must be evaluated separately (Bunnell, 1989). EUROSEM is a case in point; it has been developed by a team of scientists primarily interested in erosion process description for use by managers and policy-makers (Morgan et al., 1990; 1992b). Both groups have different needs, and EUROSEM must be evaluated separately for each group.

Often the result of a validation attempt will be that the model fails to pass the test it has been set. This may show that something in the model is incorrect, and it can be modified and, hopefully, improved. This is the most common course of action (Medawar, 1968). Models are rarely discarded completely. However, from a successful test we learn nothing about the veracity of model components.

If we are faced with an unsuccessful application we must bear in mind two things:

(1) whether or not the data collection can be trusted
(2) that not all of the model components are likely to be wrong.

In order to reject the model it must not only be tested in a variety of contexts, but tested in small sections to pin-point which area of the model is providing erroneous predictions. Alternatives can be tested (Bunnel, 1989) and the one with the best performance is accepted. Sections of the model may be rejected in some contexts, but not in others.

2.3 Measurement accuracy and precision

The validation of a physically based model requires a perfect knowledge of initial and boundary conditions (Stephenson and Freeze, 1974) as well as the model outputs. Without this it is difficult to tell whether the discrepancy between observations and predictions is due to the model, or to inaccuracies in the input.

Perfect knowledge of model outputs, initial and boundary conditions is unlikely to be available. This should not prevent validation, particularly if the model is being validated for prediction, as the model user will face the same problems in estimating the initial and boundary conditions.

The spatial variability of model inputs is one of the greatest challenges to physically based erosion modelling. Physically based models are often of a distributed nature; a small watershed may be represented by a series of cascading planes or a network of grid squares. The model inputs may vary considerably over each plane or square, but will be represented in the model as a single value. Beven (1989) gives the example of variation in hydraulic conductivity causing runoff in one part of an element, and infiltration in another. If an average value had been used it may be possible that the model, erroneously predicts small amounts of runoff. This situation is not an extreme case as many of the inputs to physically based erosion models exhibit large degrees of spatial variability. Variables identified as sensitive during the sensitivity analysis of a model (such as saturated hydraulic conductivity and surface roughness in EUROSEM), may exhibit changes of an order of magnitude over only a few metres (Beckett and Webster, 1971; Warrick and Nielson, 1980). At present no method exists for describing spatial variation of model inputs (Beven, 1989).

Measurement of erosion sub-processes, which may be difficult in the laboratory, verge on the impossible in field situations. There are no methods for measuring transport capacity in the field, or for measuring the amount of stemflow from a field of wheat. How do we assess the D50 of the entrained particles/aggregates at the point of detachment? Models such as EUROSEM use or predict this information but there is no way of checking the predictions.

3. VALIDATION OF EUROSEM

3.1 Field data collection

Eight erosion plots were established in the autumn of 1988 for the purpose of validating EUROSEM. These are located on Rothamsted Experimental Station's Woburn Experimental Farm, sited on Great Hill at the western edge of the farm. The soil is predominantly Cottenham series, a Brown Sand on Lower Greensand, although there are some patches of Stackyard series, a Brown Earth on sandy colluvium.

Each plot measures 25m × 35m and is bordered on all sides by an earth bund. At the down-slope end of the plot the bund is graded at 1:200 towards a sediment trap. The sediment trap is connected by pipe to two 2000 litre capacity tanks, holding an equivalent of 4.57mm of water running off each plot. After an erosion event the amount of water and sediment held in the sediment traps and tanks is measured. Pressure transducers have been installed on three of the plots, providing detailed hydrographs. Daily rainfall is recorded. Model inputs are measured on a regular basis.

3.2 EUROSEM as a predictive model

Total soil loss and runoff

EUROSEM was used to simulate 29 'plot' erosion events which occurred on the Woburn erosion plots on four dates during 1989 and 1990. Inputs to the model were either measured in the field, or were taken from tables in the EUROSEM user guide (Morgan et al., 1991). The erosion plots were simulated as single planes and no calibration was carried out.

If the predicted soil loss values are plotted against the observed values (Figure 1) it becomes apparent that there is a discrepancy between the prediction and the line of perfect agreement. This is particularly evident where measured soil loss is less than 60 kg.

The data can also be examined as a cumulative distribution of differences between observed and predicted soil loss (Figure 2). Here it is evident that most of the simulations lie within $0.6t.ha^{-1}$ of the observed soil loss and that all but three of the simulations lie within $1t.ha^{-1}$ of the observed values.

It is not clear from Figure 1 whether the soil loss predictions exhibit bias, with 13 points having positive residuals and 16 with negative errors. However, the sum of the positive residual exceeds that of the negative residuals indicating a tendancy for EUROSEM to over-predict by greater amounts than it under-predicts.

305

When the observed and predicted runoff are plotted (Figure 3), there is again a discrepancy between observed and predicted values. Here there appears to be a tendancy for the model to under-predict.

Figure 1. Comparison between observed and predicted soil loss (kg).

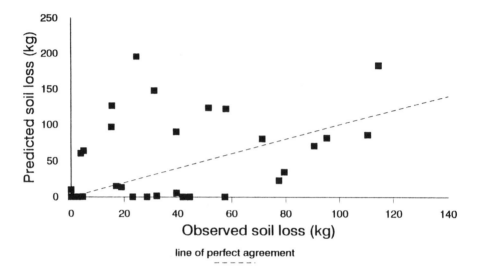

line of perfect agreement

If the cumulative distribution of the differences between predicted and observed runoff are considered (Figure 4) it can be seen that 50% of the simulations predict to within 0.8 mm of the observed runoff.

Hydrographs and sedigraphs

At present only one data set from the Woburn erosion experiment encompasses water and sediment discharges through time (23 Jan 1990). EUROSEM was used uncalibrated to simulate both hydrographs and sedigraphs.

A comparison between the observed and predicted sedigraphs is given in Figure 5. It can be seen that the shape of the two graphs and their peaks are similar. However, the peak of the sedigraph is displaced by 3 minutes and the observed sedigraph has a longer falling limb than the simulated one.

The hydrographs show a different relationship to that displayed by the sedigraphs (Figure 6). Here, there is good agreement as far as the timing of the peak is concerned, but the two graphs have different shapes; the predicted hydrograph being sharper.

Figure 2. Cumulative distribution of differences between observed and computed soil loss (kg).

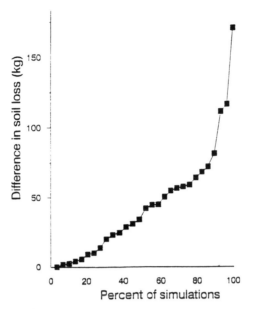

Source of errors in soil loss estimates

Errors in the estimates of soil loss by EUROSEM can come from several sources. It is interesting, particularly for the erosion modeller, to establish whether poor erosion predictions are simply a function of poor runoff predictions, or whether there is something fundamentally wrong with the erosion component of the model.

Where hydrographs are available it is possible to calibrate the model for runoff so that the observed and predicted hydrographs closely resemble each other. The predicted erosion can then be compared with the observed and assuming no measurement errors, the error can be attributed to the erosion component of the model. This approach has the disadvantage that whilst good agreement between predicted and observed hydrographs is possible using descriptor values, these have little resemblance to catchment characteristics, which may adversely affect the soil loss predictions.

A second approach which relies on the relationship between the errors in runoff and soil loss predictions will be used in this study.

Figure 7 shows the error in runoff predictions plotted against the error in soil loss predictions. Although there is considerable scatter a definite trend is evident: large negative errors of runoff are associated with negative errors in soil loss. This trend is strengthened if the circled points are discounted. Each of these points occurred on 1 February 1990, so it is possible that there was leakage in the collecting system causing the large errors in both predictions.

Although the proportion of error attributable to the erosion component of the model cannot be established, the importance of a good hydrological simulation in erosion prediction is confirmed.

Figure 3. Comparison between observed and predicted runoff (mm).

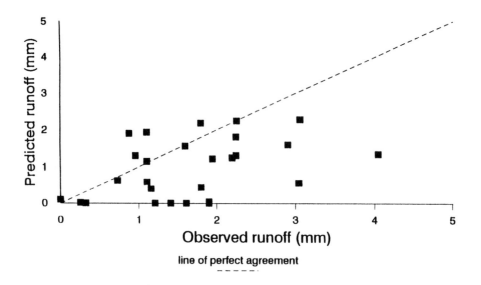

Discrepancies between observed and predicted values may not always be a due to incorrect modelling. Leakage from the erosion plots and deposition within them may help to explain EUROSEM over-predicting in some instances. Difficulties in estimating input variables and parameters may also account for some of the differences. Where timings of rainfall, runoff and sediment discharge are made independently, there may be discrepancies in the instruments, leading to the problems observed in Figure 7.

3.3 EUROSEM as a model for understanding

As EUROSEM was tested as a single model and not as individual components, it is not possible to highlight areas in which the model does not simulate the erosion processes correctly. However, from knowledge of the field site and of the model's performance in the above tests, areas which could be improved can be highlighted. These areas include surface morphology, crusting, vegetation and transport relationships.

4. DISCUSSION

Without the identification of model users and their needs, it is not possible to validate erosion models. All that can be done is to state the conditions under which the model has been tested and how well the model performed.

Figure 4. Cumulative distribution of differences between observed and computed runoff (mm).

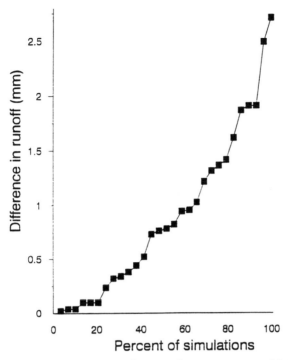

Figure 5. Observed and predicted sedigraphs for the storm on 23 Jan 1990, Woburn experimental plots.

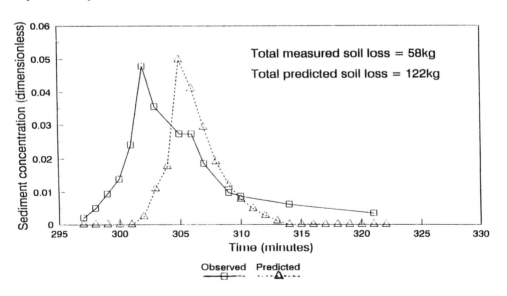

Figure 6. Observed and predicted hydrographs for the storm on 23 Jan 1990, Woburn experimental plots.

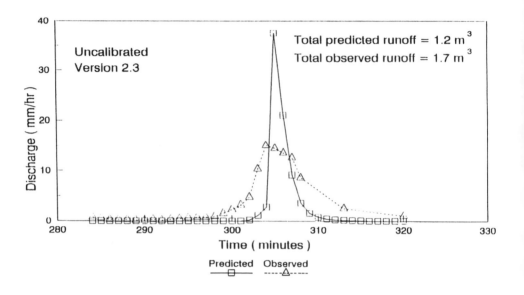

Figure 7. Errors in soil loss predictions (t.ha^{-1}) versus errors in runoff predictions (mm).

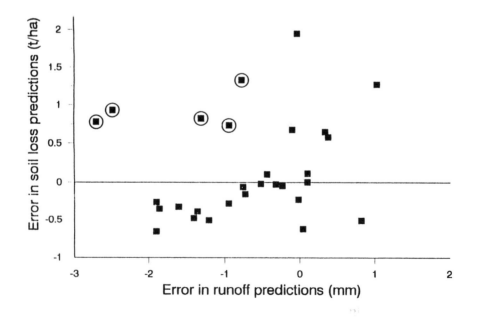

The test of an uncalibrated version of EUROSEM using data from the Woburn erosion experiment shows that EUROSEM is able to predict soil erosion to within 0.6t.ha^{-1} and runoff to within 0.8 mm in over 50% of simulations. Hydrographs and sedigraphs for the storm of 23 Jan 1990 show a reasonable predictive ability but there are problems associated with graph shape and timing.

These discrepancies may be explained by imperfect knowledge of model outputs, initial and boundary conditions, or user error. Further work is required to understand exactly where the problem lies. No decision as to whether to accept or reject the model can be made on the basis of these results.

REFERENCES

BECKETT, P.H.T. and WEBSTER, R. 1971. Soil variability: a review. Soils and Fertilizers 34,1-15.

BEVEN, K. 1989. Changing ideas in hydrology - the case of physically-based models. Journal of Hydrology 105,157-172.

BLEDSOE, J.L. and JAMIESON, D.A. 1969. Model structure of a grassland ecosystem. In: Dix and Bieldman (eds). Grassland ecosystems: a preliminary synthesis. Colorado State University press, Fort Collins, U.S.A.

BUNNELL, F.R. 1989. Alchemy and uncertainty: what good are models? USDA, Forest Service, Pacific Northwest Research Station. 27pp.

CASWELL, H. 1976. The validation problem. In: Patten, B.C. (ed.). Systems analysis and simulation in ecology. Academic Press, New York. pp. 313-325.

DANISH HYDRAULIC INSTITUTE. 1991. Validation of hydrological models, phase 1. Danish Hydraulic Institute, Horsholm, Denmark.

DE ROO, A.P.J. and WALLING, D.E. 1994. Validating the ANSWERS soil erosion model using ^{137}Cs. In: Rickson, R.J. (ed.). Conserving Soil Resources: European Perspectives. CAB International. Wallingford, Oxfordshire.

FARMER, A.H., ARMBRUSTER, M.J., TERRELL, J.W. and SCHROEDER, R.L. 1982. Habitat models for land use planning: assumptions and strategies for development. Transactions of the North American Wildlife and Natural Resources Conference 47, 47-56.

FOSTER, G.R., LANE, L.J., NOWLIN, J.M., LAFLEN, J.M. and YOUNG, R.A. 1981. Estimating erosion and sediment yield on field sized areas. Transactions of the American Society of Agricultural Engineers, pp.1253 - 1262.

GASS, S.I. 1977. Evaluation of complex models. Computer and Operations Research 4, 27-35.

HALL, C.A.S. and DAY, J.W. 1977. Systems and models: terms and basic principles. In: Hall, C.A.S. and Day, J.W. (eds). Ecosystems modelling in theory and practice. John Wiley-Interscience, New York. pp 6-36.

HOLLING, C.S. 1966. Strategy of building models of complex ecological systems. In:Watts, K. (ed.). Systems analysis in ecology. Academic Press, New York.

KANT, I. (1885) An introduction to logic (translated by Abbott T.K.). London.

KLEMES, V. 1982. The desirable degree of rigour in the testing of rainfall-runoff models. Amer. Geoph. Univer., Fall meeting, San Francisco; Eos, 63(45), Nov. 9. p. 922.

KNISEL, W.G. 1980. CREAMS: A field-scale model for Chemicals, Runoff, and Erosion from Agricultural Management Systems. U.S. Department of Agriculture, Conservation Research Report No.26. 640pp.

KONIKOW, L.F. and PATTEN Jr, E.P. 1985. Groundwater Forecasting. In: Anderson, M.G. and Burt, T.P. (eds). Hydrological Forecasting. John Wiley, Chichester. pp. 221-270.

LANE, L.J. and NEARING, M.A. 1989. USDA - water erosion prediction project: hillslope model documentation. NSERL Report No. 2. National Soil Erosion Laboratory, USDA - Agricultural Research Service, W. Lafayette, Indiana 47907.

LEGGETT, R.W. and WILLIAMS, R. 1981. A reliability index for models. Ecological modelling 13, 303-312.

LEVINS, R. 1966. The strategy of model building in population ecology. American Science 54, 421-431.

MANKIN, J.G., O'NEILL, V., SCHUGART, H.H. and RUST, B.W. 1977. The importance of validation in ecosystems analysis. In: Innis, G.S. (ed.). New directions in the analysis of ecological systems. Part 1. Simulation Councils Proceedings Series 5. pp.63-71.

MARCOT, B.G., RAPHAEL, M.G., BERRY, K.H. 1983. Monitoring wildlife habitat and validation of wildlife habitat relationships models. In: Sabol, K. (ed.). Transactions, 49th North American Wildlife and Natural Resources Conference; 1983 March 19-24 Kansas City, MO. Wildlife Management Institute, Washington, DC. pp. 315-329.

MEDAWAR, P.B. 1968. Induction and intuition in scientific thought. American Philosophical Society, Philadelphia.

MORGAN, R.P.C., QUINTON, J.N. and RICKSON, R.J. 1990. Structure of the Soil erosion prediction model for the European Community. In: Proceedings of the International Symposium on water erosion, sedimentation and resource conservation. Central Soil and Water Conservation Research and Training Institute, Dehra dun, India. pp. 49-59.

MORGAN, R.P.C., QUINTON, J.N. and RICKSON, R.J. 1991. EUROSEM a user guide. Silsoe College, Silsoe, Bedford, U.K. 56pp.

MORGAN, R.P.C., QUINTON, J.N. and RICKSON, R.J. 1992a. EUROSEM documentation manual. Silsoe College, Silsoe, Bedford, U.K. 34pp.

MORGAN, R.P.C., QUINTON, J.N. and RICKSON, R.J. 1992b. Soil erosion prediction model for the European Community. In: Hurni, H. and Tato, K. (eds). Erosion, conservation and small-scale farming. Proc. VIth ISCO Conference, Addis Abeba, Ethiopia. Geographica Bernesia.ISCO.WASWC. pp. 151-162.

POPPER, K.R. 1959. The logic of scientific discovery. Hutchinson and Co. Ltd, London.

POPPER, K.R. 1963. Conjectures and refutations: the growth of scientific knowledge. Routledge and Kegan Paul, London.

SCHELLENBERGER, R.E. 1974. Criteria for assessing model validity for managerial purposes. Decision Sciences 5,644-653.

SCHRANK, W.E. and HOLT, C.C. 1967. Critique of 'verification of computer simulation models.' Management Sciences, 14, B104-B106.

STEPHENSON., G.R. and FREEZE, R.A. 1974. Mathematical simulation of subsurface flow contributions to snowmelt runoff, Reynolds Creek Watershed, Idaho. Water Resources Research., 10, 284-298.

VAN HORN, R. 1969. Validation. In: Naylor, T.H. (ed.). The design of computer simulation experiments. pp. 232- 251.

WALTERS, C.J. 1971. Systems ecology: the systems approach and mathematical models in ecology. In: Odum, E.P. (ed.). Fundamentals of ecology, 3rd Edition.

WARRICK, A.W. and NIELSEN, D.R. 1980. Spatial variability of soil physical properties in the field. In: Hillel, D. (ed.). Applications of soil physics. Academic Press, New York and London. pp. 319-344.

WOOLHISER, D.A., SMITH, R.A. and GOODRICH, D.A. 1989. KINEROS: A kinematic runoff and erosion model: documentation and user manual. USDA, Agricultural Research Service ARS-77.

Chapter 29

EUROSEM: PRELIMINARY VALIDATION ON NON-AGRICULTURAL SOILS

J. Albaladejo, V. Castillo and M. Martinez-Mena

Centro de Edafología y Biología Aplicada del Segura. C.S.I.C.
PO Box 4195, 30080 Murcia, Spain

Summary

EUROSEM is a process-based erosion prediction model. A preliminary version of the model is now operational for individual storms at the field scale. A validation of this version is carried out on some non-agricultural soils in the Mediterranean, using data collected from two experimental plots. The results show that the predicted values over-estimate the observed values. This over-prediction may result from the modelling of concentrated flow paths which do not exist in reality. Inspite of the over-prediction, the trends in predicted and observed values are very similar.

Sensitivity analysis shows that the model is very sensitive to some parameters of land geometry such as DEPNO and DSR. The model is not sensitive to parameters dealing with soil erodibility. Further refinement in the routines dealing with runoff and transport capacity under sheet flow and soil erodibility is needed for updated versions of EUROSEM.

1. INTRODUCTION

The impact of accelerated erosion and flooding continues to be one of the main environmental problems in semi-arid south east Spain (Couchourd and Ferlosio, 1965; Lopez Bermudez, 1979). Their effects are very harmful to the socio-economic structure of the area, including loss of human life. Accurate soil loss and runoff prediction is required to control their negative effects. At the moment only imperfect tools are available for predicting erosion in non-agricultural areas (Albaladejo and Stocking, 1989). There is a need to develop physically based erosion models for conservation and planning purposes in semi-arid Mediterranean areas.

The development of EUROSEM (European Soil Erosion Model) is funded by the CEE-STEP Programme. It is a process-based prediction model for (1) evaluating soil erosion risk and (2) designing soil protection measures to combat soil erosion. It is envisaged that the model will apply to both agricultural and non-agricultural areas. Details of the proposed structure and computational procedure can be examined in Morgan et al. (1990).

The model is being developed by a multinational team involving 21 scientists from nine European countries. The model is now operational for individual storms at the field scale. A key task at this stage is to validate the model with real data collected from experimental sites, in order to assess the model performance and draw conclusions to modify the equations of the model.

The validation process is theoretically straightforward: the difference between predicted and measured values is computed and compared with the permitted error (Quinton, 1991). The specific objective of this paper is to show the first validation of the preliminary model, for non-agricultural soils in semi-arid Mediterranean areas and to indicate the possible shortcomings or deficiencies to improve the next version of the model.

2. STUDY AREA AND METHODS

Field data have been collected since 1989 from two experimental plots. One of the plots was artificially devegetated to obtain more information about the effect of the vegetal cover. Table 1 shows the environmental conditions and plot features.

Table 1: Study Area Characteristics.

Environmental Conditions:	
Soil type:	Lithic Haploxeroll
Rainfall:	300mm per year
Average temperature:	17°C
Climatic index (Thornthwaite):	Ia = 67; Im = -40.2
Vegetation:	Sparse matorral; low Mediterranean shrub with *Stipa tenacissima* and *Thymus hyemalis* as major species
Landform:	Mountainous
Soil use:	Range land
Experimental Plots	
Size:	15m long by 5m wide
Landform:	Hillside
Slope:	23%
Aspect:	North
Runoff collection:	Sediment tank with 1:5 divisor for overflow

Rainfall characteristics are measured with a rain-gauge connected to a data logger. Runoff volume is calculated from depth of water in the tanks using a pressure transducer connected to the data logger. The data logger is set up to record data every 10 seconds and calculate the mean for each minute. A detailed pluviograph and hydrograph are obtained from each storm. Sampling of the sediment in the tanks was carried out after thorough stirring. Five 1-litre samples are taken from different depths and a further five when draining the tanks. Sediment concentration is taken as the average of these.

Soil roughness is measured using a ratio of apparent to true surface length using a 1m length chain (Morgan and Rickson, 1989). Shear strength is measured at saturation with a torvane or penetrometer. Bulk density is determined using density rings. Soil porosity is calculated from bulk density and particle density as: POR = 1-(BD/PD). Moisture content is measured by drying soil samples in an oven at 105°C for 48 hours. Initial soil moisture content changes from one event to another and is determined by dividing the gravimetric moisture content at the start of the storm by the soil porosity. Maximum soil moisture content is obtained by dividing the moisture content of the soil at saturation by the soil porosity. Saturated hydraulic conductivity is determined taking undisturbed core samples for laboratory determinations. Surface crack cover, flow paths and percentage crust cover are estimated by eye or from photographs.

Textural analysis is carried out using the standard USDA method (pipette). Volume of stones is estimated by displacement and percentage stone cover from vertical photographs. Stable aggregate size distribution is measured under simulated rainfall.

Percentage canopy cover and stem angle is determined from photographs. Percentage basal cover is estimated by counting the number of plants in $1m^2$ and by measuring their stem diameter. Their stem area can then be found if all stems are assumed to be circular. Percent basal cover can then be determined by:

$$\% \text{ basal area} = \frac{\text{cumulative area} \times 100}{\text{total area}}$$

Under natural vegetation it may be impractical to make all the measurements for all the vegetation types found on the site. For those plots with a mixture of species, replicated measurements should be made for each species and then their relative proportion should be estimated, based on percentage cover.

3. PARAMETER VALUES

The EUROSEM validation database on rainfed plots comprises four sets of data sheets:

1. Soil data. This gives details of the soil measurements made every six weeks.
2. Vegetation data. This describes the vegetation measurements made every six weeks.
3. Site data. These need only be made once.
4. Event data. This gives details of the erosion event.

Table 2 shows the input variables and parameters required for EUROSEM, measured on the experimental site of Murcia for (1) the devegetated plot and (2) the natural plot.

Table 2. Values for parameters in the experimental plots of Murcia (S.E. Spain).

PARAMETER	UNITS	PLOT 1	PLOT 2
SOIL DATA			
ASR	m	0.92-0.94	0.92-0.94
(Across slope roughness ratio)			
DSR	m	0.92-0.94	0.92-0.94
(Down-slope roughness ratio)			
D50	μm	250	250
(median particle diameter of the soil)			
COH	kPa	8.0	8.0
(cohesion of the soil-root matrix as			
measured at saturation using a torvane)			
EROD	$g.J^{-1}$	7.5	7.5
(detachability of the soil particles by			
raindrop)			
FMIN	$cm.h^{-1}$	2.20	2.20
(saturated hydraulic conductivity)			
G	cm	48.5	48.5
(effective net capillary drive)			
MANN	$m^{1/6}$	0.11	0.13
(value of Manning's n, allowing for			
roughness effects of soil particles, soil			
surface micro-topography and vegetation			
cover)			
POR	%	0.23	0.23
(soil porosity)			
RECS	mm	8.40	8.40
(infiltration recession factor)			
RHOS	$Mg.m^{-3}$	2.65	2.65
(specific gravity of the sediment			
particles)			
SI		0.20-0.80	0.20-0.80
(initial relative soil saturation)			
SMAX		0.7	0.7
(maximum relative soil saturation)			
SPLTEX		2.1	2.1
(water depth exponent, affecting soil			
detachment by raindrop impact)			
VEGETATION DATA			
COV		0.0	70.0
(% canopy cover)			
DINTR	mm	0.0	1.0
(maximum interception storage)			
PBASE		0.0	0.15
(percentage basal area of the vegetation)			

Table 2 continued.

PLANGLE (average angle of the plant stems to the soil surface)		0.0	45
PLANTH (effective canopy height)	m	0.0	0.30
SHAPE (plant leaf shape factor)			1
SITE DATA			
BW (width of channel bottom)	m	0.0	0.0
DEPNO (number of concentrated flow paths (rills) per element)		0.5	0.5
PAVE (proportion of surface covered by impermeable materials)		0.0	0.0
RILLD (average depth of concentrated flow paths (rills))	m	0.01	0.01
RILLW (average width of concentrated flow paths (rills))	m	0.01	0.01
ROC (proportion of rock in the surface soil, by volume)		0.7	0.7
S (slope)	m/m	0.23	0.23
W (width of plot)	m	5.44	5.44
XL (length of plot)	m	16	16
EVENT DATA			
ACUMM.DEPTH (accumulated depth of rain)	mm	6.5-113.2	6.5-113.2
TIME (accumulated time from start of storm)	min	95-3217	95-3217
TEMP (air temperature at time of rainfall)	°C	0.30	0.30

4. RESULTS AND DISCUSSION

4.1. Runoff and Soil Loss Prediction

The predicted and observed values of runoff and soil loss from 16 events are shown in Table 3. Gentle rainfall produces very low runoff and sediment production (the model prediction is zero), but for moderate to strong rainfall, runoff and soil loss values are overestimated in relation to the observed values.

These results can be analysed in terms of how the model simulates runoff and erosion processes. Under field conditions, there were no rills or other flow paths, which is quite common over large areas of non-agricultural soils in the Mediterranean. In these areas, runoff and soil detachment occur within sheet flow. However, when a value of 0 is used for the model input parameter DEPNO (number of concentrated flow paths), no runoff or soil loss is predicted.

This is clearly in error compared with the observed values. Hence a very low value of DEPNO was used (0.5). However, the sensitivity of the model to this parameter is such that even such a low value results in an over prediction of soil detachment and transport capacity, when compared to observed values.

Figures 1-4 show the comparison of observed versus predicted values of runoff volume and soil loss. The correlation coefficient of the regression lines fitted to these data were very high (r values range from 0.93 to 0.99 with a significance level $P<0.01$). The slopes of the regression lines were much less than 1.0. This indicates a slight tendency for the model to underpredict the lower values and a marked tendency to overpredict the higher values.

These trends may be linked to a hydrological threshold, which is not exceeded in low intensity events due to high soil storage capacity. This causes underprediction for gentle rainfall. Once this threshold is exceeded, runoff production and soil loss are over-predicted because the model assumes the presence of concentrated flow paths (DEPNO). It is necessary to develop a subroutine for areas where runoff occurs only as unconcentrated sheet flow.

On the other hand, although Figures 1-4 show great differences between observed and predicted values, the trend for both is similar. These differences could reduce if calibration was undertaken to improve model performance. However, this would contradict the physically-based nature of the model and its capacity for extrapolation to other environments.

Table 3. Predicted and observed values of runoff and soil loss.

DATE	RAINFALL (mm)	I (mm.h⁻¹)	RUNOFF OBSERVED (litres)		RUNOFF PREDICTED (litres)		SOIL LOSS OBSERVED (KG)		SOIL LOSS PREDICTED (kg)	
			PLOT 1	PLOT 2	PLOT 1	PLOT 2	PLOT 1	PLOT 2	PLOT 1	PLOT 2
17.01.89	6.50	1.53	18.80	30.00	0.00	0.00	0.058	0.050	0.000	0.000
07.02.89	20.30	0.63	15.00	20.60	0.00	0.00	0.117	0.060	0.000	0.000
20.03.89	20.00	1.05	52.50	48.80	0.00	0.00	0.052	0.030	0.000	0.000
05.09.89	90.50	5.25	428.00	300.00	4000.00	3900.00	2.610	0.783	147.000	135.800
07.09.89	68.00	19.52	652.00	707.00	4500.00	4400.00	4.350	3.480	408.000	394.000
01.05.90	20.00	0.37	12.00	16.00	0.00	0.00	0.028	0.018	0.000	0.000
28.05.90	29.00	4.14	130.00	50.00	100.00	60.00	0.219	0.085	1.400	0.800
10.10.90	4.50	1.13	12.75	0.00	0.00	0.00	0.060	0.000	0.000	0.000
22.10.90	4.50	1.13	12.75	0.00	0.00	0.00	0.060	0.000	0.000	0.000
19.02.91	28.20	1.62	30.15	27.45	0.00	0.00	0.037	0.030	0.000	0.000
23.02.91	18.20	2.37	17.62	14.92	0.00	0.00	0.031	0.017	0.000	0.000
15.10.91	9.40	5.94	15.15	11.55	0.00	0.00	0.038	0.026	0.000	0.000
23.10.91	15.60	1.02	9.82	13.95	0.00	0.00	0.007	0.003	0.000	0.000
18.02.92	17.60	1.03	13.80	17.10	0.00	0.00	0.017	0.007	0.000	0.000
19.02.92	20.00	0.84	22.50	25.42	0.00	0.00	0.026	0.009	0.000	0.000
20.02.92	75.60	4.36	364.57	154.65	1100.00	900.00	0.169	0.061	23.000	17.800

Figure 1. Observed (Ro) and predicted (Rp) runoff from Plot 1.

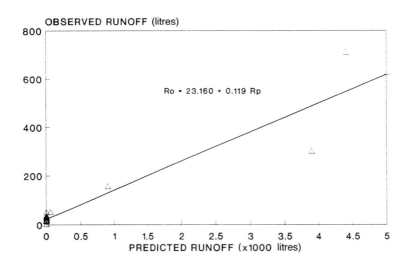

Figure 2. Observed (Ro) and predicted (Rp) runoff from Plot 2.

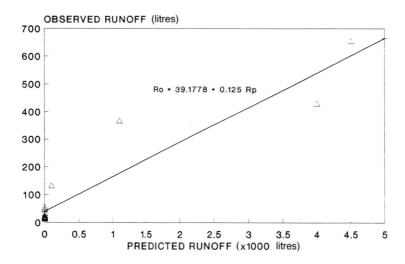

Figure 3. Observed (SLo) and predicted (SLp) soil loss from plot 1.

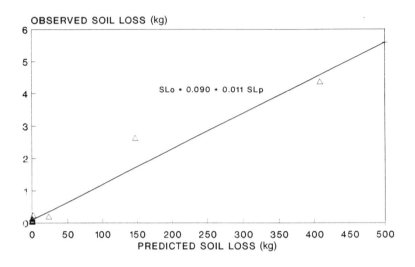

Figure 4. Observed (SLo) and predicted (SLp) soil loss from plot 2.

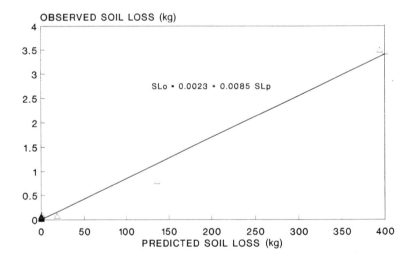

4.2. Sensitivity Analysis

The sensitivity analysis was undertaken to identify those inputs which, when modified, produce important changes to the value of the outputs. Here, only the sensitivity of selected parameters considered significant in non-agricultural lands of the Mediterranean area is studied.

Each selected input was modified in turn by increasing or decreasing its value. With some parameters of very low sensitivity the changes are exaggerated to investigate the possibility of reducing the number of inputs by the development of transfer functions to aid data input. Each model simulation is then carried out with the value of only one input changed from the original (Favis-Mortlock and Smith, 1990). The effect of this change on soil loss and runoff volume as predicted by EUROSEM was analysed. The parameters analysed were: DEPNO, DSR, SMAX, SI, FMIN, EROD, COH, MANN and D_{50}. For this analysis only the bare plot data set was used.

Runoff and soil loss are excessively sensitive to some of the land geometry parameters, such as DEPNO or DSR. An increase of 0.1 in DEPNO increases runoff from 0m^3 to 1.9m^3 and increases soil loss from 0kg to 87kg. With a DSR value of 0.91 predicted runoff is 0.2m^3 and soil loss is 8.8kg but with a value of DSR = 0.93 runoff increases to 1.5m^3 and soil loss to 67.3kg. Therefore, small changes in the value of DEPNO and DSR caused considerable changes in the values of outputs.

Changes in other parameters, such as SMAX, SI or FMIN, show important effects on model outputs. The predicted value of runoff and soil loss ranges from 0.4m^3 and 17.9kg with a SMAX value of 0.65, to 0.1m^3 and 3.7kg with a SMAX value of 0.75. In relation to SI an increase of 0.15 in the value of this input parameter produces an increase of 0.25m^3 in runoff and 8.6kg in soil loss. The predicted runoff and soil loss ranges from 0.6m^3 and 23.9kg with a FMIN value of 2.0, to 0.2m^3 and 3.6kg with a FMIN value of 2.2. These parameters are fundamental to runoff generation, but are difficult to measure accurately because there is (i) no standard method for the measurement of soil saturation, (ii) temporal variability of the SI parameter and (iii) spatial variability of these inputs.

Generally, the changes caused in runoff and soil loss are related. The only parameter that influences soil loss but not runoff is D_{50}. Changes in D_{50} produce moderate to small changes in soil loss. So, when the value of D_{50} is modified from 100 to 200µm, the soil loss value ranges from 23.2kg to 12.6kg respectively. Another important parameter of soil loss prediction is MANN. Changes in MANN produce small changes in runoff volume and moderate changes in soil loss.

The model is insensitive to parameters dealing with soil erodibility, such as EROD and COH. This is surprising as there is evidence that changes in soil erodibility cause important changes in soil loss (Albaladejo and Diaz, 1990; Wischmeier and Mannering, 1969; Rickson, 1987). In the model, a change in EROD from 7.5 to 15 and a change in COH from 1.0 to 15.0 does not affect soil loss prediction. This is a shortcoming of the model for non-agricultural soils where soil erodibility plays a key role in determining soil loss. The equations dealing with this topic need to be reviewed and other parameters which are easy and accurate to measure, such as structural stability, could be tested instead.

5. CONCLUSIONS

The validation of this preliminary version of EUROSEM enables us to propose improvements to the model for use in non-agricultural lands in semi-arid Mediterranean areas.

1. The predicted values of runoff volume and soil loss overestimate the observed values. It is suggested that a model subroutine be developed to simulate conditions in which runoff, soil detachment by runoff and transport capacity occur as sheet flow.
2. The model shows high sensitivity to the parameters dealing with soil surface characteristics. Surface roughness controls several highly sensitive inputs and it is felt that its weighting in the model is greater than in reality.
3. Variables identified as sensitive in the sensitivity analysis exhibit high spatial variability in the field. Methods for describing spatial variation of model inputs must be developed.
4. This version of the model is not sensitive to parameters dealing with soil erodibility. Further refinement of the routine dealing with this aspect is needed for future versions of EUROSEM.

ACKNOWLEDGEMENTS

This project is funded by the Commission of European Communities, Contract Nos: EV4V-0112-C(AM) & PL 900247.

REFERENCES

ALBALADEJO, J. and DIAZ, E. 1990. Degradación y regeneración del suelo en el litoral Mediterráno Español: Experiencias dn el Proyecto LUCDEME. In: Albaladejo, J., Stocking, M. and Diaz, E. (eds). Soil degradation and rehabilitation in Mediterranean environmental conditions. CEBAS-CSIC. Murcia. p. 191-214.

ALBALADEJO, J. and STOCKING, M. 1989. Comparative evaluation of two models in predicting storm soil loss from erosion plots in semi-arid Spain. Catena 16:227-236.

COUCHOURD, R. and FERLOSIO, R. 1965. Compendio cronológico de las riadas, avenidas e inundaciones. Centro de Estudios Hidrográfico. Madrid.

FAVIS-MORTLOCK, D.T. and SMITH, F.R. 1990. A sensitivity analysis of EPIC. In: EPIC-Erosion/Productivity impact calculator. 1. Model documentation. U.S. Department of Agriculture Tech. Bull. 1768. pp. 178-190.

LOPEZ BERMUDEZ, F. 1979. Inundaciones catastróficas, precipitaciones torrenciales y erosión en la Provincia de Murcia. Papeles del Departamento de Geografía No. 8.50-91. Universidad de Murcia.

MORGAN, R.P.C., QUINTON, J.N. and RICKSON, R.J. 1990. Structure of the soil erosion prediction model for the European Community. In: Proceedings of the International Symposium on water erosion, sedimentation and resource conservation. Central Soil and Water Conservation Research and Training Institute, Dehradun. India. pp. 49-59.

MORGAN, R.P.C. and RICKSON, R.J. 1989. Report to collaborating scientists (unpublished). Project number EV1V1*1591. Soil erosion in the European Community: a framework for soil protection.

QUINTON, J.N. 1991. Validation of EUROSEM: a discussion paper. (Unpublished). Presented to collaborating scientists in the meeting held in Firenze.

RICKSON, R.J. 1987. Small plot field studies of soil erodibility using a rainfall simulator. In: Pla Sentís (ed.). Soil conservation and productivity. Vol.1:339-343. Sociedad Venezolana de la Ciencia del Suelo. Maracay.

WISCHMEIER, W.H. and MANNERING, J.V. 1969. Relation of soil properties to its erodibility. Soil Sci. Soc. Amer. Proc. 31:131-137.

Chapter 30

POTENTIAL APPLICATIONS OF THE EUROPEAN SOIL EROSION MODEL (EUROSEM) FOR EVALUATING SOIL CONSERVATION MEASURES

R.J. Rickson

Silsoe College, Cranfield University, Silsoe,
Bedfordshire, MK45 4DT, UK

Summary

One of the key objectives of the European Soil Erosion Model (EUROSEM) is to evaluate the effectiveness of soil conservation measures appropriate to European conditions. This paper reviews how EUROSEM (in particular its input variables) is able to account for different soil conservation practices, and how these affect the dynamics of soil erosion processes.

Conservation practices are often classified into agronomic, mechanical and soil management measures. None of these will change the Rainfall Input File of the model, but they will influence some of the input values used in the Catchment Characteristics File. Agronomic measures such as mulching and contour grass strips can be accounted for in the model by modification of the following input variables: percentage canopy cover, maximum interception storage, Manning's n, percentage basal area of vegetation and effective canopy height. Mechanical measures such as channel or bench terracing will be accounted for in the input variables of catchment length, slope and water depth exponent. Finally, soil management practices (such as minimum tillage) can be modelled by using different input values for soil roughness ratios and Manning's n.

As a hypothetical example, the model is used to evaluate the effectiveness of a grass contour strip in reducing sediment and runoff from an agricultural field. Preliminary results indicate that EUROSEM is able to model hydrological and erosional processes observed in reality.

1. INTRODUCTION

The design requirements of the European Soil Erosion Model include the model's ability to design and evaluate soil protection measures (Chisci and Morgan, 1988). In the United States, the Department of Agriculture and the Soil Conservation Service have used predictive erosion models such as the Universal Soil Loss Equation and CREAMS to evaluate the effectiveness of soil conservation measures.

Despite the undeniable attractions of these models (not least the simplicity of the USLE) they should not be used unmodified in Europe, due to their empirical foundations being based on US data. However, European agricultural advisory services require a calibrated, validated and therefore reliable method of predicting erosion rates and evaluating soil conservation measures. It is hoped that EUROSEM will meet these requirements. Advisory services will be able to use the model for soil conservation planning at the catchment scale, by identifying areas of high soil erosion risk, and then target these areas with soil protection measures.

This paper outlines how EUROSEM models the effects of soil protection measures on erosion processes. The paper is *not* intended to assess the accuracy of the model in predicting reality (See Quinton, this volume). It is intended to illustrate how input variables can be manipulated to simulate the role of conservation practices on soil erosion processes.

2. USING EUROSEM TO EVALUATE SOIL CONSERVATION MEASURES

The detailed structure of EUROSEM has been outlined elsewhere (Morgan et al., 1990). To run the model, input variables and parameters have to be entered into the Catchment Characteristic data file, and the Rainfall data file. Details of these files can be found in the EUROSEM User Guide (Morgan et al., 1991). These input variables and parameters account for many changes in the soil/plant/water relationships which result from different land management or treatment practices. Obviously, any change in these variables influences the likelihood of soil erosion occurring for any given set of initial conditions. For example, one input variable in the catchment file is Manning's n. The value chosen would be higher where a surface mulch was used, compared with bare soil conditions, all other factors being equal. This would account for the greater degree of roughness imparted to overland flow by the mulch elements compared with the bare soil. This difference in the input data would give correspondingly different outputs of predicted soil loss and runoff.

By changing the input variables and parameters in this way, EUROSEM has the potential for evaluating agronomic and mechanical methods of soil protection and soil management practices and how these practices can affect soil erosion and runoff rates.

2.1 Mechanical measures of soil conservation

Terraces are used primarily to reduce slope lengths and catchment areas over which runoff will generate. EUROSEM accounts for this with a change in the slope length parameter in the Catchment Characteristic file (denoted as XL), and by a division of the total slope length into a number of slope sections or elements, between which sediment and runoff can be routed. The length of each element is then taken as the horizontal interval of each terrace. Each element (or terrace, in this case) has to be uniform in its catchment and rainfall characteristics.

Where terraces also change slope gradient (notably bench terraces) again the slope gradient parameter (S) in the catchment characteristic file can be changed. However, this will not account for the oversteepening on the 'riser' of the terraces, which should also be simulated in the model. This may prove difficult as the slope length of the riser (in plan) is very small, so generating many elements over one field.

The riser of bench terraces will also often have significantly different cover characteristics compared with the cultivated benches. Often, risers are grassed or supported by stones or brickwork. These covers can be accounted for in numerous input parameters such as across slope and up/downslope roughness, cohesion of soil/root matrix, percentage canopy cover, maximum interception storage, Manning's n, proportion of the surface covers by impermeable materials, percentage basal area of vegetation, angle of plant stems to the soil surface, canopy height, proportion of rock fragments in the surface soil and the plant leaf shape factor.

Whilst soil hydrology is modelled in the hydrology routine used with EUROSEM (currently the hydrological model of KINEROS), the build up of pore water pressure behind the risers which often leads to slumping and collapse of the terrace walls, is not simulated in the present version of EUROSEM.

2.2 Soil conservation through soil management practice

Tillage practices used to conserve soil and water are applied extensively in the United States, with increasing interest from European farmers. This growing interest has often come about because of the social and economic constraints of implementing and maintaining mechanical measures of soil conservation, such as terracing. 'Conservation tillage' is based on the principles of maintaining a soil surface which is resistant to erosion (by water and wind) and retaining maximum or optimum crop residues as a mulch on or just below the soil surface. The input variables and parameters of EUROSEM can be manipulated to account for these conditions. Hence, predictions of soil loss and runoff under conventional versus conservation tillage practice can be compared for any given starting condition.

Where soil management practice involves retaining crop residues, the mulch elements simulate vegetation canopy (interception of rainfall), plant stems (roughness imparted to overland flow) and roots (where the mulch has been incorporated). By drawing this analogy with 'real' vegetation EUROSEM can be applied by changing the appropriate input data of the catchment file.

2.3 Agronomic measures of soil conservation

The structure and philosophy of EUROSEM emphasise the role of vegetation in runoff generation and soil erosion processes. Hence, application of the model to the evaluation of agronomic measures of soil conservation is very straight forward. The changes in the catchment characteristics are easy to incorporate into the input data files. Table 1 shows the number of input variables and parameters in the Catchment Characteristic file that are related to vegetation and its management. In Europe, the agronomic measures of soil protection that are of interest include cover cropping, mulching (surface and sub-surface), intercropping, multiple cropping, strip cropping and contour grass strips.

Some of these practices are best evaluated on an annual or seasonal basis, as they are designed to reduce soil losses over a season as a whole. For this reason, application of EUROSEM (an event based model) is limited to test the effectiveness of these measures. Bearing this limitation in mind, however, it is possible to apply EUROSEM for any given storm and compare soil loss and runoff both with and without these soil protection measures.

Table 1. Selected input to EUROSEM's Catchment Characteristic file, related to agronomic measures of soil conservation.

VARIABLE	DESCRIPTION
ASR	Across slope roughness
COH	Cohesion of the soil-root matrix
COV	% canopy cover
DINTR	Maximum interception storage
DSR	Downslope roughness
(FMIN)	*Saturated hydraulic conductivity*
(G)	*Effective net capillary drive*
MANN	Manning's n roughness coefficient
PBASE	Percentage basal area of the vegetation
PLANGLE	Average angle of the plant stems to the soil surface
PLANTH	Effective canopy height
(POR)	*Soil porosity*
(RECS)	*Infiltration recession factor*
SHAPE	Plant leaf shape factor
(SI)	*Initial relative soil saturation*
(SMAX)	*Maximum relative soil saturation*

Note: Variables in italics may be only indirectly affected by different agronomic measures.

Table 1 also shows that there are varying degrees of confidence as to whether the use of agronomic practices will change the input variables listed. It is highly likely that in reality, interactions and indirect relationships do occur, but these require further investigation. One example of this is whether vegetation roots affect (either directly or indirectly) saturated hydraulic conductivity, effective net capillary drive, soil porosity, initial relative soil saturation, or maximum relative soil saturation. If the roots do have an effect, to what extent is this and how should the input variables of EUROSEM reflect this?

3. APPLYING EUROSEM TO EVALUATE THE EFFECTIVENESS OF A CONTOUR GRASS STRIP

The objective of this exercise was to assess how EUROSEM could be applied to evaluate the effectiveness of a contour grass strip in controlling soil erosion and runoff. This hypothetical example was chosen because contour grass strips have been used throughout Europe for soil conservation and control of sediment. They are particularly suitable in fields that have been enlarged, as a result of intensification of agriculture or where land consolidation has amalgamated formerly fragmented land holdings. Where field boundaries such as hedgerows have been removed in this way, erosion rates increase. Grass strips can replace these boundaries and reduce erosion risk, without losing the practical advantages of managing larger fields. As the contour strips do not interfere with farming operations (they can be crossed by farm machinery), they are more likely to be adopted by farmers, compared with more disruptive conservation practices. There are also ecological benefits such as creation of new habitats within the grass strip. Grass strips have also been used as 'buffers' or 'filter strips'. By intercepting sediment and runoff they may act as a sink for N, P and pesticides, thus preventing pollution of adjacent rivers and streams.

By reducing effective slope lengths, contour grass strips help to reduce runoff velocities and flow shear stresses. Roughness imparted to overland flow by the grass stems in the strip help to reduce runoff velocity further, as well as filtering out sediment eroded from upslope, which is then deposited within the grass strip.

3.1 Procedure for applying EUROSEM

Firstly, the two input files (Rainfall and Catchment Characteristics) were analysed as to how to incorporate the physical characteristics of a contour grass strip. Secondly, the indirect effects of the grass strip had to be accounted for in the input files. Thirdly, the model output in terms of predicted sediment and runoff with and without the grass strip were compared and analysed.

The rainfall data for the model simulation was taken from an event recorded at the Woburn erosion plots in Bedfordshire, UK. The storm occurred on the 25th January 1990. This typical storm was used for all the model simulations represented below. As conservation measures do not affect the rainfall input variables or parameters in this file, the same conditions were used for all the simulations, so that a true comparison could be made between the various slope configurations used. A hypothetical field of 90 metres length, 25 metres width and a slope of 12% (7°) was used. The soil type is a sandy loam, with no eroded channels or rills.

Simulation of a slope with no grass strip

The catchment characteristic file for this configuration is given in Table 2. As it is assumed that all the input variables and parameters are uniform along the total slope length (XL = 90 metres). Only one element is used in the simulation.

The vegetation cover on the hypothetical field is at an early stage of growth (5cm canopy height) and a percentage canopy cover of 2% is used, with an assumed maximum interception storage of 2mm. Percentage basal area of vegetation is low at 0.02 due to the early stage of vegetative growth. Across and downslope roughness are taken as 0.966 and 0.992 respectively, from measurements taken in the field on the Woburn erosion plots under similar conditions. Manning's n is taken as 0.05, following the guide values in the EUROSEM User Guide (Morgan et al., 1991). Soil cohesion is also taken from guide values.

Table 2. Catchment characteristic file for a single slope element, J1.

NU	NR	NL	NC1	AUXPR	NC2	NCASE	
0	0	0	0	1	0	0	
XL	W	S	ZR	ZL	BW	DIAM	MANN
90.0	25.0	0.121	0.0	0.0	0.0	0.0	0.05
FMIN	G	POR	SI	SMAX	ROC	RECS	DINTR
2.0	14.7	0.4	0.5	0.67	0.0	0.0	0.2
DEPNO	RILLW	RILLD	ZLR	IRILL	ASR	DSR	
42	0.05	0.04	0.250	0	0.966	0.992	
COVER	SHAPE	PLANGLE	PLANTBASE		PLANTH		
0.02	1	55.0	0.02		5.0		
D50	EROD	SPLTEX	COH	RHOS	PAVE	SIGMAS	
250	1.0	2	8.92	2.65	0.0	100	

These input data were fed into EUROSEM. The hypothetical resultant soil loss from the one element is given in Table 3.

Table 3. Results of sediment and runoff from 1 element.

Treatment	Element No.	Sediment (kg)	Runoff	
			mm	m^3
90 metres	1	56.417	0.554	1.2

Effectiveness of the grass strip - 2 elements

To include the effects of the grass strip, the field is subdivided into 2 elements. The upslope element (element 1) is treated as the cultivated slope, with the second, downslope element simulating the grass strip. The total slope length of the two elements is the same as for the previous, single element simulation (90 metres). Sediment and runoff are routed by EUROSEM from Element 1 (J1) downslope to Element 2 (J2).

The catchment characteristics for Element 1 are identical to those for the single element simulation, apart from the slope length which was varied according to the width of the grass strip. In all cases, the total slope length was kept at 90 metres.

Element 2 simulates the grass strip. There are a number of input variables and parameters which will have to change to reflect the change in soil/plant/water relationships within this element as compared with the cultivated slope element. The slope length of the strips was set at 10, 20 and 30 metres. It was assumed that wider strips were impractical because of the amount of cultivated land that would be taken out of production.

Manning's n was increased to 0.41 to represent a dense sward, which would impart high levels of roughness to any overland flow. Providing adequate maintenance and mowing was carried out, the short stems would not be submerged or flattened by the flow, so justifying such a high roughness coefficient. Roughness imparted by the vegetation was also accounted for in the across and up/down slope roughness ratios, although this may involve some double counting as Manning's n would also account for this roughness.

Again, the vegetation characteristics of percentage cover and basal area were increased (to 90% and 0.8 respectively) although leaf shape and plant height are assumed to be the same as for the emerging crop on the cultivated slope element. Maximum interception storage will also increase due to the dense canopy of the grass strip, hence a value of 1.2mm was used. Soil cohesion was increased to 10.92kPa, to allow for increased cohesion due to the dense grass root mat. Table 4 shows the final input data used for the two elements.

Table 4. Catchment Characteristic File for two slope elements a) J1 and b) J2.

a) **J1**

NU	NR	NL	NC1	AUXPR	NC2	NCASE	
0	0	0	0	1	0	0	
XL	W	S	ZR	ZL	BW	DIAM	MANN
80.0	25.0	0.121	0.0	0.0	0.0	0.0	0.05
FMIN	G	POR	SI	SMAX	ROC	RECS	DINTR
2.0	14.7	0.4	0.5	0.67	0.0	0.0	0.2
DEPNO	RILLW	RILLD	ZLR	IRILL	ASR	DSR	
42	0.05	0.04	0.250	0	0.966	0.992	
COVER	SHAPE	PLANGLE	PLANTBASE		PLANTH		
0.02	1	55.0	0.02		5.0		
D50	EROD	SPLTEX	COH	RHOS	PAVE	SIGMAS	
250	1.0	2	8.92	2.65	0.0	100	

b) **J2.**

NU	NR	NL	NC1	AUXPR	NC2	NCASE	
1	0	0	0	1	0	0	
XL	W	S	ZR	ZL	BW	DIAM	MANN
10.0	25.0	0.121	0.0	0.0	0.0	0.0	0.41
FMIN	G	POR	SI	SMAX	ROC	RECS	DINTR
2.0	14.7	0.4	0.5	0.67	0.0	0.0	1.2
DEPNO	RILLW	RILLD	ZLR	IRILL	ASR	DSR	
42	0.05	0.04	0.250	0	0.700	0.700	
COVER	SHAPE	PLANGLE	PLANTBASE		PLANTH		
0.90	1	55.0	0.80		5.0		
D50	EROD	SPLTEX	COH	RHOS	PAVE	SIGMAS	
250	1.0	2	10.92	2.65	0.0	100	

The results of the three simulations of 2 elements (80m:10m; 70m:20m and 60m:30m) are given in Table 5.

Table 5. Results of sediment and runoff from 2 elements.

Treatment	Element No.	Sediment (kg)	Runoff	
			mm	m ^{3}
80:10	1	48.267		
	2	0.393	0.502	1.1
70:20	1	40.359		
	2	0.056	0.465	1.0
60:30	1	32.625		
	2	0.000	0.426	1.0

NB. Sediment losses and runoff values are cumulative over the two elements.

4. DISCUSSION

The results above are purely hypothetical. They do not attempt to predict *actual* values of soil loss and runoff, as would be observed in the field. The model was run to illustrate how management practices (here a contour grass strip) can be represented by the input data of EUROSEM, and whether the model output reflects our current understanding of how grass strips function in controlling sediment and runoff production.

Comparison of the single element slope (with no grass strip) with the two element slope (with a grass strip) illustrates the ability of EUROSEM to simulate the effectiveness of a grass strip in reducing soil erosion, even though the absolute values are *not* taken to be realistic. According to the model output, soil loss from the totally cultivated slope (90m) is 56.42kg, compared with only 0.393kg when a grass strip of just 10m is applied. This dramatic reduction is only partly due to reduced slope length of the cultivated slope element, which becomes 80m where the 10m grass strip is used (soil loss from the 80m slope is 48.27kg). This implies that it is the grass strip which is limiting sediment output from the total slope length.

For the widest grass strip used in these simulations (30m), sediment production from the total slope is reduced to zero. Obviously, the correspondingly shorter cultivated slope (60m) produces less sediment than the single element slope (90m) (32.62kg versus 56.42kg respectively), yet all of this sediment from the 60m length is retained by the grass strip, according to the simulation.

In order to understand why these reductions in soil loss occur, and to highlight possible processes being simulated by the model which give such results, the input data can be analysed in relation to the model output. The increase in plant cover (represented by increases in the values of percentage cover, plant basal area, soil cohesion and maximum interception storage) would result in less sediment from the grass strip itself. However, these factors alone would not explain the more significant function of filtering out sediment eroded from the cultivated slope above. Simulation of this process comes about due to the given increases in Manning's n and surface roughness. Roughness imparted to the flow would reduce flow velocity and hence transporting capacity, so encouraging deposition within the grass strip. However, such a dramatic reduction in sediment load modelled here may be too optimistic. This may be partly due to the effective double counting of roughness imparted to the flow by modification of both soil and plant induced roughness.

Runoff volume is *not* reduced by the presence of a grass strip to the same extent. For the 80:10m configuration, 1.1m ^{3} of runoff is predicted, as opposed to 1.2m ^{3} for the single element slope. This relatively small difference in runoff totals contrasts with the relatively large difference in sediment from the different slope configurations. This fact implies that the ability of runoff to transport sediment eroded from the cultivated slope section is reduced by runoff velocity rather than runoff volume. Indeed, transport capacity of flow is known to be more sensitive to reductions in runoff velocity than runoff volume.

Thus the model output implies a reduction in runoff velocity which is sufficient to deposit sediment within the grass strip, and yet is not sufficient to retard flow velocity to such an extent that ponding and infiltration of runoff occurs in the strip.

The model output suggests that grass strips are effective at reducing soil losses from above slope elements, but not effective at reducing runoff volumes. This has also been observed experimentally by Chisci and Borschi (1988). The question then arises as to the erosivity of the flow once it leaves the grass strip. The philosophy behind EUROSEM is such that if transport capacity has not been reached, then the flow is still potentially erosive. If the grass strip has removed all or much of the sediment, then the flow's potential erosivity (i.e. transport deficit) is increased. If the model output is realistic, then this conclusion implies that the grass strip should always be placed at the bottom of the slope, rather than used as a means of breaking up a long slope length. This is an important consideration if contour grass strips are to be used as replacements or alternatives to mid-slope field boundaries which have been removed during agricultural intensification.

5. CONCLUSION

Input data to EUROSEM are easily modified to represent a variety of soil conservation and management practices. However, often guide values have to be used for important and sensitive parameters such as Manning's n, due to the difficulties in obtaining true, site specific values.

The model is able to simulate the expected effect of a grass strip in controlling sediment and runoff production from a hypothetical slope. Modifications in the input data can be used to represent possible processes simulated by the model, although interactions between the input variables and parameters require further investigation. The results are a good relative approximation of what is likely to happen in the field, as reasoned from our current knowledge of erosion processes and their control. There may be some concern as to the routing of runoff from one element to another, which may overestimate the erosivity of flow, once its sediment load has been removed by the grass strip.

Further confidence in the model output can be gained only from field validation, using a wide range of possible applications of the model. Field observations will also give the most appropriate input data. The predicted results presented in this paper have to be compared with observed values from field observations. Only this exercise will evaluate the true ability of EUROSEM to predict accurately soil erosion and runoff rates from different soil conservation measures. It is essential that EUROSEM is robustly tested in this way, in order that advisory services have confidence in using the model as a management tool.

REFERENCES

CHISCI, G. and BORSCHI, V. 1988. Runoff and erosion control with hill farming in the sub-coastal Apennines climate. Soil and Tillage Research 12(2):105-120.

CHISCI, G. and MORGAN, R.P.C. 1988. Modelling erosion by water: why and how. In: Morgan, R.P.C. and Rickson, R.J. (eds). Erosion assessment and modelling. CEC Report No. EUR10860EN. pp.237-253.

MORGAN, R.P.C., QUINTON, J.N. and RICKSON, R.J. 1990. Structure of the soil erosion prediction model for the European Community. In: Proceedings of an International Symposium on Water Erosion, Sedimentation and Resource Conservation. Central Soil and Water Conservation Research and Training Institute, Dehra Dun, India. pp.49-59.

MORGAN, R.P.C., QUINTON, J.N. and RICKSON, R.J. 1991. EUROSEM: A user guide. Silsoe College, Cranfield Institute of Technology, Bedfordshire, UK.

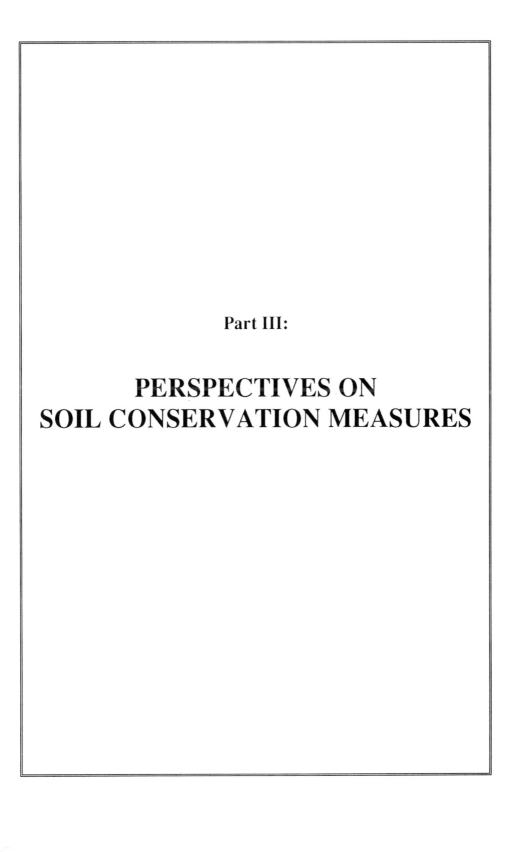

Part III:

PERSPECTIVES ON
SOIL CONSERVATION MEASURES

Chapter 31

PERSPECTIVES ON SOIL PROTECTION MEASURES IN EUROPE

G. Chisci

Dipartimento di Agronomia e Produzioni Erbacee,
Universita di Firenze, Italy

Summary

Many observations and data illustrate that recent changes in farming systems and associated infrastructure have activated a new cycle of accelerated erosion in some European landscapes, producing on-site and off-site damage.

Reconnaissance analysis of erosional systems, taking into account the occurrence and different types of erosion forms in different land units, may represent the best strategy for choosing integrated soil protection, suitable for erosion control in specific sites.

A number of soil protection measures controlling the main erosion factors, such as soil erodibility, runoff generation and runoff concentration, are known from past experience and more recent experimentation. With integrated application of these measures a consistent abatement of accelerated soil erosion processes is possible.

Unfortunately, whilst better soil protection appears attainable from a strictly technical point of view, the main problem is establishing operative conditions for the adoption of adequate soil protection measures by farmers.

1. INTRODUCTION

The historical background of soil degradation and geomorphic dynamics, due to agricultural colonisation of European slopes is deeply inscribed in the 'cultural landscapes' of the Continental and Mediterranean environments.

European cultural landscapes, having been densely settled for a long time, have been subjected to anthropologically related cycles of soil erosion for hundreds and, sometimes, for thousands of years. Consequently, actual soil erosion intensity must be seen as an amalgam of present and historical factors (Richter, 1980).

The long-term geomorphic landscape evolution is relatively well documented in various areas of the Mediterranean (Hughes and Thirgood, 1982; Susmel, 1986; Chisci, 1979, 1980, 1987) and of central Europe (Vogt, 1958; Meyer and Willerding, 1961; Bork and Rohdenburg, 1979; Bork, 1983, De Ploey, 1986).

Nevertheless, compared to the huge soil erosion phenomena in tropical and temperate areas of the world (with high climate aggressiveness and fragile soils), the European situation did not appear to be a serious problem. However, in the Mediterranean area, climate aggressiveness and geomorphology make the situation intrinsically more susceptible to accelerated erosion processes. A long history of human pressure on sloping land has produced erosion damage in the past following deforestation and upland agricultural colonisation. Many deforested upland Mediterranean areas have been reduced to matorral, degraded pastures and, sometimes to bare rocks, as is evident in the Spanish Sierras and in the Italian Apennines.

Sustainable agriculture based on grassland, long term rotation with forage crops, animal husbandry, organic matter turnover, contour tillage and terracing on steep slopes has been practised in Mediterranean hill farming, adopting agricultural systems in which water management structures (such as contour channels, field and farm roads, live fences, underground drainage and terracing), and agro-biological soil conservation practices (such as rotations with fallow and/or forage crops, organic manuring and contour tillage, etc.) were well integrated on a variety of soils.

In the past, knowledge of the evolution of the different European 'cultural landscapes' and the evaluation of the long-term role played by erosion was the main body of soil erosion studies in Europe. Only recently has a more quantitative approach and a better knowledge of the mechanics of erosion process been acquired. Indeed, only in the 1980s European scientists and agriculturists became aware by quantitative measurements that a new cycle of accelerated erosion occurred after World War I, both in the Mediterranean and in the Continental areas (Richter, 1963; Bollinne, 1982; Chisci, 1980; Morgan, 1986; Schwertmann, 1986; De Ploey, 1986).

2. GENERAL CAUSES OF THE ACTUAL SOIL EROSION CYCLE

The causes of such acceleration were pointed out by many authors for different areas (Chisci, 1980; De Ploey, 1986; Monnier and Boiffin, 1986).

At the centre of the problem is the socio-economic evolution of the rural environment, characterised by rural depopulation and ageing of agricultural operators, along with increasing markets for agricultural products, compared with the previous self sufficiency by the farmers.

Parallel to such socio-economic changes, was the increasing use of industrial products in agriculture: (a) chemical fertilisers, herbicides and pesticides; and (b) machinery substituting animal and manpower (Figure 1).

It is generally recognised that innovative agricultural technologies, interacting with improved varieties of agricultural crops and improved breeds of animals, have increased the marketable yield of agricultural products, reducing, at the same time, the costs of their production. As a consequence, better revenues were obtained for all agricultural operators. Such changes made an impact not only on farming management but also on the morphographic-infrastructural elements of the landscape.

Figure 1. Flow chart of the general trend of agricultural evolution in Europe.

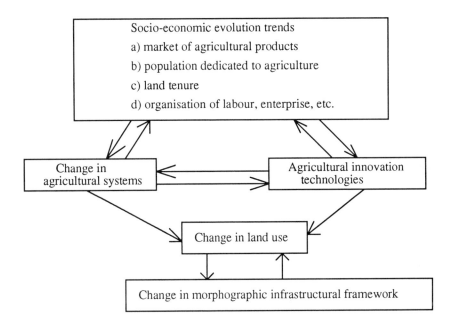

Some of the changes (Table 1) have enhanced the fragility of the land toward degradation of soil fertility, decrease of soil infiltration, disorder in soil water regime, increased runoff generation and runoff concentration, increased soil erodibility and reduction of cover and protection.

In order to apply soil protection measures, the benefit has been emphasised of identifying erosion systems that have to be controlled site by site (Papy and Boiffin, 1989).

In this way, the main erosional systems and the relative importance of different erosion processes and forms were collected from the literature for some representative European cultural landscapes (Table 2) (Chisci and Morgan, 1986; Morgan and Rickson, 1988; Schwertmann et al., 1989).

3. MORPHOGRAPHIC-STRUCTURAL CLASSIFICATION OF SOIL PROTECTION MEASURES

The erosion system activated in field, farm and catchment units is strictly related to the morphographic-infrastructural framework of each landscape, the seasonal pattern of climate and crop cover. These correspond to the sequence of agricultural practices, unique to each land-unit of a given cultural landscape (De Ploey, 1989).

Table 1. **Main changes to farming systems introduced in European cultural landscapes.**

---

a) <u>CONTINENTAL EUROPE</u>

1) Increase of arable land at the expense of grassland;
2) Shortening of rotation and reduction of rotational forage crops;
3) Continuous arable crops, sometimes in crop monoculture;
4) Animal husbandry concentration and reduction of animal breeding on small farms;
5) Tree crops in up-and-down slope plantations, tilling the soil between tree rows;
6) Decreasing use of FYM on arable land;
7) Use of heavy machinery increasing soil compaction and up-and-down slope oriented roughness (tractor tracks);
8) Increasing use and number of operations by tillage implements producing pulverisation of mechanical soil structure;
9) Reduction and/or elimination of terraces on steep hillsides;
10) Neglect of the maintenance of field and farm infrastructures;
11) Reducing ecotones and increasing field dimensions to facilitate mechanisation.

b) <u>MEDITERRANEAN EUROPE</u>

1) Crop specialisation in separate fields of arable crops and tree crops (chess-board pattern of cultivation on hillsides);
2) Shortening of rotations and sometimes crop monoculture on arable land;
3) Reduction of long term forage crops in the rotations;
4) Animal husbandry concentration and reduction of animal breeding on small farms;
5) Decreasing use of FYM on arable land and specialised tree stands;
6) Use of heavy machinery increasing soil compaction and up-and-down slope oriented roughness (tractor tracks);
7) Increased use and number of operations by tillage implements producing pulverisation of mechanical soil structure;
8) Levelling of hillside morphologies in cultivation units to facilitate machinery management;
9) Up-and-down tilling and planting;
10) Elimination of mechanical structures such as terraces, contour channels, live fences, and underground drainage;
11) Maintenance of structures is neglected;
12) Abandonment of arable marginal hilly land, followed by the development of natural vegetation exerting scarce soil protection against erosion. Abandonment of existing structures leads to collapse, runoff concentration and land disorder;
13) Use of fire for vegetation management in natural pastures and for residue management in arable land (especially straw in cereal farming);
14) Reducing ecotones and increasing field dimensions to facilitate mechanisation.

Table 2. Relative impact of main erosional forms in some European 'cultural landscapes'.

Geographic areas of Europe	Soil type	Topo-graphy	Infra-structural linearity	Land use	Relative ranking of erosional forms (*)									
					Splash	Wash	Rill	Gully	Channel-ling	Piping	Sliding	Creeping	Soli-fluction	Collu-viation
Northern central	loess loamy	gently rolling	moderate	arable	(4)	(3)	(3)	(2)	(1)	(2)	(1)	(1)	(1)	(4)
Central	skeletal-brown	steep	high	vineyard	(2)	(2)	(4)	(1)	(1)	(1)	(1)	(1)	(1)	(2)
Northern Mediterranean	sandy-loamy	steep fractional	very high	vineyard orchard arable	(1)	(2)	(4)	(1)	(1)	(1)	(1)	(1)	(1)	(2)
Northern Mediterranean	marl-arenaceous	steep incised	high	vineyard orchard arable	(3)	(2)	(3)	(2)	(1)	(1)	(1)	(1)	(2)	(2)
Central-southern Mediterranean	marls	moderately steep incised	high	vineyard olive arable	(1)	(1)	(2)	(2)	(4)	(3)	(3)	(2)	(2)	(1)
Central-southern Mediterranean	brown	moderately steep fractional	high	arable vineyard olive	(1)	(2)	(3)	(1)	(1)	(1)	(1)	(2)	(1)	(1)
Central-southern Mediterranean	skeletal-red	moderately steep	moderate	arable; olive vineyard pasture	(1)	(2)	(3)	(1)	(1)	(1)	(1)	(2)	(1)	(1)
Central-southern Mediterranean	clayey	moderately steep fractional	low	arable pastures fallow	(1)	(1)	(2)	(2)	(2)	(3)	(4)	(3)	(2)	(2)

*Relative ranking of erosional forms:
(1) = absent or small
(2) = moderate
(3) = high
(4) = very high

Table 3. General classification of morphographic-infrastructural elements, and crop and tillage management practices for soil protection on field and farm land-units.

Non point elements	Organic matter management	- organic fertiliser - crop residues
	Use of soil conditioners	
	Tillage practices	- conventional tillage (ploughing and seeding) - minimum tillage (harrowing, disking, chiselling, rotor tilling) - zero tillage (sowing in the sod with special soil preparation and sowing machines)
	Crop cover management	- perennial and rotational grass cover; cover cropping
	Mulching	- with live material - with agricultural and industrial residues
Point elements	Trees and/or shrubs Seedling and distanced crop plants Soil cloddiness	
Linear elements	Field and farm roads Tractor tracks Furrows Ditches Underground tile drainage Channels Linchets Terraces Tillage direction Plant row direction	
Areal elements	Field topography Ecotones pattern	

In classifying soil protection measures, it may be worthwhile to adopt a morphographic-structural approach which is useful for the selection of one or more soil protection measures for the control of a complex erosion system (Table 3.).

Each soil protection measure is more or less valuable for the control of one or more erosion factors. The main erosion factors are soil erodibility (soil splashability, crustability and rillability), runoff generation and runoff concentration, in various combinations and with different relative importance. The relative value of different soil protection measures can be assessed, when the erosion system of a specific cultural landscape is known (Table 4). Assessment for each land-unit of erosion forms and processes and definition of the cropping system and soil protection measures seems the best strategy for erosion control (Papy and Boiffin, 1989).

However, in establishing such a strategy, it must be remembered that when a variety of erosion processes and forms affect a specific land-unit, it is necessary to avoid adopting a protection measure which may control one process but enhance the hazard of another one at the same time (De Ploey, 1989; Chisci, 1991; Zanchi, 1989).

Table 4. Main erosion factors and point of control in agricultural management.

EROSION FACTORS	POINT OF CONTROL OF EROSION PROCESSES AND FORMS
a) Soil erodibility (splashability; crustability; rillability)	1. Soil structure and stability improvement 2. Management of soil mechanical structure 3. Crop and/or mulch cover management
b) Runoff generation	1. Soil structure and stability improvement 2. Management of soil mechanical structure 3. Surface and deep soil compaction control
c) Runoff concentration	1. Management of linear structures 2. Management of linear infrastructures 3. Management of area architecture

4. PERSPECTIVES ON SOIL PROTECTION MEASURES IN SOME REPRESENTATIVE EUROPEAN AREAS

Only a summary examination of soil protection measures is attempted, as can be extrapolated from experience and recent literature for some representative European landscapes.

Technical refinement and integrated design of soil protection measures for erosion control depends on individual site features. Successful adoption of one or more soil protection measures by the farmers depends on technical skill and socio-economic considerations.

Table 5. Perspectives on soil protection measures adoptable at field and farm scale in the loessial-loamy belt of central-northern Europe.

Main soil protection measures	Expected effects of application	Main advantages	Main constraints
Organic farming using compost and/or crop residue management in conventional tillage systems	Reduction of splashability and crusting; decreased soil erodibility; increased infiltration and control of runoff generation	Amelioration of humus content and soil structural stability	Availability of compost of adequate quality; 1) high isohumic coefficient; 2) low to zero heavy metals and synthetic organic compounds; 3) sterilisation of disease organisms and weed seeds; 4) cost of compost
Reduced tillage and mulching	Reduction of splashability and crusting; decreased soil erodibility; increased infiltration and control of runoff generation	Control of surface erosion processes	Soil compaction; difficulties in controlling weeds and diseases. Necessity of use of herbicides
Inclusion of cover crops in arable rotations	Reduction of splashability and crusting; increased soil infiltration and control of runoff	Provide soil protective cover during critical periods of rainfall aggressiveness; provide mulch material in reduced tillage; possible use as green manure	Competition for nutrient with main crop
Reduction of field traffic by use of combined machinery and low-pressure tyres; use of subsoiling	Increased soil infiltration; control of runoff generation and runoff concentration in oriented roughness	Reduction of soil compaction and oriented water pathways	Cost of the machinery and reduced manageability
Seed-bed cloddiness	Reduction of crust formation; increased soil infiltration; control of runoff generation and runoff velocity	Increase of water micro-storage on slopes and soil roughness	Possible difficulties in sowing and establishing small grain
Management and maintenance of linear and areal elements	Reduction of runoff concentration and point-erosion processes	Better water management; control of deepening and head advancement of existing gullies	Cost of operations

4.1. Loessial Belt

Soil protection measures in the loessial belt, as identified from past experience and innovative experimentation (Chisci and Morgan, 1986; Morgan and Rickson, 1988; Schwertmann et al., 1989.) are illustrated in Table 5. The expected effect of each measure on erosion control and the main advantages and constraints of their potential adoption are also shown.

It may be worthwhile to mention that adequate erosion control in Northwestern Europe could be easily attained by reforestation and/or turning critical arable fields into pasture. This operation seems rather unfeasible from a socio-economic point of view, considering the high natural fertility of the loessial, loamy areas of this region (De Ploey, 1986).

However, the possibility of reducing soil erosion within an acceptable level is not out of reach with the adoption of some agrotechnical soil protection measures and more care in the morphographic-infrastructural management of the land.

4.2. Vine growing Areas

Potential soil protection measures in the European vine growing areas are summarised in Table 6.

From experimental findings, grass-sodding in vineyards appears to be one of the most promising solutions for erosion control, both in central Europe and in the northern Mediterranean area. More difficult is to convince the farmers to spend time and money on management and maintenance of existing intensive linear structural and infrastructural elements. The building of linear structures, such as contour channels or underground tile drainage, is necessary to control runoff concentration in the Mediterranean area. The cost of such structures means the farmers may not adopt them even when vine cultivation is profitable and the vineyard could be in production for many years.

In grass-sod management, it is important to avoid the formation of up-and-down slope tractor tracks. Such tracks become perennial waterways for concentrated runoff.

The on-site and off-site impact of soil erosion in newly established vineyards is important. Here, soil erosion corresponding with high magnitude rainfall events is very common. Measured soil loss of 300-400 $t.ha^{-1}$ was assessed in north Mediterranean vineyards (Bazzoffi et al., 1989). In the Mosel area of Germany, high soil losses were also measured (Richter, 1989).

Planting in holes or deeply ploughed trenches may reduce erosion. However, up until now experimental evidence is lacking on the possible reduction of soil loss, and on the growth, yield and sustainability of vines resulting from these practices.

4.3. Clay and Marl Arable Areas

Potential soil protection measures suitable on clay and marl landscapes of central southern Mediterranean hilly areas are summarised in Table 7.

Chisci (1979, 1989) pointed out that surface and sub-surface runoff concentration management takes precedence in these soils to control a combination of landslides, solifluction and channel incision.

The use of extensive linear water management structures to protect cultivation units from upslope runoff concentrations are recommended. These include stormwater channels, contour and/or traverse ditches, contour channels and natural and/or artificial waterways (Landi, 1989). Deep up-and-down slope ploughing and/or subsoiling can drain these slopes, and underground tile drainage stabilises slopes (Chisci, 1989; Zanchi, 1989). These measures would increase the efficiency of water management within the cultivation units, especially in the control of runoff concentration.

4.4. Aspects of Soil Protection in the Mediterranean Area

While agrobiological soil protection measures may be sufficient for the control of accelerated erosion in northern central Europe, agricultural exploitation on slopes in the Mediterranean area would require the adoption of a more intensive mechanical water management system. This would operate with agrobiological measures (Chisci, 1979).

This is reflected in the traditional evolution of the cultural landscapes of these areas. The linearity index (De Ploey, 1989) of morphographic-infrastructural features is very high, as a consequence of the natural morphology on which man-made linear structures such as terraces, channels, fields and farm roads, live fences, tree boundaries, embankments and linchets have been superimposed.

Empirical experience and experimental evidence (Chisci, 1989, 1991) has clearly confirmed that agrobiological soil protection measures alone are not sufficient to cope with erosion, by reducing soil erodibility and increasing water infiltration into the soil. Runoff generation and concentration would not be sufficiently controlled in arable land during the intensive rainstorms of the Autumn-Winter season of the Mediterranean. Consequently, mechanical soil protection measures must be integrated into the cultivation system to reduce erosion to an acceptable level (Chisci, 1991).

Specialised-mechanised farming, based on arable and tree crop cultivation, has imposed chaotic morphographic-structural and infrastructural features on many Mediterranean landscapes (Table 1). Moreover, the new pattern of linear and areal structures appears to be less coherent than in the past (Chisci, 1979, 1980).

Elimination of obsolete mechanical structures, and reshaping of field boundaries and slope morphology by levelling may, if carefully designed, be feasible. However, new mechanical water management measures compatible with the mechanisation of cultivation units, must replace the obsolete ones to control runoff concentration.

Unfortunately, the farmers in these areas have shown an attitude of non-adoption of up-to-date linear water management structures in order to keep the cost of hill farming as low as possible. The maintenance of the old water management structures has also been neglected. It is not a question of insufficient technical innovation or lack of knowledge and know-how by the farmers, but the high cost of building and maintaining these mechanical structures. This is compounded by the decreasing revenues from hill farming.

Table 6. Perspectives on soil protection measures adoptable at field and farm scale in the vine growing regions of Europe.

Main soil protection measures	Expected effects of application	Main advantages	Main constraints
Permanent grass-sodding over the vineyard surface	Reduction of soil splashability and rillability	Continuous cover of the soil; increase in carrying capacity of the soil in relation to machinery	Possible increase in soil compaction and formation of permanent tractor tracks. Expensive grass-growth control; competition of grasses for water and nutrients; formation of shallow rooting of cultivated grapes; hosting of grape diseases on herbage plants
Inter-tree rows grass-sodding	Reduction of soil splashability and rillability on grass-sod strips	Provide a discontinuous soil cover; increase in carrying capacity on grass-sod strips; reduction of competition for water and nutrients with grapes; better sanitary conditions in relation to continuous grass-sodding	Increase in soil compaction in grass-sod strips; possible formation of permanent tractor tracks. Expensive grass-growth control; possible rill formation on oriented tilled strips
Cover crops and continuous or partial mulching with different mulch material (grass, straw, compost, stubble, etc.)	Reduction of soil splashability and rillability	Provide a soil cover during critical periods of rainfall aggressiveness. Increase soil organic matter by incorporating green and/or mulch material into the soil	Downslope flowing of mulch on steep slopes; shallow rooting of cultivated vines
Reduction of field traffic load using low pressure tyres	Increased soil infiltration; better control of runoff generation and concentration in oriented roughness	Reduction of soil compaction and oriented tractor tracks	Cost of machinery and reduced manageability
Redesign, management and maintenance of linear and areal structural and infrastructural elements	Control of runoff concentration and point-erosion processes	Better water management control; reduction of gullying and mass movements	Cost of operations

Table 7. Perspectives on soil protection measures adoptable at field and farm scale in clay and marl areas of central-southern Mediterranean Europe.

Main soil protection measures	Expected effect of application	Main advantages	Main constraints
Organic farming using composts and/or crop residue management in conventional tillage systems	Reduction of slaking and crusting; increased isotropic water infiltration; control of runoff generation	Amelioration of soil humus content and structure stability	Availability of compost of adequate quality: 1) high isohumic coefficient; 2) low to zero content of heavy metals; 3) sterilisation of disease organisms; 4) sterilisation of weed seeds; 5) cost of the compost
Ploughing system: contour direction on marls on moderate slopes; up-and-down slope direction on clay soils. Use of subsoiling or double-layer ploughing	Increased water infiltration and storage in the soil; reduction of landslides and solifluction	Amelioration of soil water drainage	Possible increase in piping and gullying in absence of adequate linear structures for sub-surface runoff control
Linear structures within and on boundaries of cultivation units: stormwater channels upslope; contour-channels or contour roads; contour temporal ditches; underground tile drainage	Control of runoff concentration and outflow of excess water at a non-erosive velocity. Management of drainage to prevent piping, gullying, solifluction and mass movement	Sub-optimal water management in marl to clayey arable soils. Slope stabilisation	Cost of construction and maintenance of structures

The situation is somewhat better for the more valuable specialised tree crops (grapes, olives and orchards). The costs of building and maintenance of water management structures may be more easily sustained for these crops.

On the other hand, the increasing marginality of some hilly, arable areas of the Mediterranean has led to abandonment of previously cultivated land. Quite often, abandoned fields have obsolete mechanical structures that are subject to collapse and invasion by noxious weeds, which offer scarce cover or soil protection. The abandoned fields represent a source of land disorder, sediment production and runoff concentration. To allow some use of the abandoned land for pasture, fire is often used to destroy thorny bush and weeds, thus reducing the already low soil protection further.

Abandoned fields and farms must not be left without care. Reforestation, grass-sodding and forage shrub planting (Chisci et al., 1991) must be provided, together with an extensive programme for the maintenance of existing structural and infrastructural elements of the landscape.

5. CONCLUSION

A considerable number of soil protection measures is available to confront and control the cycle of soil erosion acceleration affecting some European 'cultural landscapes'.

Erosion studies, technological innovation and experimentation on soil protection measures by European scientists have increased recently. This has enhanced knowledge of erosion phenomena and their evolution in time and space, and understanding of the implications of accelerated soil erosion on the welfare of society.

The real problem lies with the operative implementation of the acquired knowledge. The majority of the soil protection measures reviewed here depends on adoption by farmers.

On the other hand, all of society, not only the people directly involved in exploiting the land, is deeply affected by the possible soil and environmental degradation due to such exploitation, when adequate soil protection measures are not adopted.

Any soil protection measure has to be compatible with maximum possible revenues from hill farming. It may also be necessary to induce a modification of the economic attitude of farmers. Although they have to make a living from agricultural enterprise, this does not give them the right to mismanage a natural non-renewable resource, without a thought to the possible damaging effects on the environment.

It may be asserted that there is a duty for the landlord and/or agricultural entrepreneur to protect the land, and that there is also a component of responsibility for off-site damage due to land mismanagement. For instance, adequate care of morphographic-infrastructural elements is the farmer's responsibility.

351

It is also important to stimulate pride in having well-managed land. For instance, it may be suggested to promote renewed care for the land by issuing a friendly competition between neighbouring farmers, with the land authorities conferring special prizes to farmers who use adequate and effective soil protection measures.

REFERENCES

BAZZOFFI, P., CHISCI, G. and MISSERE, D. 1989. Influenza delle opere di livellamento e scasso sull'erosione del suolo nella collina cesenate. Rivista di Agronomia, XXIII, 3. pp.213-221.

BORK, H.R. 1983. Die holozane Relief und Bodenentwicklung in Lossgebieten. CATENA, Suppl 3. pp.1-93.

BORK, H.R. and ROHDENBURG, H. 1979. Beispiele für jungholozane Bodenerosion und Bodenbildung im Unter-Eichsfeld und Randgebieten. Landschaftsgenese und Landschaftsokologie. pp.115-134.

BOLLINNE, A. 1982. Etude et prevision de l'erosion des sols limoneux cultives en Moyenne Belgique. These presentee pour l'obtention du grade de Docteur en Sciences Geographiques, Université de Liege.

CHISCI, G. 1979. Considerazioni sulle trasformazioni dei sistemi di agricoltura collinare in relazione al regime idrologico e al dissesto dei versanti. Geologia Applicata e Idrogeologia, XIV, III, Bari. pp.225-249.

CHISCI, G. 1980. Physical soil degradation due to hydrological phenomena in relation to change in agricultural systems in Italy. In: Boels, D., Davies, D.B. and Johnson, E.A. (eds). Soil degradation. A.A. Balkema, Rotterdam. pp.95-103.

CHISCI, G. 1987. Effetti antropici sull'erosione dei suoli. Atti del 1° Congresso Internazionale di Geoidrologia sulla 'Autropizzazione e la Degradazione dell'Ambiente Fisico', Firenze. pp.535-552.

CHISCI, G. 1989. Measures for runoff and erosion control on clayey soils: a review of trials carried out in the Apennines hilly area. In: Schwertmann, U., Rickson, R.J. and Auerswald, K. (eds). Soil erosion protection measures in Europe. Soil Technology, Series 1. pp.53-71.

CHISCI, G. 1991. Interventi tecnici e strategi operative per la difesa del suolo in ambiente declive mediterraneo. Relazione al 'Convegno sulla Difesa del Suolo in Ambiente Mediterraneo', ERSAT, Cala Gonone (Sardegna), 12-14 Giugno.

CHISCI, G.C. and MORGAN, R.P.C. (eds). 1986. Soil erosion in the European Community. Balkema, Rotterdam.

CHISCI, G., STRINGI, L., MARGINEZ, V., AMATO, G., GRISTINA, L. 1991. Ruolo degli arbusti foraggeri nell'ambiente semi-arido siciliano. 2. Funzione protettiva contro l'erosione idrometeorica. Rivista di Agronomia, XXV, 2, pp.332-340.

DE PLOEY, J. 1986. Soil erosion and possible conservation measures in loess loamy areas. In: Chisci, G. and Morgan, R.P.C. (eds). Soil erosion in the European Community. Balkema, Rotterdam. pp.157-163.

DE PLOEY, J. 1989. Erosional systems and perspectives for erosion control in European loess areas. In: Schwertmann, U., Rickson, R.J. and Auerswald, K. (eds). Soil erosion protection measures in Europe. Soil Technology, Series 1. pp.93-102.

HUGHES, J.D. and THIRGOOD, J.V. 1982. Deforestation in ancient Greece and Rome: a cause of collapse. The Journal of Forest History, 26,1. pp.196-208.

LANDI, R. 1982. Revision of land management systems in Italian hilly areas. In: Schwertmann, U., Rickson, R.J. and Auerswald, K. (eds). Soil erosion protection measures in Europe. Soil Technology, Series 1. pp. 175- 88.

MEYER, B. and WILLERDING, U. 1961. Bodenprofile, Pflanzenrests und Fundmaterial von neuerschlossenen neolithischen und eisenzeitlichen Siedlungs-stellen in Gottinger Stadtgebiet. Gottinger Jahrbuch. pp.21-38.

MONNIER, G. and BOIFFIN, J. 1986. Effect of the agricultural use of soils on water erosion: The case of cropping systems in Western Europe. In: Chisci, G. and Morgan, R.P.C. (eds). Soil erosion in the European Community. Balkema, Rotterdam. pp.17-31.

MORGAN, R.P.C 1986. Soil degradation and soil erosion in the loamy belt of northern Europe. In: Chisci, G. and Morgan, R.P.C. (eds). Soil erosion in the European Community. Balkema, Rotterdam. pp.165-172.

MORGAN, R.P.C. and RICKSON, R.J. (eds). 1988. Erosion assessment and modelling. EC Report EUR 10860. EN.

PAPY, F. and BOIFFIN, J. 1989. The use of farming systems for the control of runoff and erosion (Example from a given country with thalweg erosion). In: Schwertmann, U., Rickson, R.J. and Auerswald, K. (eds). Soil erosion protection measures in Europe. Soil Technology Series 1. pp.29-38.

RICHTER, G. 1963. Bodenerosion. Schaden und gefahrdete Gebiete in der Bundesrepublik Deutschland. Forschungen zur dt. Landeskunde 152. Bad Godesberg.

RICHTER, G. 1980. On the soil erosion problem in the temperate humid area of Central Europe. Geojournal 4.3. pp.279-287.

RICHTER, G. 1989. Erosion control in vineyards of the Mosel region, FRG. In: Schwertmann, U., Rickson, R.J. and Auerswald, K. (eds). Soil erosion protection measures in Europe. Soil Technology, Series 1. pp. 149-156.

SCHWERTMANN, U. 1986. Soil erosion: extent, prediction and protection in Bavaria. In: Chisci, G. and Morgan, R.P.C. (eds). Soil erosion in the European Community. Balkema, Rotterdam. pp.185-200.

SCHWERTMANN, U., RICKSON, R.J. AND AUERSWALD, K. (eds). 1989. Soil erosion protection measures in Europe. Soil Technology, Series 1.

SUSMEL, L. 1986. Principi di ecologia nella pianificazione del verde. In: Viola, F. (ed.). Criteri forestali nella pianificazione territoriale. INVET, F. Angeli Ed., Milano. pp.13-44.

VOGT, J. 1958. Zur historischen Bodenerosion in Mitteldeutschland. Petermanns Geographische, Mitt. 102,199-203.

ZANCHI, C. 1989. Drainage as soil conservation and soil stabilizing practice on hilly slopes. In: Chisci, G. and Morgan, R.P.C. (eds). Soil erosion in the European Community. Soil Technology, Series 1. pp.73-82.

Chapter 32

CROPPING SYSTEMS OF FODDER MAIZE TO REDUCE EROSION OF CULTIVATED LOESS SOILS

F.J.P.M. Kwaad
Laboratory of Physical Geography and Soil Science, University of Amsterdam, Nieuwe Prinsengracht 130, 1018 VZ Amsterdam, the Netherlands

Summary

In 1985 a plot study was set up on sloping cultivated loess soils in south Limbourg (the Netherlands) to evaluate the effects of various cropping systems of fodder maize on runoff, soil loss and crop yield under natural rainfall. Plot length was 22m. The results of the growing seasons and winters of 1987, 1988 and 1989 are presented. The following cropping systems of fodder maize were compared:

(a) conservation cropping system A; maize with winter rye as winter cover crop and direct drilling of maize,
(b) conservation cropping system B; maize with summer barley as spring cover crop,
(c) conventional cropping system C; maize with stubble field in winter,
(d) permanently bare soil (system C'), tilled as the conventional system C and wheelings formed at seeding and harvest time of maize.

For the whole study period, total soil losses of systems C', C, B and A were the equivalent of 16.0, 10.8, 3.4 and 1.7 $t.ha^{-1}yr^{-1}$ respectively. High winter runoff on system C was strongly reduced by systems A and B. Maize crop yields of system A varied from 86 to 105% of system C and those of system B from 91 to 102% of system C.

1. INTRODUCTION

During the last two or three decades, damage by rainfall induced accelerated erosion, with associated off-site effects, is reported to have increased in south Limbourg (Schouten et al., 1985). Damage affects a hilly area of 40,000ha with loess soils, which is part of the west European loess belt. In 1987, regulations were issued by the Provincial authorities to the effect that measures must be taken to reduce the on-site and off-site effects of soil erosion (Provinciale Staten, 1987).

A possible cause of the increased rate of soil erosion in this area is the expansion in recent years of the area of row crops (sugar beets, fodder maize, potatoes and onions) at the cost of small grain crops (Schouten and Rang, 1987). Since 1960 the area of erosion prone crops has increased from about 4,000ha to 12,000ha.

One way to reduce runoff and soil loss from cropped land is to adopt certain farming practices, such as winter cover cropping, plant residue mulching and conservation tillage systems. In south Limbourg, however, farmers have no experience of conservation farming systems. Also in other west European countries data are limited concerning the effectiveness of conservation cropping systems in erodible crops like sugar beets and maize (Schwertmann et al., 1989; Boardman et al., 1990). In Germany and Switzerland various studies have been carried out recently, on the agronomic feasibility and crop yield aspects of conservation cropping systems in sugar beets and maize (Ammon et al., 1990; Diez et al., 1988; Röper and Lütke-Entrup, 1987; Sturny and Meerstetter, 1989, 1990; Wunderlich and Schwerdtle, 1982; Wolfgarten et al., 1987).

In 1985, a plot study was started at the Experimental Farm Wijnandsrade, in the south of the Netherlands, with the aim to develop cropping systems of fodder maize and sugar beets, which effectively reduce runoff and erosion without reducing crop yield, and which are technically feasible. Four phases were envisaged in the development and introduction of conservation cropping systems: trial or exploration, optimisation, extrapolation and implementation. The trial phase was ended after the winter of 1989/90. In this paper, results of the runoff and soil loss measurements of the trial phase for fodder maize are presented. The objectives of the trial phase were: (a) to provide quantitative data on the rate of soil erosion on arable land in south Limbourg, and (b) to test some cropping system alternatives regarding runoff, erosion and crop yields.

For maize the trial phase included measurement of runoff, soil loss and crop yield of two newly devised conservation cropping systems as well as the conventional cropping system and permanently bare soil. For sugar beets, only the effect on crop yield was evaluated during the trial phase.

2. MATERIALS AND METHODS

Twelve plots with three replications of the three cropping systems of maize and permanently bare soil were laid out in a randomised block design (Quenouille, 1953) on a 6% sloping field (Figure 1). Plot length was 25m and plot width was 10m (cropped plots), and 5m (fallow plots). Direction of tillage and sowing was up and down slope. On each plot, a segment of 22m long and 1.80m wide, including two maize rows and two tractor wheelings, was fenced off for erosion measurements (Mutchler et al., 1988). Measurements were carried out under natural rainfall. Runoff and sediment was collected and stored in three storage tanks per plot, fitted with modified Geib multislot divisors (Brakensiek et al., 1979). Divisors with ten slots were used on 100 litre tanks. The total storage capacity was the equivalent of 10,825 litres of runoff. The interval between visits was four weeks, in which 200mm of rainfall can be expected with a recurrence interval of 100 years. Plots were laid out in October 1985. Maize was grown continuously on the plots between 1986 and 1989. Systematic measurements of runoff and soil loss were carried out from May 1987 until March 1990, both in summer and winter.

Soils at the experimental site were imperfectly drained, truncated gleyic luvisols with less than 0.80m overlying colluvium on the whole plot length. During part of the year a (perched) water table was present within one metre of the soil surface. Parent material was decalcified loess. Textural classes of the plough layer were silt and silt loam. Average particle size distribution was 13% clay, 81% silt (2-50μm) and 6% sand. Organic matter

Figure 1. Layout of experimental plots.

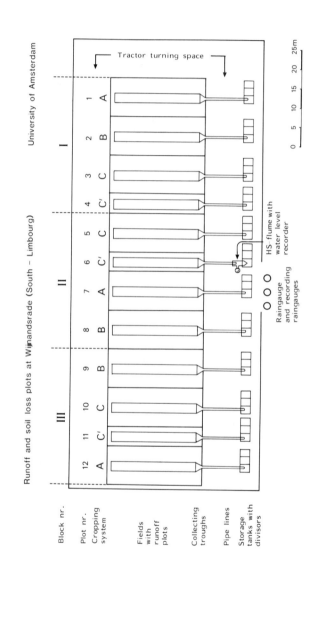

Runoff and soil loss plots at Wijnandsrade (South – Limbourg)   University of Amsterdam

content of the plough layer was 1.8%. Loess is a highly erodible soil material due to its sensitivity to soil tructure breakdown, resulting in sealing and crust formation at the soil surface (Imeson and Kwaad, 1990). Soil cover and soil tillage are important controls of slaking, infiltration and generation of Hortonian overland flow.

Average yearly precipitation is 750mm, with rain in all seasons and 23 days with snow. High intensity rains only occur in the summer, beginning in April. The 30-minute intensity that is exceeded once a year is 24mm.h^{-1} (Buishand and Velds, 1980). Rainfall erosivity (Wischmeier and Smith, 1978) is relatively low, with an R-factor value of 75 (metric units) according to Bergsma (1980). Risk of erosion is highest in April-June, when cover of crops like maize and sugar beet is still incomplete and high intensity rainfall may occur. Rapid snowmelt can be a cause of surface runoff in winter.

Fodder maize (*Zea mais* L., cultivar Sonia) was grown as a continuous culture on the experimental plots. The whole maize plant is harvested, cut up in small pieces and stored in silos for fodder. Experience with cover crops for wind erosion control in the Netherlands was used in devising the conservation cropping system. The characteristics of the maize cropping systems were as follows (Geelen, 1987):

(a) Conservation cropping system A consisted of ploughing and seedbed preparation by rotary harrowing after harvest of the preceding maize crop in October, directly followed by drilling of winter rye. The winter rye was killed by spraying in late April. Maize was sown by direct drilling in the surface mulch of winter rye. Only winter rye could be used as the winter cover crop in continuous culture of maize, in view of the late harvest time of the maize.

(b) Conservation cropping system B involved coarse loosening of the soil with a point tine cultivator after harvest of the preceding maize crop in October. Ploughing and seedbed preparation occurred in late winter/early spring, followed by sowing of summer barley. Maize was drilled directly into the summer barley, which was killed by spraying before emergence of the maize.

(c) Conventional cropping system C comprised stubble left from the time of harvest of the preceding maize crop until the next spring. Ploughing and seedbed preparation occurred in late April/early May, immediately followed by maize drilling.

(d) Permanently bare fallow system C' involved ploughing and seedbed preparation as for system C. The plots of C' were passed over with sowing and harvesting machinery at the time of sowing and harvesting of system C.

Plough depth was 30cm. Seedbed preparation in autumn and spring on all systems included loosening of the top 5cm of soil with duckfoot tines, rotary harrowing of the uplifted clods and light rolling of the broken clods, all in one combined operation. Details on the use of fertilisers, herbicides and pesticides on the plots are given by Geelen (1987) and Geelen and Kwaad (1988; 1989; 1990).

Besides runoff and soil loss, the following variables were observed during each four weekly visit: (a) degree of surface slaking of plots by comparison with a series of ten reference photos (Boekel, 1973); (b) random and oriented microrelief, measured as vertical distance between highest and lowest point of the soil surface within a horizontal distance of fifteen centimetres; (c) degree of surface cover by plants and/or plant residues including weeds; and (d) presence and nature of erosion features.

Differences between cropping systems have been statistically tested by means of analysis of variance. Significance level was 0.05.

3. RESULTS

For the figures in this section, mean values of the three replications per cropping system were used. Because of missing data for the winter 1989/90, the results only refer to the summers of 1987, 1988 and 1989 and the winters of 1987/88 and 1988/89. The summer season extends from the time of maize drilling to maize harvest. In 1987, 1988 and 1989, this was roughly from 1st May until 15th October. Results of statistical significance tests are given in Table 1. It appeared that interaction of treatment with date of observation was significant. This interaction effect has been taken into account in evaluating significance of differences between treatments.

Table 1. Significance of differences between cropping systems at p=0.05, including interaction effects of treatment with date of observation.

	Period 1987-90	Summer 1987	Winter 1987/88	Summer 1988	Winter 1988/89	Summer 1989	Winter 1989/90	Number of months*
Runoff								
A and C	sign.		sign.		sign.		sign.	9
B and C	sign.		sign.		sign.		sign.	10
A and B	sign.				sign			2
Runoff %								
A and C	sign		sign.				sign	12
B and C	sign.		sign.		sign.		sign.	15
A and B	sign.				sign.			2
Soil loss								
A and C	sign.	sign.		sign.			sign.	6
B and C	sign.		sign.	sign.	sign.		sign.	6
A and B		sign.			sign.			3

* Number of months with significant difference between treatments (total is 36 months)

Precipitation amounts were 873.2mm in 1987, 780.4mm in 1988 and 695.8mm in 1989. Monthly precipitation amounts are given in Figure 2. They ranged from 18.0 to 126.0mm. No snow fell on the experimental site during the winters of 1987/88 and 1988/89.

Figure 2. Monthly precipitation at Wijnandsrade, May 1987 - September 1989.

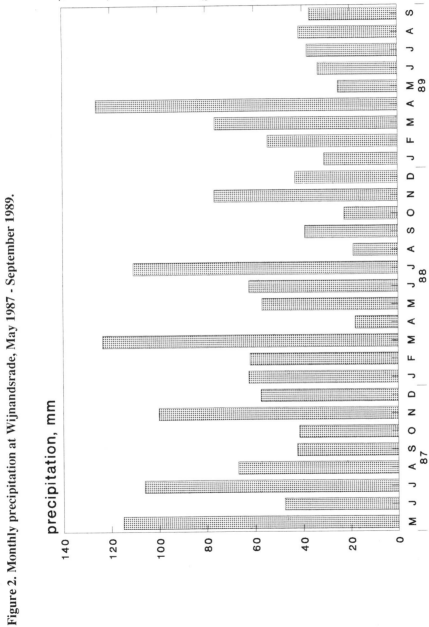

From Figure 3 it appears that most runoff has occurred in the winter season on all cropping systems. Runoff was reduced on both cropping systems A and B, as compared to system C. Reduction was most effective in winter and less effective to nil in summer. Over the three summers in the period of study, runoff was reduced to 77.7% of that of system C on system A and to 54.7% on system B. During the two winters runoff was reduced to 16.9% on system A and to 5.0% on system B. Over the whole period of study, runoff on system A was 30.9% of that of system C and runoff on system B was 16.5% of that of system C. During summer 1987, winter 1987/88 and summer 1988, runoff volumes of systems A and B were not significantly different. During winter 1988/89, runoff was significantly lower on system B than on system A. During summer 1989 runoff volumes were very low with no significant differences between treatments.

From Figure 4 it is apparent that summer soil losses were higher than winter soil losses on all cropping systems. Sediment concentration of runoff was higher in summer than in winter (Fig. 6). Total soil loss on permanently bare soil (system C') was 4kg.m^{-2} in three summer and two winter seasons, of which 64.2% was lost in the three summers and 35.8% in the two winters. Average yearly soil loss on bare soil during the period of study was 16 t.ha^{-1}. Soil loss was reduced on both cropping systems A and B as compared to system C (Fig. 7). Over the three summers of study, soil loss was reduced to 13.6% of that of system C on system A and to 41.2% on system B. During the two winters of study, soil loss was reduced to 21.9% on system A and to 8.8% on system B. Over the whole period of study, soil loss on system A was 16.1% of that of system C and soil loss on system B was 31.6% of that of system C. During summer 1987 soil loss on system A was significantly lower than that on system B. In winter 1987/88 and summer 1988 soil losses were not significantly different on systems A and B. In winter 1988/89 significantly less soil was lost on system B than on system A. Soil losses in summer 1989 were very low with no significant differences between treatments.

In 1986, 1987 and 1988, maize yield (dry matter) was 8 to 14% lower on cropping system A, and 2 to 9% lower on system B than on system C. In 1989, yield was 5% higher on system A, and 2% higher on system B than on system C (Geelen, 1987; Geelen and Kwaad, 1988; 1989;1990).

4. DISCUSSION

The runoff amounts were unexpectedly high in winter on the fallow plots and on the stubble fields of the conventional maize cropping system. Monthly runoff percentages reached values of 100 or higher in some winter months on some plots of system C'. Winter runoff did not cause much erosion within the bounds of the plots. Sediment concentration of winter runoff was much lower than that of summer runoff. On longer bare slopes near the plots, however, rill erosion was observed during winter. It may well be that winter runoff originated (partly) as saturation overland flow (Kwaad, 1991). Shallow (perched) groundwater tables were present on and near the plots in the winter half of the year. On some occasions in winter, water flow in rills was observed up to 38 hours after the last rainfall. In most winter months, runoff was reduced very effectively on systems A and B. In some winter months, however, runoff was also high on some plots of system A.

Figure 3. Half yearly runoff amounts.

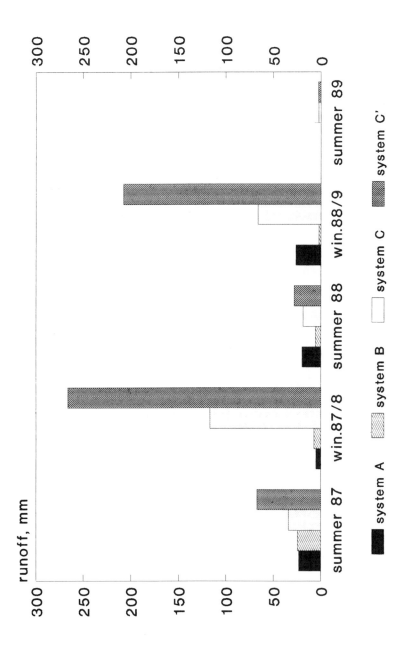

Figure 4. Half yearly soil loss amounts.

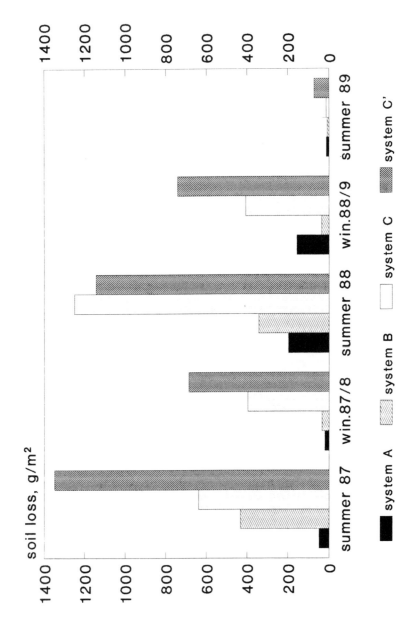

This can be explained by the saturation overland flow hypothesis. Apparently, soil moisture storage capacity of system B with its large clods in winter was high enough to prevent saturation overland flow generation.

4.1 Permanently bare fallow (C')

The soil surface on the permanently bare plots rapidly slaked by the first rains after seedbed preparation in late April-early May, and stayed that way until the following spring. Tractor wheelings were present, both in summer and winter, formed by passes over the plots with sowing and harvesting machinery. Presumably, runoff was generated in the wheelings. Also the wheelings acted as surface drainage ways for the plots of system C'. System C' gives an indication of the erodibility of the loess soils in south Limbourg. A soil loss of 16.0 t.ha^{-1}yr^{-1} on a 6% and 22m long slope, together with an annual R-factor value of 75 in metric units for the middle of south Limbourg (Bergsma, 1980), gives a K-factor value of 0.37 in metric units, adjusted for a 9% slope angle (Wischmeier and Smith, 1978). This can be converted to a value of 0.28 in English units by dividing by 1.317 (Foster et al., 1981). With the K-factor nomogram (Wischmeier and Smith, 1978) a first approximation (valid for agricultural soils with a fine granular topsoil and moderate permeability) of 0.47 (English units) is found. According to Bollinne et al. (1979) the average annual R-factor value of central south Limbourg is about 60 (metric units). Together with the measured soil loss of 16 t.ha^{-1}yr^{-1} this yields a K-factor value of 0.47 (metric units) or 0.35 (English units), adjusted for a 9% slope angle. So, a discrepancy exists between the measured and the calculated value of the K-factor, for which no explanation can be offered. The metric values of 0.37 and 0.47, however, are not outside the range of 0.29 to 0.49, published by Schwertmann et al. (1987) for loess soils in south Germany. Bollinne and Rosseau (1978) mention a K-factor of >0.45 (English units) for the loamy soils of Belgium.

4.2 System C (conventional maize cropping)

System C was characterised by a fine seedbed in May that rapidly slaked by the first rains after maize drilling. The winter condition was a slaked stubble field. Tractor wheelings were formed during maize drilling, which were further deepened during harvest of maize. As on system C', runoff originated and developed in the wheelings. Average yearly soil loss during the period of study on the 22m long plots of the conventional system C was the equivalent of 10.8 t.ha^{-1}. On longer slopes higher soil losses may be expected. 10.8 t.ha^{-1} is about the maximum rate of soil loss that is considered by some to be acceptable under most favourable conditions in the USA (Mannering, 1981; McCormack and Young, 1981; Williams, 1981; Crosson and Stout, 1983; Schertz, 1983). No soil loss tolerance level has been set for south Limbourg (or elsewhere in Europe) so far.

4.3 System A (maize cropping with winter rye as cover crop)

System A was characterised by a fine and smooth winter rye seedbed in autumn, which rapidly slaked by the first rains after the sowing of winter rye. No tractor wheelings were present on system A, either in winter, or in summer. The winter rye remained very low (<5% soil cover) from the time of sowing (late October/early November) until April. Total soil loss on system A was 1.7 t.ha^{-1}yr^{-1} during the period of study. System A effectively reduced winter runoff and erosion, inspite of strong surface slaking. At the time of maize drilling the surface had become very compact and even. In spite of this, runoff on system A was lower than on the freshly tilled system C during May and June. This is tentatively ascribed to the presence of continuous vertical biogenic macropores on system A (root channels and worm holes), formed since the sowing of winter rye, and which were not disrupted by spring tillage (Kwaad and Van Mulligen, 1991). Surface storage capacity was increased by the presence of winter rye remnants. Summer runoff was slightly reduced and summer soil loss was very strongly reduced on system A. As a result, sediment concentration of summer runoff was very low on system A (Figure 5). This constitutes a hazard, if runoff on system A is allowed to concentrate on longer slopes or to run onto adjacent downslope fields with an erodible topsoil. Low summer soil losses on system A are ascribed to a high mechanical resistance of the compact soil surface and to a decreased velocity of overland flow due to the winter rye remains lying on the surface.

4.4 System B (maize cropping with summer barley as spring cover crop)

System B was characterised by the presence of coarse clods (10cm) and a high random surface roughness from the time of cultivation with point tines in October-November until the sowing of summer barley in February-March. The clods of the barley seedbed were also relatively coarse and hard, and did not slake as easily as the seedbed that was prepared late April on system C, when the soil was drier. Tractor wheelings were present in summer after maize drilling, but not in winter after fall cultivation and in early spring after sowing of summer barley. Total soil loss on system B was 3.4 $t.ha^{-1}yr^{-1}$ during the period of study. System B effectively reduced winter runoff and winter soil loss. Summer runoff was also strongly reduced by system B. Summer soil loss was less effectively reduced by system B than by system A.

5. CONCLUSIONS

The data show that runoff and soil loss from fodder maize can be reduced substantially (on a plot scale) by introducing changes in the conventional way of growing crops. Winter runoff and winter soil loss was practically eliminated by not leaving stubble behind after maize harvest, but tilling the field or tilling and sowing a cover crop. Measures like loosening the soil and cover crops are only effective, however, if surface sealing is the main cause of runoff generation. There is evidence, however, that winter runoff on the experimental plots may, at least partly, have originated as saturation overland flow. Summer soil loss was very effectively reduced by cropping system A inspite of only a slight or no reduction of runoff. System B reduced summer runoff slightly better than system A, but did not reduce erosion as well. Further research should aim at reducing summer runoff, as runoff concentration on longer slopes may lead to erosion.

Figure 5. Sediment concentration in runoff.

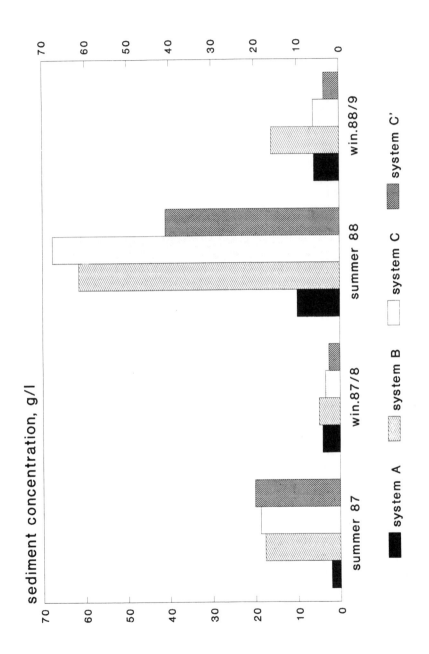

ACKNOWLEDGEMENTS

Financial support was received from the Foundation Experimental Farm Wijnandsrade, the Landscape Engineering Service at Roermond and the Provincial Government of Limbourg. The plots were located on the property of the Experimental Farm Wijnandsrade, and all agricultural operations were carried out by the Experimental Farm with Mr. Q. Kerckhoffs as acting manager. The cropping systems of maize were devised in large part by Ing. P. Geelen, research worker at the Experimental Farm. Dr. H.J. Mücher helped with mapping and classifying the soils at the experimental site. Analysis of variance was performed by W. v.d. Berg at the PAGV Institute, Lelystad. Mrs. A. Bolt, E. van Mulligen and C. Zeegers assisted in various ways with the field work.

REFERENCES

AMMON, H.U., BOHREN, Ch., and ANKEN, Th. 1990. Breitband-Frässaat von Mais in Wiesen- und Gründüngungsbestände mit Mulch-Schnitt zwischen den Reihen. Zeitschr. Pflanzenkrankh. Pflanzenschutz, Sonderheft 12:229-235.

BERGSMA, E. 1980. Provisional rain-erosivity map of the Netherlands. In: de Boodt, M. and Gabriels, D. (eds). Assessment of erosion. John Wiley, Chichester.

BOARDMAN, J., FOSTER, I.D.L. and DEARING, J.A. (eds). 1990. Soil erosion on agricultural land. Wiley, Chichester.

BOEKEL, P. 1973. De betekenis van de ontwatering voor de bodemstructuur op de zavel- lichte kleigronden. Instituut voor Bodemvruchtbaarheid, Haren. Rapport 5.

BOLLINNE, A., LAURANT, A. and BOON, W. 1979. L'érosivité de précipitations à Florennes. Révision de la carte des isohyètes et de la carte d'érosivité de la Belgique. Bulletin Société Géographique de Liége. 15:77-99.

BOLLINNE, A. and ROSEAU, P. 1978. L'érodibilité des sols de Moyenne et Haute Belgique. Bulletin Société Géographique de Liége. pp. 127-140

BRAKENSIEK, D.L., OSBORN, H.B. and RAWLS, W.J. (eds). 1979. Field manual for research in agricultural hydrology. US Department of Agriculture. Agriculture Handbook 224.

BUISHAND, T.A. and VELDS, C.A. 1980. Neerslag en verdamping. KNMI, de Bilt.

CROSSON, P.R. and STOUT, A.T. 1983. Productivity effects of cropland erosion in the United States. Resources for the Future, Washington DC.

DIEZ, Th., KREITMAYR, J. and WEIGELT, H.B. 1988 Erosionsschutzmassnahmen im Mais- und Zuckerrübenanbau. Die Landtechnische Zeitschrift, 1/88. pp.1-19.

FOSTER, G.R., McCOOL, D.K., RENARD, K.G. and MOLDENHAUER, W.C. 1981. Conversion of the Universal Soil Loss Equation to SI metric units. Journal of Soil and Water Conservation 36:355-359.

GEELEN, P. 1987. Bestrijding van watererosie. Jaarverslag over 1986 van de Proefboerderij Wijnandsrade. pp. 83-85.

GEELEN, P. and KWAAD, F.J.P.M. 1988. Bestrijding van watererosie. Jaarverslag over 1987 van de Proefboerderij Wijnandsrade. pp.95-99.

GEELEN, P. and KWAAD, F.J.P.M. 1989. De bestrijding van watererosie bij de continuteelt snijmais. Jaarverslag over 1988 van de Proefboerderij Wijnandsrade. pp.143-150.

GEELEN, P. and KWAAD, F.J.P.M. 1990. De bestrijding van watererosie bij de continuteelt snijmais. Jaarverslag over 1989 van de Proefboerderij Wijnandsrade. pp.133-140.

IMESON, A.C. and KWAAD, F.J.P.M. 1990. The response of tilled soils to wetting by rainfall and the dynamic character of soil erodibility. In: Boardman, J., Foster, I.D. and Dearing, J.A.(eds). Soil erosion on agricultural lands. John Wiley, Chichester.

KWAAD, F.J.P.M. 1991. Summer and winter regimes of runoff generation and soil erosion on cultivated loess soils (The Netherlands). Earth Surface Processes and Landforms 16:653-662.

KWAAD, F.J.P.M. and VAN MULLIGEN, E.J. 1991. Cropping system effects of maize on infiltration, runoff and erosion on loess soils in south Limbourg (The Netherlands): a comparison of two rainfall events. Soil Technology 4:281-295.

MANNERING, J.V. 1981. The use of soil loss tolerances as a strategy for soil conservation. In: Morgan, R.P.C. (ed.). Soil conservation. Problems and prospects. Wiley, Chichester.

McCORMACK, D.E. and YOUNG, K.K. 1981. Technical and societal implications of soil loss tolerance. In: Morgan, R.P.C. (ed.). Soil conservation. Problems and prospects. Wiley, Chichester.

MUTCHLER, C.K., MURPHREE, C.E. and McGREGOR, K.C. 1988. Laboratory and field plots for soil erosion studies. In: Lal, R. (ed.). Soil erosion research methods. Soil and Water Conservation Society, Ankeny.

PROVINCIALE STATEN VAN LIMBURG. 1987. Streekplan Zuid-Limburg (Algehele Herziening). Maastricht.

QUENOUILLE, M.H. 1953. The design and analysis of experiment. Griffin, London.

RÖPER, W. and LÜTKE-ENTRUP, N. 1987. Mulchsaatverfahren im Zuckerrübenanbau. Rationalisierungs-Kuratorium für Landwirtschaft (RKL), Kiel. pp.632-669.

SCHERTZ, D.L. 1983. The basis for soil loss tolerances. Journal of Soil and Water Conservation 38:10-14.

SCHOUTEN, C.J. and RANG, M. 1987. Bodemerosie in Zuid-Limburg. Natuur en Milieu 11:9-13

SCHOUTEN, C.J., RANG, M.C. and HUIGEN, P.M.J. 1985. Erosie en wateroverlast in Zuid-Limburg. Landschap 2:118-132.

SCHWERTMANN, U., RICKSON, R.J. and AUERSWALD, K. (eds.). 1989. Soil erosion protection measures in Europe. Soil Technology Series 1, Catena Verlag, Cremlingen-Destedt.

SCHWERTMANN, U., VOGL, W. and KAINZ, M. 1987. Bodenerosion durch Wasser. Vorhersage des Abtrags und Bewertung von Gegenmassnahmen. Ulmer Verlag, Stuttgart.

STURNY, W.G. and MEERSTETTER, A. 1989. Konservierende Bodenbearbeitung und Mulchsaat im Zuckerrübenanbau - ein Bestandteil integrierter Produktion. Forschungsanstalt für Betriebswirtschaft und Landtechnik, Tänikon, Schweiz, FAT-Berichte, Nr. 363.

STURNY, W.G. and MEERSTETTER, A. 1990. Mulchsaat von Mais in Gründüngungsbestände. Forschungsanstalt für Betriebswirtschaft und Landtechnik, Tänikon, Schweiz, FAT-Berichte, Nr. 376.

WILLIAMS, J.R. 1981. Soil erosion effects on soil productivity: a research perspective. Journal of Soil and Water Conservation 36:82-90.

WISCHMEIER, W.H. and SMITH, D.D. 1978. Predicting rainfall erosion losses. A guide to conservation planning. US Department of Agriculture, Agriculture Handbook No. 537.

WOLFGARTEN, H.J., FRANKEN, H. and ALTENDORF, W. 1987. Mulchsaat oder Direksaat? DLG-Mitteilungen 5/87:1-3.

WUNDERLICH, K.H. and SCHWERDTLE, J.G. 1982. Erosion im Zuckerrübenanbau. KWS, Einbeck.

Chapter 33

MECHANICAL MEASURES FOR RUNOFF MANAGEMENT AND EROSION CONTROL IN THE VINEYARDS OF NORTH EAST SPAIN

J. Porta, J.C. Ramos and J. Boixadera
Department of Meteorology and Soil Science
ETSEA, Lleida, Spain

Summary

The Penedés and Anoia regions of north east Spain are important areas for the production of grapes for high quality wine. In these areas, soil and water conservation is an important issue because the climatic and the soil characteristics necessitate the implementation of measures to control runoff and soil erosion.

Local measures are usually implemented by farmers without any technical advice. The type, design and geometry of these measures are discussed, analysing their function and efficiency in relation to some empirical criteria.

1. INTRODUCTION

Decision makers and farmers have different points of view as far as conservation measures are concerned. In the first case a small scale approach can be sufficient, but this is not appropriate in advising farmers. In this case it is not sufficient to evaluate potential soil losses, but measures are needed for efficient control of these losses. Because farmers implement their own measures it will be interesting to know whether a) they have a scientific basis and b) what can be learnt from them.

If such measures fail, scientific and technical analysis would allow improvements to be proposed for the conservation of natural resources.

1.1 The erosion problem

Soil and water conservation is one of the major farming concerns in the Anoia-Penedés vineyard areas of north east Spain. The farmers systematically remove weeds to avoid moisture competition with the vines. The bare soil may be easily eroded by rain splash, causing detachment of soil particles, and by runoff creating a rill network after each erosive rainfall event. The area is dissected by a deep gully network whose initiation may have been a complex process. Over an area of 20,000ha, the gully network represents about 20% of the area. These gullies have vertical side walls, and are ten or more metres deep in many places. The gully system grows at points where surplus runoff is concentrated, and also by mass movements on the gully slopes.

In this paper some local measures implemented by farmers to control runoff are analysed with a discussion of their functions and efficiency for soil and water conservation. It would be interesting to use these local methods to demonstrate possible future erosion control methods. Present technical solutions used in designing new, sustainable vineyard cultivation should take into account local experience and learn from the past.

1.2 The study area

The study area is located about 30km south west of Barcelona between the Anoia and the Llobregat rivers (Ordnance Survey Grid Reference 31T DF 020915) on alluvial sediments. The topography consists of gentle slopes dissected by a network of deep gullies. Inter-gully areas are rolling, with complex slopes.

The area lies within the temperate to maritime Mediterranean climate. The mean annual rainfall ranges from 471 to 670mm, and is very irregularly distributed throughout the year, with high intensity rainstorms. Torrential rain falls in the autumn following the dry season, with a large erosive potential (Ramos et al., 1991). The soil moisture regime is xeric and the mean annual temperature is 21°C.

The lithology of the area is complex, although only Miocene sedimentary rocks are present. These include calcilutites (marls) with occasional sandstones and conglomerates. Quaternary deposits of variable thickness cover the miocene materials which outcrop in some places. The calcilutites have very poor rooting properties. Detailed descriptions of the behaviour of the calcilutites and soils developed from these materials are provided by Porta et al. (1991). The soils are highly calcareous and may be classified according to Soil Taxonomy (SSS, 1990) as subgroups of Xerorthents, Xerofluvents and Xerochrepts. They have been described and mapped by Danés (1977), Boixadera (1983) and Porta et al. (1991).

This area is important in the production of grapes for high quality wine, especially 'cava', a wine produced by the champagne method. Arable land use is mainly vineyards, but some fields have peach trees and cereals.

2. METHODS

Rainfall has been measured with a pluviograph linked to a data logger, recording rainfall at five minute intervals. After every occurrence of erosion, reference was made to the rainfall record, from which the total rainfall, maximum intensity during the rainstorm (mm.h^{-1}) and kinetic energy were calculated. The rainfall kinetic energy was computed by summing the values of the following relationship over the total storm duration:

$$KE = (11.9 + 8.73 \ \log I_i) \ P_i, \ \text{if I (intensity)} < 76mm.h^{-1}; \text{ and}$$

$$KE = 28.3 \ P_i, \text{if I} > 76mm.h^{-1} \text{ (Foster et al., 1981)}.$$

Replicate measurements of hydraulic conductivity were obtained using the Porchet method. The particle size analyses were made by the pipette method without removing the calcium carbonate. Moisture retention was determined after equilibration on a ceramic pressure plate. Textural class, calcium carbonate equivalent and organic matter were also determined.

The type, geometry and efficiency of local soil conservation practices were studied. Sites were identified by photo-interpretation and by field surveys. Efficiency was evaluated by the presence of rills on the protected field, and by the presence of breaks in the ditches, measured by the chain method over segments of 25m length.

3. RESULTS AND DISCUSSION

3.1 Rainfall characteristics

The intensity of the rainfall has been measured since 1990. The characteristics of the most important rainstorms are shown in Table 1. All these rainstorms belong to periods of maximum precipitation. During the autumn months, short duration and high intensity rainstorms are frequent, registering maximum intensities over 30 minutes which range from 18 to 45mm.h^{-1}. The values of kinetic energy calculated for these rainstorms range from 345 to 858 J.m^{-2}.

Table 1.Rainfall characteristics of some rainstorms in the area during 1990-91.

Date	Total amount (mm)	Duration	Intensity I(mm.h^{-1})	Maximum intensity during 30 min (mm.h^{-1})	Maximum intensity during 5 min (mm.h^{-1})	Kinetic energy KE(J.m^{-2})
900915	30.6	2h.40min	38.0	38	112.8	772
900925	29.4	1h.45min	2.1	7.6	16.8	461.9
901015	19.4	14h	7.2	14.0	26.4	345
901022	22	2h.40min	33.0	43.2	108.0	567.7
901025	29.2	15h	2.0	8.4	14.4	555.7
901109	25	15h.25min	5.2	6.8	7.2	347.7
910301	31.9	12h.15min	2.6	9.2	9.6	624.2
910505	18.2	3h	2.7	10.4	14.4	262.6
910508	29.8	16h.50min	2	4	4.8	460.6
910509	38.2	14h.35min	2.6	11.2	14.4	673.3
910925	36	3h.25min	10.5	45.2	93.6	857.8
910926	14.4	2h.40min	5.6	18.4	31.2	308.5
911005	31.1	8h.10min	6.0	30.6	72.0	674.1
911126	39.4	7h	5.6	24.8	45.6	856.7

Duration of rainfall and maximum intensities are used to calculate water discharge in hillside ditch channels and waterways, to evaluate the performance of the existing soil conservation measures (Table 2).

Table 2. Field characteristics, geometry and efficiency of local soil conservation measures in the Anoia-Penedés area.

Ditch profile (Fig.1)	Field slope (%)	Hillside ditch characteristics				Ditch XSA (m²)	RD (m)	Geometry parameters (m) (See Figure 1)				Efficiency		
		VI (m)	HD (m)	S (%)	L (m)			a	b	c	d	R	CE	BR
A	9.8	2.55	22.4	1.9	124	0.53	3	1.0	0.65	3.1	4.3	H	L	H
	5.7	2.60	44.9	1.9	326	0.65	3	0.5	3.0	3.	6.3			
	4.2	1.90	45.9	2.3	285	1.11	3	0.4	3.2	2.6	6.0			
	4.0	2.55	47.0		78	0.69	3	0.5	0.6	2.9	3.4	VL	L	VL
B	7.5	2.60	19.4	1.0	123	0.79	3	0.6	0.9	2.4	4.3	H	L	H
C	4.2		45.9	2.3	285	0.35	3	0.5	3.1	0.3	6.4	H	L	H

VL = Very Low; L = Low; H = High;

VI = Vertical interval between ditches   HD = Horizontal distance between ditches
 S = Slope of the ditch channel   L = Length of the channel
RD = Distance between vine rows   R = Rills between ditches
CE = Channel eroded   BR = Breaks in the ridge

3.2 Soil characteristics

Soil characteristics affecting soil water movement are summarised in Table 3. Hydraulic conductivity ranges from 0.001m.day^{-1} (very low) to 0.34m.day^{-1} (low to medium). This is controlled by underlying calcilutites. The low values explain the low infiltration rates where calcilutites are near the soil surface.

Calcium carbonate is a common component of the soils. It ranges from 30% to 50%. Silt ranges from 20% to 45%, and organic matter content is low, from 0.7% to 1.8%. All these data refer to the topsoil.

3.3 Types of local soil conservation measures

The main soil conservation measures used in the area are hillside ditches (local name 'rases'). Their function is to intercept surface runoff and convey it out of the field. The frontslope of the terrace also provides access for farm machinery. In fact, it is a broadbase, channel terrace (Schwab et al., 1981) commonly used by farmers. The design, geometry and efficiency of the hillside ditches are described in Table 2. The traditional designs are the result of the farmers' empirical knowledge.

In the rather large farms of the area where conservation measures are commonly found, the old vineyards (>20 years) sought to convert the hillside ditches into contour lines as much as possible. At present, this is not the case in the new plantations if they are very intensive and mechanised. According to the farmers' criteria, row vineyards must be in straight lines, regardless of landform and rows must be as long as possible to decrease the time involved in ploughing and other farm operations.

Figure 1. Cross section of specific hillside ditches.

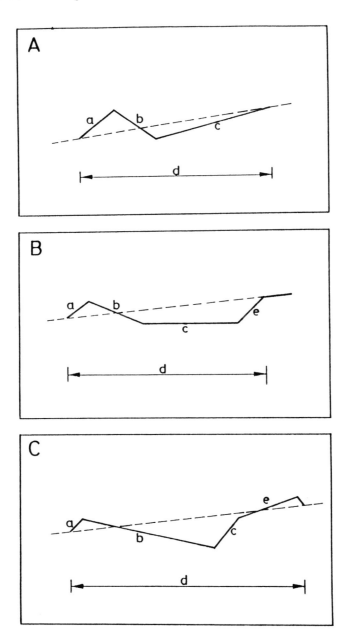

Table 3. Soil characteristics of benchmark soils of the area affecting water movement (adapted from Danés, 1977; Boixadera, 1983; Porta et al., 1991)

Classification (SSS,1990)	Rooting depth (cm)	Top soil characteristics				Hydraulic conductivity (Porchet method) (m.day^{-1})	
		Texture (USDA)	CaCO$_3$ content (%)	Silt content (2-50μm) (%)	o.m (%)		
Calcixerollic Xerochrept, fine silty, mixed, thermic	>120	Loam	33-37	30-45	1.1	0.06-0.150	low
Typic Xerorthent, fine silty, mixed, thermic	30-45	Loam, Sandy loam	30-50	45	0.8 to 1.4	0.015-0.036	low
Calcixerollic Xerochrept, fine silty, carbonatic, thermic	>120	Loam	43	29	1.4	0.024-0.002	low
Typic Xerofluvent, fine silty, mixed, thermic	>120	Sandy clay loam	35-58	22-27	0.7 to 1.2	0.025-0.042	low
Petrocalcic Xerochrept, loamy, carbonatic, mixed, shallow	35	Sandy clay loam	50	23	1.8	0.343	low to med
Typic Xerorthent, fine silty, carbonatic, thermic	0.45	Loam	40	47	0.8 to 1.3	0.042-0.008	low
Aquic Xerorthent, fine silty, mixed, thermic	0.8	Sandy clay loam	35	50	1.0	0.017-0.003	low

3.4 Hillside ditch design

Farmers construct ditches according to their own experience without any technical advice. Usually they construct them prior to or at the same time as they plant the new vineyards; after that it is very difficult to change the location of the ditches without disturbing the vines which remain in the field for decades - some of them for half a century. Technical advice has to rely on many empirically-based calculations, but these may be far removed from the experience gained by local research.

The data from the field survey are used to assess the efficiency of the farmers' measures. The slopes in the area range from 4% to 10%; if the slope of the area is higher, usually bench terraces exist, but this is unusual in this area. The vertical interval of the ditches ranges from 2m to 2.6m. The horizontal distance ranges from 20m to 50m and the length of the ditches ranges from 80m to 325m.

The efficiency of the measures of soil erosion control (Table 2) is evaluated in relation to:
- rill erosion between ditches (R)
- erosion of the channel (hillside ditches or waterways) (CE)
- breaks in the banks of the hillside ditches (BR).

In order to try to explain any failures of the erosion control system, the design criteria commonly used by Schwab et al. (1981) and Magister (1973) have been applied. Concentration time and runoff have been computed according to Frevert and Ramser, cited in Schwab et al. (1981).

The results of the calculation of hillside ditch geometry according to the standard design procedure show that Bennet's criteria give results quite different from the farmers' practical solutions. The results of the other methods (Saccardy, Israel, Schwab and Spanish S.C.S. in Magister, 1973) give similar results for vertical distance on slopes of 7%, from 2.3m to 2.7m. Farmers use from 2m to 2.6m. Values for the horizontal distances between ditches can be considered to be inside the recommended values, but in some places they exceed this: 20 to 47m in the study area (Table 2), compared with 33 to 39m for the calculated design. The length of ditches recommended by the Ministerio de Agricultura of Spain ranges from 330m to 400m. In the area of study, the length does not exceed these values.

The total amount of water to be removed by each ditch is computed with the Rational formula (Schwab et al., 1981) using the maximum rainfall intensity ($mm.s^{-1}$) (Table 1) during the concentration time, for a return period of ten years.

The maximum allowable velocity in the channel ditch should not exceed $0.45m.s^{-1}$ for erodible soils, using a Manning roughness coefficient of 0.04. Quantitative data on velocity in the channels are not available, but Table 2 shows a low efficiency (L) in channels, because erosion was identified in the lower sections. This is a general problem in the area due to the empirical design criteria applied by the farmers.

The factors responsible for failures of the system (Table 4) are deduced from observations as farmers have insufficient knowledge of rainfall and soil characteristics. Rill formation is observed where ditches are more than 40m apart, with the field slope not steeper than 10%. At present, these observed rills result from overflow in some cases, and from excessive slope length between ditches in others. The steep gradients of the channels may explain the erosion observed in some channel ditches. Breaks in the ditches seem more related to the size of the ridge than to the channel capacity. This is most probably due to the fact that the ridges have two functions - drainage and as a road. Small ridges are prone to damage by machinery traffic.

Concentration times are always shorter than five minutes, computed with the Kirpich's formula. The observed rainfall during the study period (Table 1) did not exceed $100mm.h^{-1}$ over five minute periods. Design rainfall for the conditions in Table 2 are well above $250mm.h^{-1}$ for periods between 1.5 to 3 minutes. This information, although still preliminary, seems to confirm that the failures observed are related to the ridge size, and not to the channel of the ditch.

Table 4. Factors responsible for the failure of local measures for soil and water conservation in Penedés-Anoia areas (NE Spain).

FAILURES OF THE SYSTEM	FACTORS RESPONSIBLE
Rill formation in the field between ditches	Distance between ditches insufficient. Low infiltration rate
Ditch channel erosion	Steep channel slope gradient
Breaks in the ridges	Channel slope locally too gentle Channel cross section too small; too much land has been planted Traffic damage to channels

Table 5 displays calculated values of ditch geometry for several channel lengths. It is clear from the Manning formula that farmers construct over-sized channel ditches (Table 2). It points out that farmers are more worried about excess water than about water conservation. Failure can be explained by poor construction of the ditches, insufficient ridge size or insufficient gradient on some sections of the channels (Table 4), not by the size of the channels.

3.5 Hillside ditch maintenance

Farmers clean the ditches every year using specially designed farm machinery. In the cleaning of the channels, vegetation is removed as well as evidence of erosion. The farmers also reshape the ditches as they clean them, ensuring the continued operation of the soil erosion control measures, providing cleaning is done properly.

Table 5. Characteristics of hillside ditches according to standard design calculations.

Length (m)	Ditch Slope (%)	Riser Slope (%)	Section (m ^{2})	Maximum Speed (m.s^{-1})	Ditch depth (m)	Ditch width (m)
250	0.4	15	0.47	0.38	0.30	4.88
250	0.7	15	0.42	0.48	0.29	4.72
100	1.0	15	0.26	0.50	0.24	4.00

3.6 Ditch outlets

In the study area, most hillside ditches discharge freely into a gully, usually without any physical protection. In some cases, the outlets run into a waterway. Some waterways may be vegetated with wild vegetation including *Cynodon dactylon* and *Arundo donax*.

Often the head of a gully begins at the end of a waterway. The network of gullies grows at the head as well as laterally, where waterways or hillside ditches discharge runoff water. In a very few cases where hillside ditches or waterways join gullies, there are protective concrete or masonry structures. These control measures are very efficient in the protection of gully sides, but water can erode the gully wall or bed, because often no apron is constructed. Therefore, the concrete structure collapses as runoff undermines it.

Until now, farmers have been unable to curb the gully extension which is enhanced by these conservation measures. According to the survey, farmers are aware of the causes of gully growth, because they use loppings of lupins and *Arundo donax* to control gully development, but these measures are inadequate. Other techniques used to convey water to the bottom of the gullies, as used in highway engineering, are too expensive and cannot be used by the farmers.

4. CONCLUSIONS

When the older vineyards were planted much more consideration was given to erosion problems. Hillside ditches ('rases') were converted to contour lines and large ridges were used to control erosion. These vineyards may be considered sustainable with stable soils.

New plantations aim for maximum planted area. This results in narrower ditches and ridges which are not efficient and cannot cope with the disposal of road and drainage water.

Farmers tend to make their hillside ditches oversized compared with the dimensions given by standard design calculations. Channel erosion results from steep channel gradients even where the soil is not erodible. For given hydrological conditions and for the same channel cross section, ridge cross section is a critical factor in ridge failure.

Although the farmers' conservation measures generally give acceptable results in removing excess water from the fields and allowing farm operations, the concentration of runoff by the hillside ditches greatly enhances gully growth, because the farmers do not take into account the outlets of the ditches into the gully. From this point of view, the vineyard system cannot be considered sustainable.

REFERENCES

BOIXADERA, J. 1983. Proyecto de un área modelo de conservación de suelos en Piera-Masquefa (Barcelona). Proyecto final de carrera, ETSEAL.

DANÉS, R. 1977. Estudi de sòls de la finca Can Massana. Masquefa. Special Report, DMCS, Lleida (Spain).

FOSTER, G.R., McCOOL, D.K., RENARD, K.G. and MOLDENHAUER, W.C.1981. Conversion of the USLE to SI metric units. Journal of Soil and Water Conservation. Nov-Dec. pp.355-359.

MAGISTER, M. 1973. Conservación de suelos. ETSIAM, Madrid.

PORTA, J.J., BOIXADERA, J. and POCH, R.M. 1991. Soil conservation measures in new vineyards in NE Spain: Physical conditions and social constraints. ESSC Seminar on Combating Soil Erosion in Vineyards, Trier, Germany.

RAMOS, M.C., PORTA, J. and BOIXADERA, J. 1991. Rainfall characteristics and soil losses in vineyards in the NE Spain. ESSC Seminar on Combating Soil Erosion in Vineyards, Trier, Germany.

SCHWAB, G.O., FREVERT, R.K., EDMINSTER, T.W. and BARNES, K.K. 1981. Soil and water conservation engineering. John Wiley, New York.

S.S.S. 1990. Soil taxonomy. Agriculture Handbook No. 436. Soil Conservation Service, USDA.

Chapter 34

EFFECT OF DIFFERENT SITE PREPARATION TECHNIQUES ON RUNOFF AND EROSION IN PLANTATION FORESTRY

S. Lucci and S. Della Lena

Centro di Sperimentazione Agricola e Forestale - S.A.F.,
Rome (Gruppo E.N.C.C.), Italy

1. INTRODUCTION

Plantation forestry in central southern Italy over the last two decades has entailed the use of intensive site preparation treatments (such as brush clearing, soil preparation and post-planting cultivation) and a high degree of mechanisation. In several cases, these techniques are vital to the success of plantation management and for overcoming socio-economic restraints (Lucci, 1990a; Minotta, 1987). However, land availability in hill and mountain areas has caused these techniques to be extended to steep slopes and to zones of delicate environmental balance without due investigation. This has given rise to various questions as to the amount and form of soil degradation in terms of erosion, reduction of organic matter, disruption to microbiological activity and nutrient cycling and acidification, and their effect on water quality and long term productivity (Lucci, 1987; Utzig and Walmsley, 1988).

It is suggested that where favourable physical, chemical and biological soil characteristics and/or limiting, fragile conditions (such as steep slopes, erosion hazard, thin soil profiles and poor soils) are present the highest productive capacity and/or protection against erosion may be achieved with less intensive measures (Susmel, 1983; Shepherd, 1986; Lucci, 1990b).

It is therefore urgent to make thorough analysis of actual or proposed land use and to seek possible ways of reconciling productive requirements with those of conservation.

2. OBJECTIVES OF THE STUDY

The primary aim of the study is to measure runoff, erosion and the removal of soil organic matter and nutrients where preparation operations are carried out. These operations involve deep ploughing and ripping downslope, which is commonly used for plantation forestry in southern Italy, and furrowing along the contours, which is feasible for slopes less than 40-45%. These operations have been compared with those carried out on land where the bush vegetation has remained undisturbed and where the maquis has been burnt. The three treatments were:

(a) total brush clearance by dozer blade, and ripping to 1m;
(b) brush clearing as above combined with complete deep ploughing, using a single mouldboard plough;

(c) brush clearing as above, forming windrows (brushpiles) with the removed material on the contours. Parallel to these, furrows were created every 3m by turning the sod downhill. The brushpiles are arranged with a spacing of approximately 30m, depending on the slope (Donmez, 1984) in order to reduce slope length, so creating a barrier to runoff and minimising erosion.

Observations on the physical, chemical and biological changes in the soil and on the evolution of the vegetation were made for all three treatments (Valenziano, 1990; Angeloni, 1991). The productivity of the plantation for each treatment was also observed (Fusaro and Lucci, 1989).

3. EXPERIMENTAL AREA AND METHODS

The experimental area, the Marganai Regional Public Forest, is representative of the hill and mountain environment of south western Sardinia (Table 1). It is situated south of Mount Linas, in the upper basin of the Leni torrent, at altitudes of between 570 and 800m. The area forms part of the middle and medium-cold subzone of the *Lauretum* (type I - dry summers).

Table 1. Site characteristics of the experimental blocks (both south-facing).

Block	Altitude (m)	Slope (%)		Soil	Vegetation
		m	max		
I	650 - 700	36	40	Shallow soil (31-60 cm), sandy loam, frequent rock outcrops, on weathered granite.	Partially degraded Mediterranean maquis with *Arbutus unedo, Erica arborea, Rosmarinus officinalis, Genista angustifolia, Cistus* spp. (Ø 2-6 cm, mean height < 2m)
II	610 - 650	30	35	Moderately deep to deep soil (61-120 cm), sandy loam, locally very stony, rare large quartz boulders, on weathered granite.	Mediterranean maquis with *Arbutus unedo, Erica arborea, Genista corsica, Phillyrea angustifolia* (Ø 2-6 cm , mean height 2 - 2.5 m)

The different treatments were compared in two randomised blocks, each of 5 plots, measuring 40m width × 100m length downslope. The single experimental blocks, although not far from each other, have fairly different soil, vegetational and topographic characteristics. The plots represent conditions close to the extreme limits of workability using ordinary mechanical means.

Erosion measurements were carried out on areas of 12.5 × 80m, bounded within subplots (20 × 100m). Instrumentation included runoff and sediment collectors, a sedimentation tank, an H flume and ultrasonic water level gauge and a Coshocton type runoff sampler (1/200 sampling rate). The plots were planted, in all three cases, with *Pinus radiata* D. Don. Observations on the development and evolution of the vegetation were carried out on four blocks with an experimental split-plot arrangement and two subplots of 20 × 100m planted with *Pinus pinaster* Aiton. and *P. radiata* D. Don, respectively.

Sampling for sediment and physical-chemical analyses was carried out for each storm. In the case of small rainfall events, where runoff was less than one cubic metre (i.e. the sedimentation tank volume) and no sediment was produced, cumulative sampling was performed. Sampling started in December 1986, after the requisite instruments had been installed.

4. PRELIMINARY RESULTS AND DISCUSSION

During the first four years of observation 58 runoff events were sampled, 48 of which are presented, from November 1987 - December 1990. Previous data contained errors caused by inaccurate calibration of the instruments.

With the exception of 1990, the period examined was characterised by low intensity, frequency and amount of rainfall, compared with previous years. The mean annual rainfall recorded at the Monti Mannu weather station (2km downstream) between 1925 and 1950 was 1131mm (min 700-800mm; max 1400-1500mm). During the trials, the rainfall measured at Marganai was 398mm in the period September - December 1986; 587mm in 1987; 552mm in 1988; 630mm in 1989; and 940mm in the period January - December 1990. These quantities correspond to little more than half the average value mentioned for the Monti Mannu station. In any case, events of such intensity and duration only rarely generate any appreciable runoff and erosion.

The practical difficulties in applying the treatment to the two replications, together with the existing variability among the treatment plots (within the block), made interpretation of the data difficult. The analysis of variance applied on treatment means and on total (cumulative for the observation period) and yearly data rejects the hypothesis of any significant difference among the treatments. Future work on data analysis and interpretation will be based on comparison of obtained results with estimated data using models (e.g. USLE) which will allow for the specific differences among the replications and treatment plots (Bayesian analysis). Data observations, supported by simple statistical means (such as the t test among replicates and treatments) permit some preliminary considerations.

Compared with the 'control' plot (undisturbed vegetation), all the other treatments produced, to differing degrees, a clear increase in runoff and, to a certain extent, in sediment (Figures 1 and 2; Tables 2 and 3) during the study period. This phenomenon is more accentuated in Block I, probably because the vegetation and soil were already appreciably degraded before the treatments were imposed, with sparse, stunted vegetation, lack of organic horizons over a large part of the area and the outcropping of horizon C or of fairly weathered parent rock.

Table 2. Annual runoff (mm).

Year	Events No.	Rainfall mm	Control		Contour furrowing		Ripping		Deep ploughing		Burnt	
			1	2	1	2	1	2	1	2	1	2
1987*	4	156	6.0	8.1	6.6	9.4	28.6	13.4	6.9	21.4	8.5	8.5
1988	14	552	20.4	29.9	115.6	122.6	156.6	95.4	48.8	144.4	114.8	42.2
1989	13	630	4.3	3.4	14.8	7.1	46.2	39.0	43.4	184.0	12.0	6.2
1990	17	940	46.3	4.9	93.8	98.3	71.3	90.1	52.6	169.4	82.4	49.3

* From 18th November

Table 3. Annual sediment (kg.ha^{-1}).

Year	Events No.	Rainfall mm	Control		Contour furrowing		Ripping		Deep ploughing		Burnt	
			1	2	1	2	1	2	1	2	1	2
1987*	4	156	13	11	27	36	244	23	77	2521	14	22
1988	14	552	69	58	348	340	4124	611	814	9065	324	161
1989	13	630	16	15	40	15	432	325	247	4158	29	18
1990	17	940	51	16	233	302	388	170	176	2495	311	166

* From 18th November

Figure 1. Total runoff (mm) from 18/11/87 to 19/12/1990 (48 events).

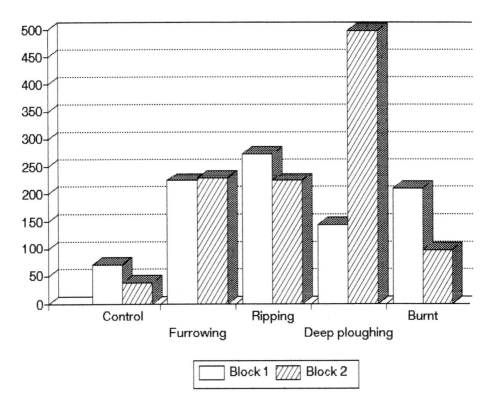

Very high runoff and sediment values occurred in just one of the two replications of the 'complete deep ploughing' and 'ripping' treatments. The clear difference observed between the replications of these two treatments can be explained by differences in site characteristics, such as soil conditions and conformation of the slope profile, and by other incidental causes such as:

(a) accentuation of soil diversity by the land treatment (e.g. in Block I, ploughing brought coarse detritic material to surface, and this partly blocked runoff and erosion);
(b) runoff concentration and rill erosion caused initially by the placement of galvanized iron sheets and flumes required small-scale adjustments;
(c) a ridge created by working in the ploughed plot in Block I, made reshaping of the plot surface necessary.

Within the replications for the 'control' and for the 'contour furrowing', a greater uniformity in behaviour is noted, with no significant statistical difference. In the latter, there appears to be a clear relationship between runoff and the amount and intensity of rainfall. Readjustment capacity seems to remain high in the majority of cases, but is insufficient for a number of more intense, abundant events.

Figure 2. Total sediment (kg.ha^{-1}) from 18/11/87 to 19/12/1990 (48 events).

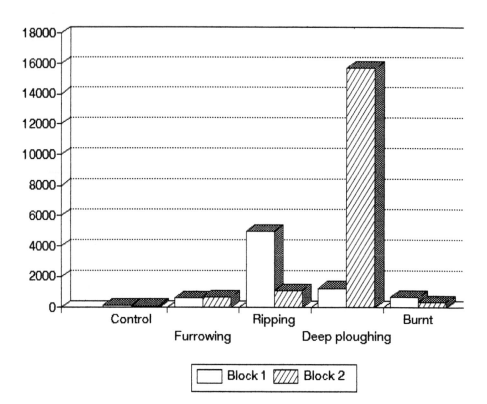

An increase in runoff is not followed by any appreciable worsening of erosion. It may be assumed that the furrow and ridge created by the treatments are undersized or that they are discontinuous or are not perfectly aligned on the contour so they are unable to fully control rill erosion. A different choice of installation method and equipment should improve the control and water storage capacity of the soil. The abundant regrowth of the bush vegetation has not yet demonstrated any appreciable erosion control.

In the 'burnt' treatment a difference was found between the replications, which is only statistically significant for the runoff. In comparison with the other treatments, the effect of the fire has been remarkable, particularly in regard to the runoff and, to a much lesser extent, to sediment (Figures 1 and 2; Tables 2 and 3).

During the first four years of observation, a clear attenuation of erosion has only been noted for the intense treatments (ripping and ploughing). For the contour furrow, burnt and control treatments, the phenomena followed an irregular trend that, apart from specific causes mentioned above, can be attributed to the variation in rainfall rates and, in particular, to higher precipitation in 1990. However, erosion is still high in one replication of the ripping (Block I) and ploughing (Block II) treatments. This can be partially explained by the poor cover development of herbaceous and bush vegetation (Douglass and Goodwin, 1980).

In another part of Sardinia, researchers have found that even a 30% herbaceous cover reduced erosion by around 90%. In less than 2 years between 60 and 100% of the cover can re-establish itself and erosion can be practically stopped. The very intense nature of the treatments probably delayed the control of the erosion.

5. CONCLUSIONS

A drastic reduction in erosion was observed for the more intensive treatments (ripped and ploughed plots). Even in the absence of cultural operations, erosion still remains active and considerable after four years of data collection. This is due to the severity of the mentioned treatments, the scanty evolution of the herbaceous vegetation and the increase in rainfall rates in the last years of observation, particularly in 1990.

Although it is too early to draw any final conclusions, a number of indications can be made:

1. Disturbing vegetation and soil causes an increase in runoff and sediment, even four years after treatment.
2. Treatment following the contours seems to guarantee substantial erosion control. There is, however, no corresponding water control capacity. This could be improved by improving the installation of the treatment.
3. Treatments such as deep ploughing and ripping up and downslope can, at least temporarily, cause an appreciable increase in runoff and, in particular, erosion. The extent of the phenomenon seems to vary appreciably even over a short distance, due to variations in soil and specific morphological characteristics of the slope.
4. The accidental fire (wildfire) that burnt the vegetation caused a temporary increase in runoff, but only a marginal increase in erosion.

REFERENCES

ANGELONI, A. 1991. Comparisons of different methods of mechanized land preparation for reforestation: environmental impacts. '10ème Congres Forestier Mondial: La Forêt, Patrimoine de l'Avenir', Paris, 17-26 September 1991.

DONMEZ, E. 1984. Methods and machines for the rehabilitation of low productivity forests. ECE/FAO/ILO - Joint Committee on forest working techniques and training of forest workers. Seminar on techniques and machines for the rehabilitation of low productivity forests, Izmir (Ginevra).

DOUGLASS, J.E. and GOODWIN, O.C. 1980. Runoff and soil erosion from forest site preparation practices. USDA Forest Service, Coweeta Hydrologic Laboratory Report.

FUSARO, E. and LUCCI, S. 1989. Contribution to: 'Tavola rotonda - Il piano forestale nazionale: un ponte tra passato e futuro'. Atti Euroforestalegno 1989.

LUCCI, S. 1987. Meccanizzazione delle attivatà forestali in relazione alle esigenze di conservazione del suolo. Atti Prima Conferenza Regionale: Conservazione del Suolo e Forestazione in Calabria, Bovalino. Laruffa Editore, Reggio Calabria. pp. 143-58.

LUCCI, S. 1990a. Studi parcellari su metodi di preparazione del terreno per rimboschimenti produttivi in aree con limitazioni ambientali. I. Descrizione delle prove e delle installazioni di campagna per la misura dell'erosione. Quaderni di Ricerca SAF. 30.

LUCCI, S. 1990b. Considerazioni su una prova di preparazione del terreno secondo le curve di livello per attività di rimboschimento. Quaderni di Ricerca SAF 29.

MINOTTA, G. 1987. Confronto tra differenti tecniche di lavorazione del suolo nel rimboschimento di terreni ex-agricoli su formazioni argillose dell'Appennino settentrionale. Istituto di Coltivazioni Arboree, Università degli Studi di Bologna (PhD Dissertation).

SHEPHERD, K.R. 1986. Plantation silviculture. Martinus Nijhoff Publishers, Dordrecht.

SUSMEL, L. 1983. Quando lavorare il suolo per prepararlo al rimboschimento? Economia Montana 15(6):23-6.

UTZIG, G.F. and WALMSLEY, M.E. 1988. Evaluation of soil degradation as a factor affecting forest productivity in British Columbia. A problem analysis - Phase I. FRDA Report 025.

VALENZIANO, S. 1990. Spontaneous vegetation: possible variations as a result of different methods of site preparation for planting. International Symposium on agroecology and conservation issues in temperate and tropical regions. University of Padova, Department of Biology, September 1990.

Chapter 35

THE INFLUENCE OF SOIL SURFACE CONFIGURATION ON DEPRESSION STORAGE, RUNOFF AND SOIL LOSS

G.M. Edwards, N.C. Taylor and R.J. Godwin
Silsoe College, Cranfield University, Silsoe,
Bedfordshire, MK45 4DT, UK

Summary

The effect of surface roughness on depression storage, runoff and soil loss was investigated using simulated rainfall on a sandy loam. Surface roughness and roughness orientation played a major role in the infiltration and storage of water on the soil surface. The roughness index and depression storage influenced the time to initial runoff and runoff total. The combination of all these effects gave a substantial reduction in soil loss with increasing roughness. In addition to the experimental work, a simple model was developed and confirmed by laboratory work, which shows that slope steepness reduces depression storage in a linear manner between 0° and 20°. Porosity, bulk density, surface sealing, surface storage and orientation of surface roughness were all found to play a role in runoff production. Water harvesting techniques on these sandy loams will be more effective where the surface has been compacted and where surface storage capacity has been reduced to a minimum.

1. INTRODUCTION

Tillage practices that modify soil surface configuration to retain precipitation and reduce nutrient and soil loss are vitally important. Although the importance of surface roughness is well known in soil management, only a few studies involving quantitative descriptions of surface roughness and its effects on runoff have been conducted. The majority of the work carried out in this area has been on the finer textured soils with low hydraulic conductivities. Some work has been carried out as to the measurement and description of soil roughness, with regard to its role in depression storage. Models to describe both roughness and depression storage have been developed (Barron, 1971; Mitchell and Jones, 1978) but do not include the effects of infiltration and slope.

Tillage induced roughness can be described as the surface configuration of the soil caused by equipment, and the size and orientation of the clods at the soil surface after cultivation. The roughness of a cultivated surface has two components: random roughness and oriented roughness. The random roughness is most pronounced after some forms of primary tillage, and oriented roughness after ridgers and drills (Barron, 1971).

Some work has been carried out to describe the degree of roughness and cloddiness created by various tillage implements and operations (Evans, 1980; Godwin, 1990). Evans (1980) describes ploughing as producing a rough cloddy surface with local variations in height of

120 to 160mm. Secondary cultivation is seen to reduce these height variations from 30 to 40mm, whilst rolling and drilling decreases it further.

Depression storage is directly linked to surface roughness; the rougher the soil surface the greater the potential depression storage. Work carried out by Larson (1964) indicates that deep furrows resulting from operations such as contour listing have a potential depression storage of more than 76mm of water on the surface, compared to about 25mm of water after smooth conventional tillage.

Infiltration (Burwell and Larson, 1969; Burwell et al., 1968), runoff retardation (Mitchell and Jones, 1978; Moore and Larson, 1979; Monteith, 1974) and erosion (Cogo et al., 1984) have all been shown to be closely related to soil roughness. The most significant changes in soil properties due to tillage except the effects on roughness and depression storage, were seen to be a loosening/compaction of the tilled layer, and an increase/decrease in pore space.

Burwell et al. (1968) found that infiltration before ponding, on freshly tilled surfaces, increased as tillage-induced random roughness increased. It was found that the average infiltration of rainfall, prior to runoff on freshly ploughed surfaces, was at least twice that on freshly ploughed, disked-harrowed surfaces. Once runoff began infiltration was only slightly greater for the ploughed surfaces than for other less rough surfaces, due to soil capping. The effect of tillage operations on roughness, runoff and soil loss was investigated by Cogo et al. (1984) on a Barnes loam soil. An increase in roughness was seen to decrease soil loss much more than it did runoff. This was due to the deposition of suspended sediments as a result of the increased drag caused by the surface roughness. Tillage is seen to cause simultaneous changes in both roughness and bulk density/macroporosity (Currie, 1966).

2. MATERIALS AND METHODS

In this paper the effects of soil surface roughness and depression storage on runoff and soil loss are reported from a series of field experiments with the following objectives:

i.   to quantify the relationships between roughness, water storage, runoff and soil loss and to quantify the change in roughness and storage depth after a severe rainfall event;
ii.   to develop a simple model to calculate roughness and total available depression storage from point data;
iii. to develop a mathematical model to quantify the effect of slope on depressional storage.

All the experiments described have been carried out on an erodible sandy loam soil of the Cottenham series (King, 1969) (Table 1). This soil behaves in a similar fashion to many drought prone soils throughout the world and is highly susceptible to both wind and water erosion.

Table 1. Particle size analysis.

% Sand fine	% Sand medium	% Sand coarse	% Silt	% Clay
21.48	42.69	10.32	12.86	12.65

Five soil surface treatments were produced to provide varying degrees and types of soil surface roughness. The five treatments were: (i) ploughed; (ii) smooth and compacted; (iii) seedbed conditions; (iv) compacted ridges across slope; and (v) compacted ridges up/down slope. In order to obtain the ridged surface, a steel plate (1m × 1m) with ridges was placed on the soil surface at the required orientation and a one ton weight placed on it to print the ridges into the soil.

Each plot was 3 × 3m with the data collected from a 0.5 × 2.5m strip down the centre of the plot. The strip was sectioned off on three sides using strips of metal sheeting inserted in the ground to about 200mm to prevent seepage or undercutting by any runoff. A Gerlach trough was placed at the bottom of each plot. Care was taken to ensure good contact between the trough lip and the soil surface. The trough was connected to a collecting bucket by a pipe, For the collection of runoff and sediment.

Each treatment was replicated a minimum of three times to give a total of fifteen experimental plots. Using a randomised experimental block design, the plots were divided among three blocks. Each block consisted of one plot of each different treatment and was treated independently of the other two.

The experimental area was first ploughed using a mould board plough to a depth of approximately 200mm. All plots were irrigated to have a similar initial soil moisture content. Where necessary the tillage treatments on each plot were carried out by hand. On each plot the following measurements were taken: (i) gravimetric soil moisture content; (ii) soil surface roughness before and after the design storm; (iii) surface depression storage; (iv) total runoff/soil loss; and (v) soil porosity.

Surface roughness was measured using a soil profile meter, which consisted of 50 needles, 0.5m long, mounted on a frame, 20mm apart. The frame was oriented across the slope and placed on runners parallel to the soil surface in the down slope direction so that the needles rested on the soil surface.

All height readings were taken from a marked datum point. The results were then analysed and the average of the standard deviation obtained for each set of results was taken as the roughness index (Monteith, 1974).

The results obtained from the profile meter were used to calculate the depressional storage. The storage volume was estimated by integrating the relative height of the pin above the datum point times the area of the vertical columns for all needles. The profile meter readings were analysed for the presence of large clods on the soil surface. These were then removed from the analysis as they are of little or no storage value. The adjusted height readings were then used to determine the total storage depth over the sample area. Both the roughness index and total storage depth were measured before and after the design storm (75mm.h^{-1} for 60 min). This rainfall was produced using a pressurised nozzle rainfall simulator.

The effect of slope was thought to affect the amount of depressional storage. The literature, however, does not state at what angle this starts to be significant. Hence, two models were developed to determine the effect of slope. These were derived from taking simple geometrical shapes of cubes and inverted pyramids. An almost one-to-one relationship was found between slope and percentage volume reduction, using the cubes/columns to represent

389

the soil surface (Figure 1). A much larger reduction in volume with slope was observed when modelling the effect of slope on volume for a square based inverted pyramid model (Figure 1). This indicates that although the cube and inverted pyramid model give similar depression storage volumes on the horizontal, their depression storage capacity differ greatly on sloped surfaces.

In this study the average slope of the field plots was 4°. Theoretically, the corresponding reduction in storage volume due to this slope is 3.49%, using the cube model. This figure is so small that it lies within the boundaries of experimental error and was therefore disregarded as far as any corrections to the experimental work and in its analysis was concerned. The cube model was evaluated against laboratory measurements using an egg box to represent the soil surface depressions. The results obtained are as shown in Figure 1 and show that the model realistically predicts the effect of slope on storage values.

Figure 1. Models used. ☐ **Cube model;** + **Pyramid model;** ■ **Lab results**

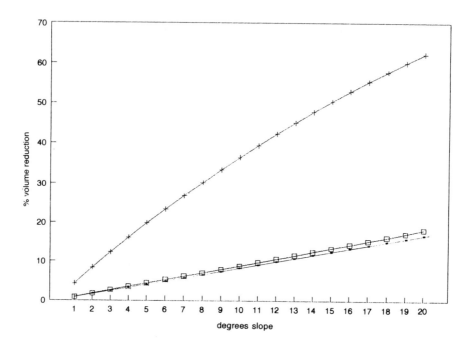

3. RESULTS

There were significant differences at the 5% probability level in the roughness index between the treatments, both before and after the storm, apart from those for the compact and seedbed surfaces (Figure 2). It was found that roughness index values followed the sequence: compact/seedbed<ridges<ploughed, for values measured both before and after the storm. These findings are in line with those found by Larson (1964). Figure 2 also shows the relative change in roughness caused by the design storm. No statistically significant differences between the treatments were obtained with respect to the effect of the storm.

3.1 Effect of Roughness on Storage

Only the treatments of ploughed, seedbed and cross-slope ridges were of any storage value. The storage depth on both compact-smooth and down-slope ridges was effectively zero. Each of the treatments possesses a characteristic roughness index and minimum storage depth, as shown in Table 2. Figure 3 shows the total available storage both before and after the storm. The observed relationship between roughness and storage is shown in Figure 4. This shows that the rougher the soil surface, the greater the storage depth. Statistical analysis showed the ploughed and cross-slope ridges were not statistically different from each other but both differed significantly from the seedbed that has the lowest available storage (Table 3).

As discussed by Barron (1971), some of the variation of storage with random roughness may be explained by the fact that the same storage depth values may be obtained for surfaces with different surface configurations and orientations. This is clearly demonstrated in Figure 4 between the ploughed and cross-slope ridges treatments. The storage does not differ significantly between these two treatments, but the roughness indices are significantly different - the ploughed being rougher than the cross-slope ridges due to the more irregular surface.

The ratio of storage depth after the storm to that before the storm is shown in Table 2. Statistically the three treatments involved differ significantly from each other. The ploughed treatment shows a decrease in storage. This may be explained by the breakdown of the clods, and the depressions being filled by the resulting sediment. An increase in both storage and roughness is also seen on the seedbed treatment, due to the localised compaction caused by raindrop impact on the light tilth. Little change is observed for the cross-slope ridges.

3.2 Effect of Roughness on Runoff/Soil Loss

As expected, runoff volume declined with increasing storage depth and increasing roughness. A non linear regression analysis was performed and from this an equation describing the relationship between storage depth and runoff was derived.

$$\text{Runoff depth (mm)} = 10.2 + 17.645 \times 0.731^a \tag{1}$$

where a = storage depth in mm after the storm.

Figure 2. Surface roughness: before and after storm.

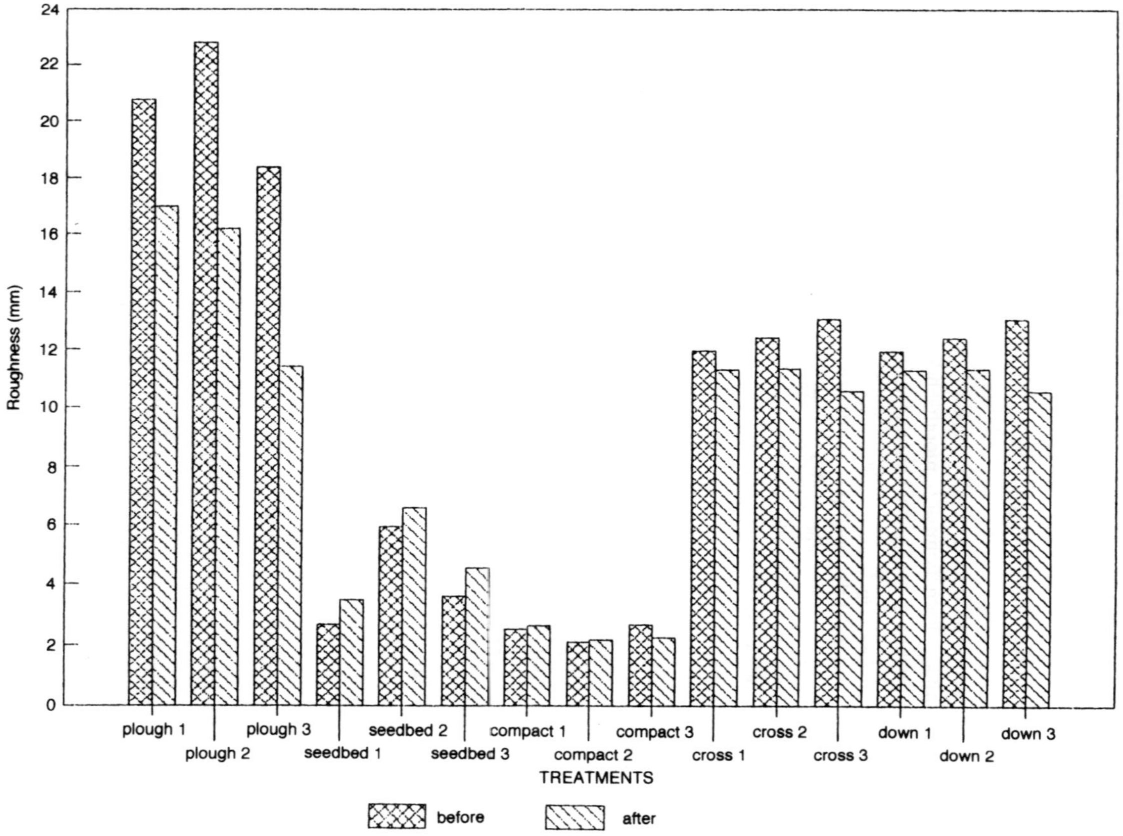

Table 2. Results obtained (3 replications).

Treatment	RB	RA	SB	SA	SL	RV	MC
Ploughed	20.77	17.01	21.43	21.63	0.00	0.00	11.89
	22.83	16.22	22.77	16.20	0.00	0.00	11.51
	18.38	11.45	20.52	12.24	0.00	0.00	11.27
Seedbed	2.73	3.52	5.17	5.57	1.07	3.27	10.99
	5.98	6.63	7.18	9.01	0.56	0.92	9.43
	3.62	4.57	5.39	7.04	0.92	1.48	10.87
Compact	2.55	2.69	0.00	0.00	11.70	20.68	10.42
	2.13	2.23	0.00	0.00	20.54	22.61	10.33
	2.70	2.30	0.00	0.00	8.05	8.65	8.71
Cross-ridges	12.01	11.36	18.10	21.32	0.00	0.00	8.20
	12.49	11.39	18.10	20.98	0.00	0.00	8.51
	13.14	10.63	23.70	19.38	0.00	0.00	9.13
Down-ridges	12.01	11.36	0.00	0.00	*	*	8.45
	12.49	11.39	0.00	0.00	15.93	11.51	8.57
	13.14	10.63	0.00	0.00	18.08	23.34	8.01

Treatment	Applied treatments
MC	% moisture content
RB/RA	Surface roughness (mm) before/after storm.
SB/SA	Storage depth (mm) before/after storm
SL	Soil loss (g.m^{-2})
RV	Runoff volume (mm)

Table 3. Statistical analysis.

Treatment	P	S	C	CR	DR	Least squares difference at 5%	Standard error	Coeff. of variation %
RB	20.66	4.11	2.46	12.55	12.55	1.93	1.27	12.20
RA	14.89	4.91	2.41	11.13	11.04	2.16	1.42	16.00
SB	21.57	5.91	0.00	19.98	0.00	2.61	1.70	18.0
SA	16.69	7.21	0.00	20.56	0.00	3.47	2.28	25.70
SL	0.00	0.61	2.61	0.00	2.89	0.36	0.23	19.60
RV	0.00	7.41	9.68	0.00	9.82	0.66	0.43	8.00
MC	11.56	10.4	9.82	8.61	8.30	1.03	0.60	7.00
RA/RB	0.72	1.22	0.98	0.89	0.88	1.02	0.07	7.00
SA/SB	0.77	1.21	0.00	1.05	0.00	0.22	0.15	24.00

P	Ploughed	S	Seedbed
C	Compact	CR	Cross-slope ridges
DR	Down-slope ridges		

Figure 3. Storage depth: before and after storm.

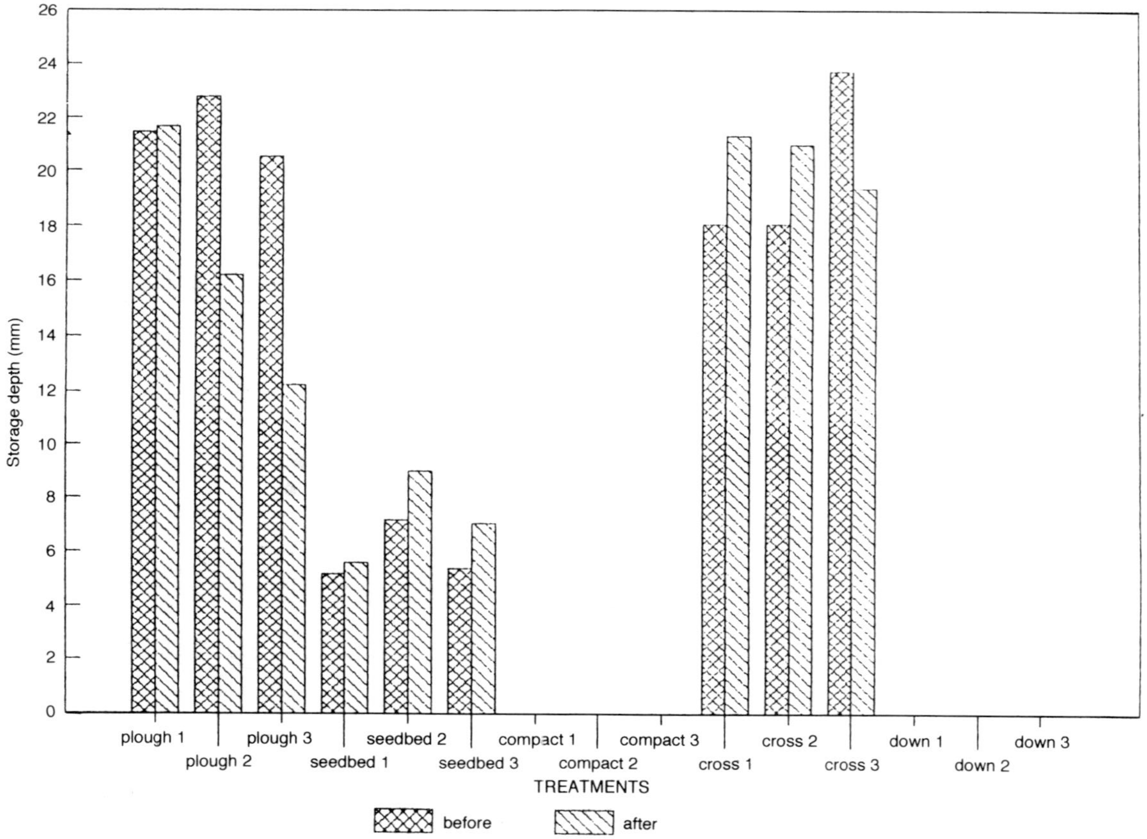

Figure 4. Roughness/storage: before storm.

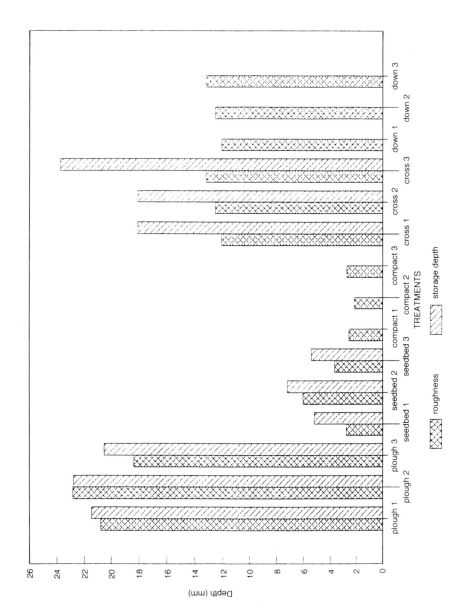

The highest runoff volume occurred with the down-slope ridges. For the compact treatment an average of 17.42 $l.m^{-2}$ of runoff was collected. This compares with an average runoff volume of 1.89 $l.m^{-2}$ from the seedbed treatment, and indicates that a slight increase in roughness can have a dramatic effect on runoff volumes. No runoff was produced from either the ploughed or the cross-slope ridges treatment due to their large storage depths and roughness indices.

On both compact and down-slope ridges the high runoff volumes obtained can be explained by the surface sealing and reduced infiltration capacity (Burwell et al., 1968). The sandy loam soil is very prone to capping due to the breakdown of soil aggregates on impact from raindrops. As described by Larson (1964), the detached particles from the down-slope ridges are eroded into the depressions between the ridges, forming a surface seal, which reduces the infiltration capacity considerably. This, combined with the up/down slope orientation, means water is readily transported down slope. This has the potential of being a useful water harvesting technique. On both the compact and down-slope ridges the surface was compacted to a depth of approximately 30-40mm. This compaction was seen to reduce the infiltration capacity of the surface layers dramatically, promoting runoff (Taylor, 1991).

As runoff is seen to decrease with increasing roughness, it is safe to assume that soil loss also decreases as roughness increases. As a linear relationship was found to exist between runoff and soil loss, we would expect the relationship between soil loss and storage to be similar to that between runoff and storage. Surface roughness is seen to reduce soil loss in three ways. First, it increases the water storage capacity of the soil surface. Second it decreases the velocity of the runoff that in turn is seen to decrease both detachment and transport capacities of the runoff. Lastly, soil loss is reduced by the trapping of sediment in the depressions created by the roughness.

4. DISCUSSION AND CONCLUSIONS

The experimental work showed that even a slight increase in surface roughness has a dramatic effect on runoff. The higher the roughness index, the higher the surface storage, and thus the greater the time to initial runoff and the lower the total runoff depth. Equation (1) was derived to describe the relationship between storage depth and runoff. A high surface storage is seen to retain water from short duration storms, reducing the erosion risk. However, the amount of storage is seen to reduce when aggregates are broken down by raindrop impact. Similarly, soil loss was seen to decrease as roughness increases. The following equation was derived to describe the relationship between runoff and soil loss:

$$\text{Soil loss}(g.m^{-2}) = 0.182 + 0.8035 \, (\text{Runoff(mm)}) \tag{2}$$

Across slope ridges gave protection against both soil and water losses, preventing runoff by retaining the water, allowing it to infiltrate slowly. This is of great importance in areas of low rainfall on soils low in water holding capacity. The down-slope ridges were seen to be of potential use for water harvesting. This treatment yielded large total runoff depths (75mm), equal to that obtained on the compact smooth plots (72.5mm). The advantage of the up/down slope ridges is that the water can be easily directed into suitable storage containers compared with smooth and compact plots. The effect of slope on storage volume is seen to depend on the geometrical shape assumed for the soil surface.

In order to reduce soil loss over high erosion risk periods, the soil surface should be made as rough as possible. This will have the dual effect of reducing both total runoff and soil loss. As the depressional storage reduces after successive rainfall events, it would be desirable to re-roughen the surface midway through this period, if required. The tillage operation used should provide a soil surface with a minimum roughness index of around 8mm to ensure adequate protection.

REFERENCES

BARRON, N.A. 1971. Soil surface depression storage by geometrical models. MSc Thesis, University of Illinois.

BURWELL, R.E. and LARSON, W.E. 1969. Infiltration as influenced by tillage-induced random roughness and pore-space. Proceedings of the Soil Science Society of America 33:449-452.

BURWELL, R.E., SLONEKER, L.L. and NELSON, W.W. 1968. Tillage influences on water intake. Journal of Soil and Water Conservation 23:185-186.

COGO, N.P., MOLDENHAUER, W.C. and FOSTER, G.R. 1984. Soil-loss reductions from conservation tillage practices. Proceedings of the Soil Science Society of America 48:368-373.

CURRIE, J.A. 1966. The volume and porosity of soil crumbs. Journal of Soil Science 17:24-35.

EVANS, R. 1980. Mechanics of water erosion and their spatial and temporal controls: an empirical viewpoint. In: Kirkby, M.J. and Morgan, R.P.C. (eds). Soil erosion. John Wiley, Chichester. pp.109-128.

GODWIN, R.J. 1990. Agricultural engineering in development: Tillage for crop production in areas of low rainfall. FAO Agricultural Services Bulletin no. 83.

KING, D.W. 1969. Soils of the Luton and Bedford district: a reconnaissance survey. Soil Survey of England and Wales.

LARSON, W.E. 1964. Soil parameters for evaluating tillage needs and operations. Proceedings of the Soil Science Society of America, 28:118-122.

MITCHELL, J.K. and JONES, B.A. 1978. Micro-relief surface depression storage: changes during rainfall events and their applications to rainfall runoff models. American Water Resources Authority Water Resource Bulletin 12:1205-1222.

MONTEITH, N.H. 1974. The role of surface roughness in runoff. Journal of Soil Conservation of New South Wales 30:42-45.

MOORE, I.D. and LARSON, C.L. 1979. Micro relief surface storage from point data. Transactions of the American Society of Agricultural Engineering 22:1073.

TAYLOR, N.C. 1991. An investigation into the effect of surface tillage treatments on rainfall runoff on water harvesting. Unpublished MSc Thesis, Silsoe College, Bedfordshire, UK.

Chapter 36

EFFECT OF SOIL CONDITIONERS ON THE INFILTRATION RATE OF AN ALLUVIAL SOIL

Alfredo Gonçalves Ferreira, Carlos Alexandre and Seco Cassamá
Dept. of Agr. Eng., University of Évora, Apart 94,
70001 Évora, Portugal

Summary

Several samples of a Xerofluvent derived from river and marine deposits were treated with different combinations of 4000kg.ha^{-1} Phosphogypsum and 10kg.ha^{-1} anionic polycrilamide (PAM). The treated samples were then subjected to simulated rainfall. The objective of this work was to evaluate the effect of these applications on the soil's infiltration rate.

The combined effect of the Phosphogypsum and the PAM was the most efficient, with a final infiltration rate triple that of the control. Regarding the sequencing of application, the low concentration of PAM in a simple application or in a split application with a month of ageing before the experiment, gave a final infiltration rate double that of the control.

For field use, the most effective technique is Phosphogypsum broadcasting, row seeding, split application of low concentration PAM, with four applications of 20ppm.

For future research it would be useful to look at the effect of the time lag between the PAM application and the first rainfall event.

1. INTRODUCTION

Infiltration rate depends on the soil texture, structure, mineralogy, exchangeable sodium percent (ESP) and the salt concentration of the percolating solution (Frenkel et al., 1987; Shainberg and Letey, 1984).

A soil crust results from the dispersion of clays, due to the kinetic energy of raindrop impact and the difference in electrolyte concentration between the natural or simulated rain and the soil solution. The crust is characterised by a higher density, finer pores and lower saturated conductivity, compared with the underlying soil (Shainberg and Singer, 1985).

The crusting process during a simulated or natural rainfall event can be the result of:
(a) Breakdown of soil aggregates caused by the impact of raindrops and compaction of the uppermost soil layer (Shainberg and Singer, 1985);
(b) Physio-chemical dispersion of clay particles which can migrate into the soil along with the infiltrating water, and clog the pores immediately beneath the surface (Shainberg and Singer, 1985);

(c) Slaking of the soil aggregates resulting from air compression inside the soil pores by water during wetting (Le Bissonnais et al., 1989).

In order to prevent crust formation the impact energy of the drops should be reduced. This can be done by providing cover to the soil during the rainfall event, or by decreasing the drop size, and/or height of fall.

The physio-chemical dispersion of clays can be prevented, or at least decreased, by increasing the electrolyte concentration of the water that reaches the soil, and by decreasing the exchangeable sodium percent of the soil (Keren and Shainberg, 1981; Shainberg et al., 1990). The aggregate stability can be improved by increasing the aggregate strength with a bonding agent and/or preventing the shrinking and swelling of the clay minerals, by decreasing the difference in electrolyte concentration between the incoming water and the soil solution (Shainberg et al., 1990).

In the Mediterranean climate and for irrigated crops improving the aggregate stability by increasing soil organic matter content is limited by the high rate of mineralization, and the high value of agricultural residues such as straw. This cost limits the use of crop residues and artifical covers to small areas of highly intensive agricultural production.

For these reasons, there is increasing interest in materials that are able to improve soil physical characteristics such as aggregate formation and stability. For example, water soluble soil conditioners like the anionic polycrylamide (PAM), 100% modified polycrylamide can bind with clay through divalent calcium (Wallace and Wallace, 1990). Another soil conditioner, phosphogypsum, is a subproduct of the fertilizer industry. It enhances the effect of the PAM by increasing aggregation and the electrolitic concentration of the incoming water (Shainberg et al., 1990).

The objective of this study is to evaluate the effect of the application technique of soil conditioners, on the infiltration rate of an alluvial soil developed on marine and river sediments.

2. MATERIALS AND METHODS

The soil is a Xerofluvent, from the river Tejo floodplain near Lisboa, whose infiltration rate was studied in the laboratory under simulated rainfall (Table 1).

Soil samples were packed 3cm thick inside acrylic boxes, over a bed of sand and gravel. The lower edge of each box has a weir to allow runoff and the bottom was connected to an outlet, to collect water that passes through the soil sample. Every sample was saturated from the bottom with a saline solution (electrical conductivity (EC) = 6 dS.m^{-1} and sodium adsorbtion ratio (SAR) = 10), drained and dried for 72 hours at 45°C.

Before placing the samples under the rainfall simulator (Munn, 1976), they were saturated with the same solution and then subjected to rainfall with drops of 3.2mm in diameter and a fall height of 2.7m, at a rate of 31.2mm.h^{-1} (standard deviation = 8.91mm.h^{-1}). The amount of water that drained out of the bottom outlet was recorded every 5 minutes.

The soil samples were treated with Phosphogypsum and PAM at a rate of 4000kg.ha[-1] and 10kg.ha[-1] respectively, as follows:

C   - Control soil with no application.
G   - As above, with Phosphogypsum spread over the soil surface before the first saturation.
P20   - PAM solution at 20ppm sprayed over the soil and dried at 45°C for 48 hours.
4P20   - PAM solution at 20ppm sprayed over the soil, one quarter of the amount each time, and dried at 45°C for 24 hours, after each application.
4P20(II)   - As above, with one month between the last application and the rainfall event.
P500   - PAM solution at 500ppm sprayed over the soil and dried at 45°C for 24 hours.
P20-G   - As P20, with Phosphogypsum spread over the surface before the rainfall event.
4P20-G   - As 4P20, with Phosphogypsum spread over the surface before the rainfall event.
P500-G   - As P500, with Phosphogypsum spread over the surface before the rainfall event.
G4P20   - As 4P20-G, with Phosphogypsum spread over the surface before the first saturation.
G-P500   - As P500-G, with Phosphogypsum spread over the surface before the first saturation.
Gm-P500 - As above, with the Phosphogypsum incorporated into the soil to 1cm depth

Table 1. Soil texture and chemical properties.

Coarse sand	(%)	1.6
Fine sand	(%)	6.5
Silt	(%)	45.5
Clay	(%)	46.4
Ca^{++}	(meq/100g)	23.3
Mg^{++}	(meq/100g)	8.5
K^+	(meq/100g)	1.2
Na^+	(meq/100g)	8.2
C.E.C.	(meq/100g)	41.4
Na/C.E.C.	(%)	19.8
Conductivity	$(dS.m^{-1})$	0.6
pH		7.7

3. RESULTS AND DISCUSSION

The infiltration rate was plotted against the cumulative rainfall. It is evident that after 20mm of cumulative rainfall the high PAM concentration (P500) and the Phosphogypsum (G) did not differ substantially from the control (Figure 1). This may result from the fairly high amount of sodium in the exchange complex in a neutral environment (Table 1). The clays remain dispersed and, due to the lack of a divalent cation, the bridging between the anionic polymer and the anionic clay is not possible (Aly and Letey, 1988).

The Phosphogypsum (G) was slightly more effective than the P500. It seems that the Phosphogypsum has some efficiency in preventing crust formation (Figure 1).

Figure 1. Infiltration rate vs. cumulative rainfall for C, G, P20, 4P20, 4P20(II) and P500 treatments.

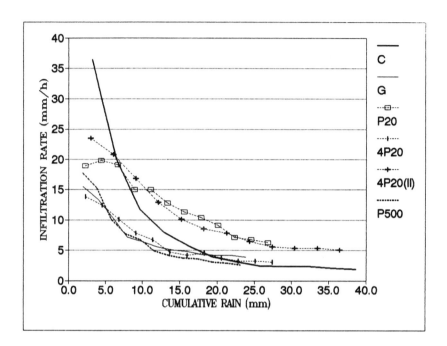

With the lower PAM concentration and when all the conditioner was applied at one time (P20) the infiltration rate was generally about twice that of the control. It seems that the larger amount of solution that corresponds to the lower PAM concentration, P20, gives a better distribution of the conditioner within the soil layer, so reaching and stabilising more soil aggregates (Figure 1). The split application (4P20) did not differ from the control after 20mm of cumulative rainfall. A reason for this could be that only the soil surface is treated and thus only surface aggregates are stabilised (Figure 1).

It is interesting to point out that the split application which was tested one month after treatment, 4P20(II), gave similar results to the P20, which shows that the effect of the conditioner is time dependent (Figure 1). It should be noted that in these two treatments the infiltration rate, after 25mm of cumulative rainfall, doubled that of the control, after 20mm of cumulative rainfall.

When the Phosphogypsum is applied together with the PAM, a synergetic effect takes place, due to the double effect of flocculation and binding and bridging (Wallace and Wallace, 1990; Shainberg et al., 1990) (Figures 2 and 3).

Figure 2. Infiltration rate vs. cumulative rainfall for C, P20-G, 4P20-G and P500-G treatments.

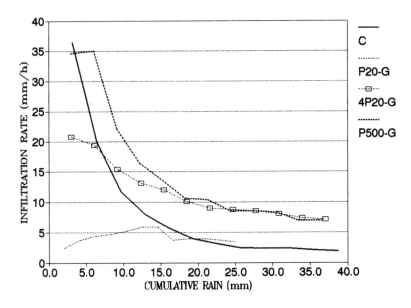

Figure 3. Infiltration rate vs. cumulative rainfall for C, G-4P20 and Gm-P500 treatments.

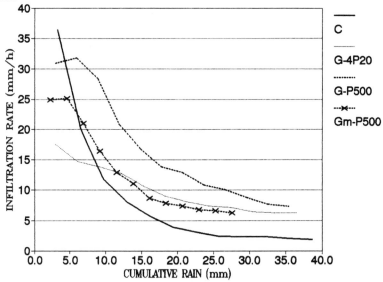

When the Phosphogypsum is applied after (a) the PAM split application (Figure 2), 4P20-G, and (b) the one time application, P500-G, both samples showed an almost similar final infiltration rate. However, initial infiltration rate is higher in the P500-G treatment than for the 4P20-G, due to the higher aggregation at the surface. However, this did not last during the rainfall event (Figure 2). Both treatments, however, increased the final infiltration rate by three times that of the control.

The one time application, P20-G, behaved strangely, probably due to the large amount of liquid applied, which distributed the polycrylamide throughout the soil sample depth (Figure 2). Compared to the control, the spreading of the Phosphogypsum slightly improved the infiltration rate, after 20mm of cumulative rainfall (Figure 2).

When the Phosphogypsum is applied before the PAM (Figure 3), the G-P500 had the best performance. The effect of the gypsum on preventing crusting was evident. The result was similar to the treatment 4P20-G (Figure 2).

The performance of the split treatment, G-4P20, however, did not reach the same final infiltration rate as the G-P500 treatment, and stability was reached after a lesser amount of cumulative rainfall (Figure 3).

When the Phosphogypsum was incorporated in the sample, Gm-P500, the effect on the electrolytic concentration of the infiltrating water was not as effective as the treatment where the Phosphogypsum was spread over the soil, G-P500 (Figure 3). However, in these treatments the final measured infiltration rate was three times higher than that of the control for the G-P500 treatment and more than twice the control rate for the G-4P20 and Gm-P500 treatments.

4. CONCLUSIONS

The experiments show that PAM application should be coupled with a divalent cation, which favours aggregation and flocculation, and allows bridging between the polymer and the soil clay minerals. Applications at low concentrations (P20 and 4P20(II)) raised the infiltration rate, after 20mm of cumulative rainfall, from 2.5mm.h^{-1}, for the control, to 5mm.h^{-1}.

The split application of low PAM concentration, or the application of larger concentrations of PAM have the same effect when applied prior to the Phosphogypsum, increasing the infiltration from 2.5mm.h^{-1}, for the control, to 7.5mm.h^{-1}, after 25mm of cumulative rainfall. In the field this technique is not easy to perform, because the PAM is applied after seeding, and so, for the Phosphogypsum spreading, another field operation is needed, which is undesirable.

It is difficult to use more than 10ppm of PAM concentration in irrigation water, which limits the use of higher concentrations. It is also difficult to spray 10kg.ha^{-1} of PAM at 20ppm (P20) at one time due to the large amount of water required.

The broadcasting of Phosphogypsum, row seeding and split application of PAM (G-4P20) appears to be the easiest and the most efficient technique. This increases the infiltration rate from 2.5mm.h^{-1} (control) to 6mm.h^{-1}, after 20mm of cumulative rainfall.

In future, it would be interesting to look at the effect of the time lag between the application of the conditioner and the first irrigation, and how the intensity and frequency of precipitation interacts with the infiltration rate of soils treated with the anionic polymer.

REFERENCES

ALY, S.M. and LETEY, J. 1988. Polymer and water quality effects on flocculation of montmorillonite. Soil Sci. Soc. Am. J. 52:1453-1458.

FRENKEL, H., GOETZER, J.O. and RHOADES, J.D. 1987. Effect of clay type and content, exchangeable sodium percentage and electrolyte concentration on clay dispersion and soil hydraulic conductivity. Soil Sci. Soc. Am. J. 42:32-39.

KEREN, R. and SHAINBERG, I. 1981. Effect of dissolution rate on the efficiency of industrial and mined gypsum in improving infiltration of a sodic soil. Soil Sci. Soc. Am. J. 45:103-107.

LE BISSONNAIS, Y., BRUAND, A. and JAMAGNE, M. 1989. Laboratory experimental study of soil crusting: relations between aggregate breakdown mechanisms and crust structure. Catena Verlag.

MUNN, J. 1976. Tahoe Lake rainfall simulator. University of California, Davis.

SHAINBERG, I. and LETEY, J. 1984. Response of soils to sodic and saline conditions. Hidegardia 52 No. 2.

SHAINBERG, I., GAL, M. and FERREIRA, A.G. 1990. Effect of water quality and amendments on the hydraulic properties and erosion from several Mediterranean soils. Proceedings of the seminar on 'Interactions between agricultural systems and soil conservation in the Mediterranean belt'. European Society for Soil Conservation, Oeiras, Portugal.

SHAINBERG, I. and SINGER, M.J. 1985. Effects of electrolyte concentration on the hydraulic properties of a depositional crust. Soil Sci. Soc. Am. J. 49:1260-1263.

WALLACE, A. and WALLACE, G.A. 1990. Soil and crop improvement with water-soluble polymers. Soil Technology 3:1-8.

Chapter 37

SHEAR STRENGTH OF THE SOIL ROOT SYSTEM: IN SITU SHEAR TESTS

S. Tobias

Amt für Gewässerschutz und Wasserbau,
Fachstelle Bodenschutz, CH-8090 Zürich, Switzerland

Summary

It is well known that vegetation is effective at erosion control. Bioengineering techniques have been developing for a long time and recently have become more popular. However, there is no appropriate means of judging the effect of vegetation. Botanical and ecological studies are being made to find species and plant communities that can endure mechanical strain. Soil mechanical approaches are applied to quantify the stabilising effect of vegetation. This contribution discusses whether the theory of reinforced soil can be applied to describe the soil root interaction. Shear tests were carried out in situ in herbaceous vegetation. The results show that the reinforcing effect of roots increases with saturation. The lower the shear strength of the soil, so the reinforcement by roots for stabilisation becomes more significant. Differences in species composition created variation in results. Species which naturally occur on sites with frequent mass movement are most suitable for stabilising purposes. Ecological observations can be quantified and proved by a soil mechanical parameter.

1. INTRODUCTION

The use of living plants is one of the oldest techniques in construction engineering. In 1591 Pan, Minister of River Flooding Control in the Ming Dynasty, wrote about the stabilisation of an earth dam by willow branches (Lee, 1985). In central Europe, early texts about bioengineering date from the 18th Century (Schlüter, 1986). In the economic crisis of the 1930s, the development of bioengineering techniques was intensified. These techniques represented cheap methods for flood control and an opportunity for employment (Grubinger, 1983).

Nowadays, alongside industrial development, there is a new and strong environmental awareness, which has made bioengineering more popular. At the same time the responsibility of engineers has increased, because of the high standard of living and the demand for a safety guarantee for any construction project. With bioengineering methods this is difficult, as there are no standards, because living plants develop very distinctly according to their habitat. Choosing the right species in the right quantity at the right time depends on the personal experience of each bioengineer.

Therefore, the scientific investigation of the stabilising effect of vegetation has become more necessary and more publications have appeared in the second half of the 20th Century (Schiechtl, 1973; Lee, 1985; Schlüter, 1986).

2. SCIENTIFIC INVESTIGATION OF THE SOIL ROOT INTERACTION

Based on botanical studies of pioneer vegetation on erodible sites, experiments in the Swiss Alps on revegetation of erosion damage areas on ski tracks have been made. They showed that the species used had to be typical for these alpine sites, and that the individuals that had germinated at that altitude were the most resistant. Individuals grown in the test garden in Zürich (400m) and then transplanted in the Alps showed weak reproduction, and soon perished. So Urbanska et al. (1987) tried to reproduce the plants by splitting them into their single ramets (single shoots of one plant), that regenerated and produced more ramets on their own. The most appropriate were plants which spread by runners (Gasser, 1989).

Other authors illustrate the effect of vegetation by the use of soil mechanical parameters. Generally, roots are considered to reinforce the soil, which causes a certain shear strength increase. After the introduction of the basic soil root interaction model by Wu (1976), Waldron (1977) and Brenner and James (1977) (all in Gray and Leiser, 1982), Wu et al. (1988) created further models that compare roots with cables or piles. The shear strength of the soil root system has been measured in a number of studies. Kobashi (1984) grew his test grass in the laboratory. Other tests were done in situ (Kaibori and Sassa, 1984; Abe and Iwamoto, 1985; Wu et al., 1988; O'Loughlin and Ziemer, 1983).

The research project discussed here was conducted at the Federal Institute of Technology in Zürich (Tobias, 1991). The principal idea was to combine botanical and ecological experience with soil mechanical quantification methods. This work is based on the theory of reinforced soil. For quantification, in situ shear tests were carried out. In addition, the following questions were considered: what are the characteristics of species that are most apt for soil stabilisation, and which ecological role do they play?

3. THE PRINCIPLES OF REINFORCED SOIL

Structures of reinforced soil and root layers show similarities in their ductile reaction to strain (Figure 1). They can deform to a great extent before they fail. They follow deflections in their foundations and subsidence of the slope without losing their retaining capacity.

According to the theory of reinforced soil (Vidal, 1966) friction along the reinforcing component (e.g. roots) holds the soil particles together. Stresses on the reinforcement component cause a continuous variation of tension in the reinforcement. The point of maximum tension in the reinforcement divides the structure into an 'active' zone, where soil particles would be displaced, but are held back by friction, and a 'resistant' zone, where the reinforcement tends to slip out, which again causes friction. The points of maximum tension in the reinforcement represent the position of maximum strain of the whole system. In retaining walls they are located in potential slip planes (Figure 2).

Figure 1. Similar structure of a retaining wall of reinforced soil (a), brush layers (b) and the root system of a natural grass strip (c).

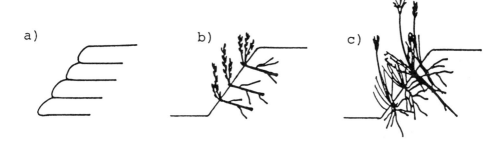

Figure 2. Distribution of tension in one reinforcement structure and location of maximum tension (T_{max}) in a retaining wall (Schlosser and Guilloux, 1981).

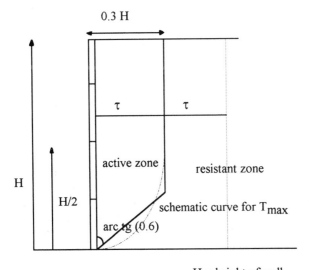

H = height of wall

τ = tension in reinforcement (kN/m^2)

T_{max} = peak tension in one reinforcement (kN/m^2)

Triaxial shear tests (Schlosser and Long, 1972, in Schlosser and Guilloux, 1981) showed that under large surcharge, reinforced soil fails by the breaking of the reinforcement at the point of maximum strain. Then, reinforcement increases apparent cohesion, c (Figure 3). Under a small surcharge, the system fails by the reinforcement slipping out of the loose soil structure. Shear test results also show an increase of the friction angle due to friction along the reinforcement component.

Artificial structures of reinforced soil, however, differ from natural soil root systems. The two parameters relevant to design are the reinforcement's tensile strength and the weight (bulk density) of the soil above the reinforcement. Orientation and distribution of the reinforcement can be selected, which allow exact calculation of the stress inside the structure. Root spreading, on the other hand, is not predictable, and the stresses are thus uncertain.

Figure 3. Failure of reinforced and non reinforced sand (Schlosser and Guilloux, 1981).

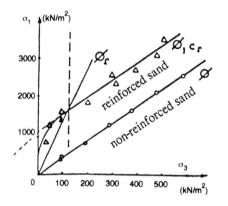

σ_1:   surcharge [kN/m^2]

σ_3:   side pressure [kN/m^2]

ϕ: friction angle of non-reinforced sand [°]

ϕ_r: friction angle of reinforced sand [°]

c_r: apparent cohesion of reinforced sand [kN/m^2]

4. SHEAR STRENGTH MEASUREMENT OF THE SOIL ROOT SYSTEM

The study aimed to reproduce natural conditions. Therefore shear tests were made in situ. Quantification of the integrity of the soil root system is difficult. The root horizon is usually not thicker than 1.5 or 2.0 metres. At the soil surface there is no surcharge and stresses are extremely small in relation to deeper layers. This fact not only demands instruments that measure small stresses precisely, but also requires special analysis of data. In the stress diagram, in which the Mohr envelope defines the function of shear strength, the measured values are close to the origin. There, the envelope is not a linear function of ϕ and c.

The size of the shear box and the test sites were selected on the basis of the problems mentioned above. The simple supposition of homogenous and isotropic distribution of the roots inside the sample is only acceptable for an extremely large shear box, or for the main root area of herbaceous vegetation. Near the soil surface, where the tests were performed, the weather has an immediate effect on soil moisture. As many tests as possible had to be made in a short period of time to ensure the same climatic conditions.

For the reasons given above, the test sites were on grassland: A monoculture of *Poa pratensis*, three sites on a forage meadow near Zürich (400m) and a fifth position on a fill-slope in the Alps (Reschenpass, 1500m), where species that were recommended for soil stabilisation had been sown.

The shear box measured 0.5m × 0.5m, with a height of 0.15m. The shearing in the main root layer took place at a depth of 0.07 to 0.08m. Reference shear tests were made underneath the root layer at 0.2m depth, as non vegetated sites were not available.

5. RESULTS

5.1. Failure of the soil root system

The shear plane was never horizontal at the bottom of the shear box in any of the tests. Soil always broke in passive rupture. A typical failure occurred along the inclined shear plane.

This phenomenon is mainly due to the small stresses near the soil surface and to the large size of the shear box. Also, in laboratory shear tests, rupture spreads progressively. The usually small shear boxes and great surcharges prevent the sample from failing in a passive way.

The points of maximum strain on the whole system formed the shear plane, which divided it into an 'active' zone (edge) and a 'resistant' zone (underneath the edge). The roots failed in that they broke and slipped out. They did not break in the shear zone, but at their ends where their cross sections were too small to absorb the applied tension. After this, they pulled out (Figure 4), as the soil at this depth is not very dense. So, contrary to artificial reinforcement, the critical cross section of roots need not be the one actually in the shear zone (Figure 4).

Figure 4. Root breakage and failure.

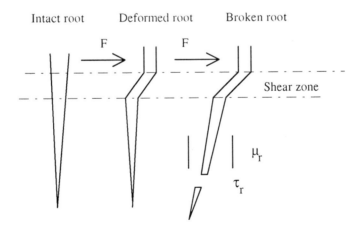

F = applied force (kN) μ_r = friction along the root surface (kN.m^{-2})
τ_r = tension in the root (kN.m^{-2})

5.2. Shear strength increase by vegetation

The shear test results are listed in Table 1. The observed failure mechanism of the roots suggests that roots effect both an increase of friction angle as well as an increase in cohesion. These parameters differ from case to case (Table 1), so that comparison is not possible by one of the shear parameters alone. A reference value $\tau(20)$ is necessary. $\tau(20)$ is the shear strength at a surcharge of 20kN.m^{-2}, calculated from the Coulomb equation. $\Delta\tau$ (20) is the difference between the shear strength of the root layer and that of the unrooted soil. The ratio of $\Delta\tau(20)$ to $\tau(20)$ of the rootless soil is expressed as a percentage.

Table 1. Measurements in the root layer.

S_r = saturation of the soil; ϕ_u = friction angle under undrained conditions, related to total stresses; c_u = section of the y-axis in the Coulomb equation under undrained conditions, related to total stresses; $\tau(20)$ = shear strength at a surcharge of 20kN.m^{-2} (reference value); $\Delta\tau(20)$ = shear strength difference between root layer and rootless soil, in (%) relation of $\Delta\tau$ (20) and $\tau(20)$ of the rootless soil.

Test field date	Main grasses	S_r (%)	ϕ_u (°)	c_u (kN.m^{-2})	$\tau(20)$ (kN.m^{-2})	$\Delta\tau(20)$ (kN.m^{-2})
Ch-C Veg 25.5.89	Alopecurus geniculatus	63	40.5	31.6	48.7	9.0 23%
WiRi 87 16.8.88	Poa pratensis	63	31.0	31.7	43.7	
Ch-A Veg 26.7.89	Agrostis stolonifera	61	27.8	28.0	38.5	5.2 16%
PB-Veg 18.10.88	Festuca pratensis Festuca rubra Poa pratensis	84	41.6	20.0	37.8	13.4 55%
WiRi 86 5.10.88	Poa pratensis	74	31.0	25.0	37.0	7.5 25%
Ch-A Veg 25.4.89	Agrostis stolonifera	100	27.8	25.2	35.7	4.8 16%
Ch-B Veg 18.5.89	Lolium multiflorum Agrostis stolonifera Poa annua	39	36.1	19.0	30.7	2.9 9%
Ch-B Veg 23.8.89	Lolium multiflorum Agrostis stolonifera Poa annua	65	36.1	15.8	30.4	-0.6 -2%
Ch-C Veg 11.10.89	Alopecurus geniculatus	100	40.5	13.0	30.1	

Especially under wet conditions (high saturation S_r) roots cause a remarkable increase in relative shear strength, and their reinforcing effect is important. This means that when soil shear strength is low, roots are effective at stabilisation.

Differences in increase in shear strength were related to the individual species. The greatest effect on soil integrity was observed in the alpine test field near Reschenpass (PB-Veg, 18-10-88), which was covered by a local pioneer grass, mainly *Festuca pratensis, Festuca rubra* and *Poa pratensis*. *Poa pratensis* as a monoculture has a significant stabilising capacity due to its rhizomes. On the wet areas in the forage meadow where *Agrostis stolonifera*, a pioneer of wet and erodible sites, had invaded, the shear strength increase was larger than where the forage species (mainly *Lolium multiflorum* and *Trifolium repens*) occurred.

6. CONCLUSIONS

6.1. Applicability of the theory of reinforced soil to the soil root system

Root tensile strength is not the only parameter indicating the stability of the soil root system. Even after breaking, the roots still keep a certain anchoring capacity due to friction along their surface. By this process the soil root system deforms to a great extent as can be seen in natural terraces where grass strips have been applied. If strain decreases (because of declining surcharge or increasing shear strength due to suction) before the roots are pulled out completely, they regenerate and so heal damage on slopes.

Qualitatively the mechanism of soil reinforcement by roots can be explained by the theory of reinforced soil. Models of reduced scale can be helpful in studying the behaviour of slopes covered by vegetation. Useful methods for quantification are not easily available. Information about the roots' tensile strength is insufficient, especially because it is not constant over the whole length of the roots. Woody parts of the root probably have a different tensile strength from smooth sections of the same root.

6.2. The effect of species composition on shear strength

Higher values were generally measured for short grass species which survive for several years, and tend to create dense swards. Some show a high resistance against foot traffic (*Poa pratensis* and *Agrostis stolonifera*), and are therefore often used on football fields. *Lolium multiflorum*, a typical forage grass, is not suitable for stabilising purposes. Basically, the measurements agree with qualitative experience. Relatively high values were measured with well known soil stabilisers like *Poa pratensis, Agrostis stolonifera, Festuca pratensis* and *Festuca rubra*. In their natural habitats these species often resist soil movement. This observation led to the preference of pioneer species for bioengineering. There is a natural selection after mechanical strain because only strong species can survive on sites with frequent mass movement. So for technical reasons the plant community that would naturally occur on a particular site at a given time should be selected. This knowledge gained by botanical studies can be proved by measurement of shear strength, a parameter which is common in engineering techniques.

REFERENCES

ABE, K. and IWAMOTO, M. 1985. Effect of tree roots on soil shearing strength. Int. Symp. on erosion, debris flow and disaster prevention, Tsukuba Japan. pp.341-346.

GASSER, M. 1989. Bedeutung der vegetativen Phase bei alpinen Pflanzen für die biologische Erosionsbekämpfung in der alpinen Stufe. Ber. Geobot. Inst. ETH Stift Rübel Zürich, 55:151-176.

GRAY, D.H. and LEISER, A.T. 1982. Biotechnical slope protection and erosion control. Van Nostrand Reinhold Comp., New York. pp.37-82

GRUBINGER, H. 1983. Bodenverfestigung durch Grünverbau. Mitt. Schweiz. Ges. Boden- u. Felsmechanik 109

KAIBORI, M. and SASSA, K. 1984. Tragbares Geländerahmenschergerät - Einige Versuchsergebnisse und Vergleich mit gewöhnlichen Scherversuchen. Int. Symp. Interpraevent 1984 Villach, 2:263-274. Forsch. Ges. vorbeug. Hochwasserbekämpfung.

KOBASHI, S. 1984. The role of vegetation to slope stability. Int. Symp. Interpraevent 1984 Villach 1:45-56. Forsch. Ges. vorbeug. Hochwasserbekämpfung.

LEE, I.W.Y. 1985. A review of vegetative slope stabilisation. Hong Kong Inst. Eng. 1985(7), Hong Kong.

O'LOUGHLIN, C. and ZIEMER, R.R. 1983. The importance of root strength and deterioration rates upon edaphic stability in steep land forests. New Zealand Forestry Service, 1570, ODC 18136:11.6.

SCHIECHTL, H.M. 1973. Sicherungsarbeiten im Landschaftsbau. Callwey, München.

SCHLOSSER, F. and GUILLOUX, A. 1981. Prinzipien und Theorie der bewehrten Erde. Mitt. Schweiz. Ges. Boden- u. Felsmech. Zürich 103:65-84.

SCHLÜTER, U.I. 1986. Pflanze als Baustoff. Ingenieurbiologie in Praxis und Umwelt. Patzer, Berlin.

TOBIAS, S. 1991. Bautechnisch nutzbare Verbundfestigkeit von Boden und Wurzel. Diss. 9483 ETH, Zürich.

URBANSKA, K.M., HEFTI-HOLENSTEIN, B. and ELMER, G. 1987. Performance of some alpine grasses in single-tiller cloning experiments and in the subsequent revegetation trials above the timberline. Ber. Geobot. Inst. ETH Stift. Rübel Zürich, 55: 64-90.

VIDAL, H. 1966. La terre armée. Suppl. Ann. Inst. tech. bâtiment et trav. pub., Paris, 223/224:887-938.

WU, T.H., McOMBER, R.M., ERB, R.T. and BEAL, Ph.E. 1988. Study of soil root interaction. Journal Geotechn. Eng. 114-12:1351-1375.

Chapter 38

EXPERIMENTAL EVALUATION OF THE EROSION ON BARE AND GEOSYNTHETICALLY PROTECTED SLOPES

D. Cazzuffi[1], F. Monferino[2], R. Monti[2] and P. Rimoldi[3]
[1]Enel/Cris, Milano, Italy
[2]Politecnico di Milano University, Milano, Italy
[3]Tenax Spa, Via Industria 3, 22060 Vigano (CO), Italy

Summary

A laboratory investigation studied the ability of some prefabricated products (such as synthetic and natural mats, and honeycomb geocells) to reduce erosion by artificial rainfall. The results are presented in the form of histograms, showing the quantity of eroded soil versus the total input flow, and with diagrams showing the solid transport flow versus the rainfall input.

The results help establish limits for the use of each product, thus giving a base for the design of erosion control applications.

1. INTRODUCTION

Since the beginning of the 1960s many different types of synthetic materials or 'geosynthetics' have increasing use, substituting for or in combination with, natural materials for geotechnical engineering. Erosion control represents one of the main applications of geosynthetics. These products, laid on a slope, are able to decrease the eroding action of rainfall, wind and runoff.

The present study deals with laboratory tests of erosion on slopes caused by rainfall and consequent runoff. Vegetative cover has an important role in erosion control. The main problem is in protecting bare slopes until the grass has established, since newly constructed slopes are highly erodible (Kirkby and Morgan, 1980).

The present study is focused on the most severe situation (i.e. bare soil). Vegetative cover, evapotranspiration and surface detention by vegetation are not taken into account. Erosion depends on the speed of runoff, becoming particularly intense when water velocity is greater than $30mm.s^{-1}$ (Rauws and Govers, 1988). One of the most effective ways to study erosion is with a large-scale laboratory model simulating slopes subjected to rainfall and runoff.

2. MATERIALS USED

The experiment is limited to geosynthetics and related products (such as biomats) used for erosion control. Geosynthetics can be classified by their physical characteristics and their function (Cazzuffi and Rimoldi, 1992). Geosynthetics for erosion control comprise geomats and geocells, while related products comprise mainly biomats. These products are characterised by a tensile resistance of 1-15 kN.m^{-1} and by their effectiveness in preventing soil movement on slopes and facilitating the growth of vegetation.

The behaviour of these products is related to containment and surface reinforcement of topsoil, protection against raindrop impact and sheet wash and high saturation capacity and weight.

Geocells are geosynthetics with a honeycomb, tri-dimensional structure, 50-200mm thick, made with cells of 80-400mm (corresponding to the diameter of the inscribed circle, hereafter referred to as the 'diameter'). They confine a certain thickness of top soil placed within the cells, avoiding mass sliding of the surface layer. Geocells are placed on the smoothened slope, opened up like an accordion and fixed to the ground with staples. The cells are slightly overfilled with soil which is then compacted. Therefore, they stabilise the soil by confinement. These products can be distinguished by the polymer used, the method of jointing and the cell diameter.

Within this category the following products were tested:
-    Tenweb 80 (produced by Tenax, Italy). This has a cell diameter of about 80mm and a thickness of 75mm. Water can flow through the different elements through a slot in every junction between adjacent cells (see Fig. 1).

Figure 1. Tenweb product.

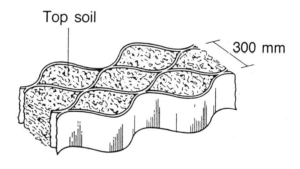

-    Tenweb 300 (produced by Tenax, Italy). This has the same characteristics as Tenweb 80, but with a cell diameter of 300mm.

-    Armater (produced by Akzo, Netherlands). This product has an hexagonal structure (see Fig. 2) with sides of 200mm length, a cell diameter of about 400mm and a thickness of 100mm. It is made of a polyester, nonwoven geotextile, cut in 100mm width strips and is jointed together in the hexagonal pattern with a special adhesive.

Figure 2. Armater product.

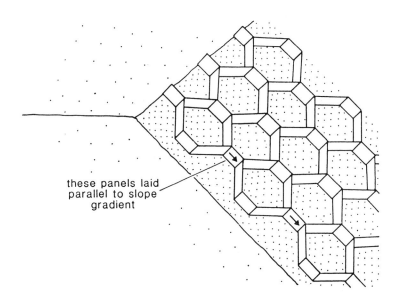

these panels laid
parallel to slope
gradient

Both the Tenweb products can be placed in a 'drainage' pattern, with the line of apertures in the junction placed along the maximum slope line; or in an 'non-drainage' pattern, with apertures at 90° to the previous direction (Fig. 1).

Geomats are tri-dimensional geosynthetics made of polyethylene, nylon or polypropylene filaments, tangled in such a way as to form a flexible mat, with a very open structure. They are 5-20mm thick and are used to stabilise a thin layer of topsoil and to improve root anchorage. Geomats are placed on a smoothened slope, fixed to the ground with staples and filled with topsoil. The soil is partially trapped within the geomat structure. Geomats avoid the formation of a soil crust and partially avoid soil loss during runoff. These products have no reinforcing action.

The products tested within this category are:

-    Enkamat (produced by Akzo, the Netherlands). This is a tangle of nylon filaments, with a constant thickness of 20mm and a flat bottom.

-    Tensarmat (produced by Netlon, UK). This product is made from lightweight grids. The two bottom grids are flat and stretched, while the two top grids are unstretched, undulated and welded to the flat bottom grids at regular points (Figure 3).

-    Multimat (produced by Tenax, Italy). This comprises three layers of a lightweight, bi-oriented grid. The central layer is mechanically and permanently folded (to provide the required thickness), while the two external layers are flat, providing high tensile strength.

415

Figure 3. Tensarmat.

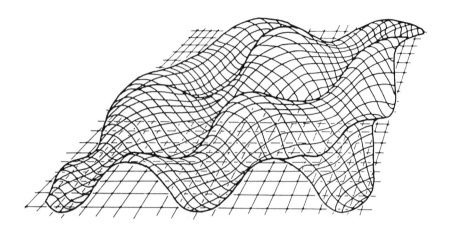

Biomats comprise either layers of natural fibres (for example straw, coconut or jute), contained within one or two layers of a lightweight polymeric grid (Figure 4), or are woven in an open mesh to form a net (Figure 5). For example, the jute biomat has a mesh size of 10-40mm and a thickness of about 5mm. The natural fibres decompose in 1-5 years, according to the raw material used. They protect the soil surface against raindrop impact during vegetation establishment. Biomats are usually characterised by a very high water retention capacity, which allows them to increase in weight during rainfall and thus adhere to the soil surface. The retained water is released in vapour form, allowing suitable conditions for the growth of vegetation.

Figure 4. Non-woven biomat.

Figure 5. Woven biomat.

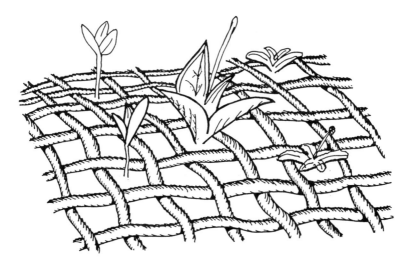

The products tested within this category are:

-   Geojute. This is made with untwisted ropes of jute, woven to give an open mesh of 25mm.

-   Bonterra K (produced by Bonterra, Germany). This biomat is made with coconut fibres held together by a lightweight polypropylene grid. It was placed on top of a Tenweb 300 geocell, in order to study the combined effect of the two types of products.

3. TEST FACILITIES

The test facility comprises an artificial slope with a rainfall and runoff simulator (Cazzuffi et al., 1991). Figure 6 shows the testing apparatus.

The slope simulator is a steel box with a width of 1.50m, slope length of 1.00m, and a depth of 200mm. It is divided into two equal parts, so that tests on two different materials can be carried out at the same time. The box can be inclined to different slopes. The bottom of the box has a steel grid with fabric to keep the soil in the box, while allowing free drainage of infiltrating water. Four measuring tanks are placed at the lower end of the box, to collect and measure runoff and infiltration water. On top of the tanks a steel grid, covered with a suitable nonwoven fabric, is placed to collect the eroded soil, whilst the runoff and infiltration water discharge into the measuring tanks.

The rainfall simulator has to simulate natural raindrop sizes, which range from 1.5 to 5mm according to the rainfall intensity. Nozzles, spraying upwards, were used on an oscillating tube standing on a 1.00m × 2.00m screen. Under this screen another three screens were placed, spaced at 200mm. These screens support three layers of plastic net, held taught on the frames (see Fig. 6). The screens break up the drops, giving a final drop diameter distribution ranging from 1.8 to 4mm, depending on the rainfall intensity used. The drops have a random distribution as with natural rainfall of the same intensity.

Figure 6. Experimental apparatus.

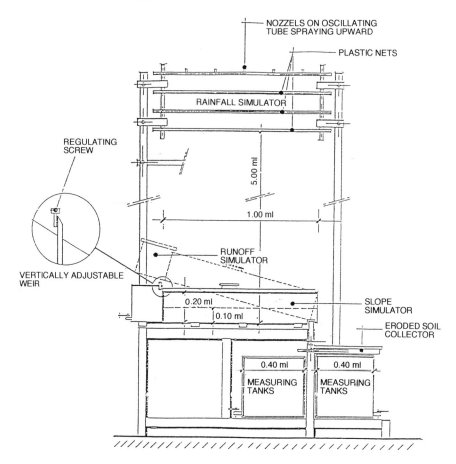

The final impact velocity of the drops must be similar to the terminal velocity of natural raindrops. For the calculation of the terminal velocity of the raindrops, reference can be made to Cazzuffi et al. (1991). Over the fall distance (5.0m) drops obtain a terminal velocity ranging between 80 and 97% of the natural terminal velocity of drops of the same diameter. Table 1 shows the limit velocity (V) and the terminal velocity (v) over 5.0m freefall for drops of different diameters, calculated according to Cazzuffi et al. (1991).

The soil used for all the tests is a silty sand, prepared by mixing 50% of medium sand, 30% of very fine sand and 20% of clay silt. This soil could be considered representative of a naturally highly erodible soil. The soil was placed in the steel box and compacted in 30mm thick layers, to a wet unit weight of about 20kN.m^{-3}. This high value was chosen in order to obtain low rates of infiltration and a high level of runoff. The soil was replaced for each test run.

Table 1. Limit velocity V, final velocity v over a freefall of 5.0m.

d[mm]	V[m.s^{-1}]	v[m.s^{-1}]	v[%V]	t[s]
1.5	5.6	5.47	97	1.28
2.0	6.5	6.17	95	1.21
2.5	7.3	6.70	92	1.17
3.0	8.0	7.10	88	1.14
3.5	8.6	7.40	85	1.12
4.0	9.2	7.60	82	1.10
4.5	9.8	7.80	80	1.09

4. TEST METHODOLOGY AND RESULTS

A control test on bare soil was run at the beginning and end of the experiments. The slope simulator was set at 26.5° inclination (1V:2H). The rainfall intensities of 50, 65 and 85mm.h^{-1} were used. During each test the following parameters were measured:

- slope angle β;
- rainfall intensity i (from which rainfall input per unit width Q_p was calculated:
  $Q_p = i \cdot 1 \cdot \cos\beta$;
- runoff flow at the downstream edge Q_u;
- flow velocity. This was measured during each test, by injecting a blue solution and measuring the time taken to flow 1.0m down the slope. This value can have large variations from point to point and from small to large rills.
- eroded soil. The soil collected was placed in an oven at 105°C for 48 hours and then weighed together with the pre-weighed collecting fabric. It is therefore possible to obtain the weight of eroded soil (W_c). The clayey silt fraction passed through the fabric and remained in the measuring tank. This fraction (W_s) was measured by weighing the volume of water containing any suspended soil particles and comparing this with the weight of the equivalent volume of clean water. The difference gave the total weight of soil particles (W_c). Thus the total weight of eroded soil (W_e) is: $W_e = (W_c + W_s)$.
  The measured weight of eroded soil was expressed as the weight of soil loss per unit area per unit of time.

The main quantitative results are summarised in Figure 7. The results show the efficiency of the tested products, based on the ratio P_e/Q_p and qualitative observations. The best sediment reduction is obtained by the geocells (Tenweb 300) for soil confinement overlaid by a biomat (Bonterra K). The erosion produced by the rainfall was in this case almost equal to zero.

The second best result was from a natural fibre product (Geojute). The jute has a high water absorption capacity, so that the runoff produced by direct rainfall is lower than for the other products. The jute, when wet, adheres to the soil, creating micro-terraces which slow down the average runoff velocity and distribute the flow uniformly. These two mechanisms reduce the formation of rills.

Figure 7. Test Results.

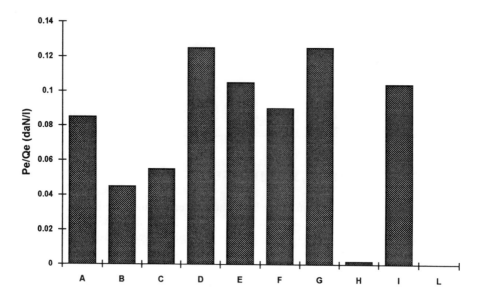

Key:

A	Bare soil	F	Enkamat
B	Tenweb 80'draining'	G	Tensarmat
C	Tenweb 'not draining'	H	Geojute
D	Armater	I	Multimat
E	Tenweb 300	L	Tenweb 300 + Bonterra

Geomats do not have the capacity of absorption nor adhere to the ground. Although they distribute the flow uniformly, they do not reduce the average flow velocity.

Of the geocells tested, the best results were given by Tenweb 80D (in the 'drainage' configuration). Compared to the other geocells, this gives a higher confining action, due to the reduced cell dimension (80mm rather than 300mm).

The poor results generally obtained by the honeycomb products are because they confine the soil, there is no real reinforcing action, and they give no protection to the surface from rainfall action. Moreover, the presence of artificial discontinuities in the soil prevents normal drainage, which results in quick saturation of the soil surface layer, and a faster and more intense surface flow.

5. CONCLUSIONS

Laboratory tests were performed with a rainfall simulator to study the ability of different prefabricated products to prevent soil erosion. The best protection against erosion was obtained with geocells overlaid by geomats. Geojute is quite effective in decreasing soil erosion over a large range of rainfall intensities. Geomats are capable of greatly reducing erosion, compared to an unprotected slope, especially under high rainfall intensities.

REFERENCES

CAZZUFFI, D., MONTI, R. and RIMOLDI, P. 1991. Geosynthetics subjected to different conditions of rain and runoff in erosion control applications: a laboratory investigation. Proc. XXII IECA Conf., Orlando, FL, USA.

CAZZUFFI, D. and RIMOLDI, P. 1992. European proposal for standardization of geosynthetic materials in erosion-control applications. Proc. XXIII IECA Conference, Reno, NV, USA.

KIRKBY, M.J. and MORGAN, R.P.C. (eds). 1980. Soil erosion. John Wiley, London, UK.

RAUWS, G. and GOVERS, G. 1988. Hydraulic and soil mechanical aspects of rill generation on agricultural soils. J. Soil Sci. 39:111-124.

INDEX